LANDSCAPES, GENOMICS AND TRANSGENIC CONIFERS

Managing Forest Ecosystems

Volume 9

Series Editors:

Klaus von Gadow
Georg-August-University,
Göttingen, Germany

Timo Pukkala
University of Joensuu,
Joensuu, Finland

and

Margarida Tomé
Instituto Superior de Agronomía,
Lisbon, Portugal

Aims & Scope:

Well-managed forests and woodlands are a renewable resource, producing essential raw material with minimum waste and energy use. Rich in habitat and species diversity, forests may contribute to increased ecosystem stability. They can absorb the effects of unwanted deposition and other disturbances and protect neighbouring ecosystems by maintaining stable nutrient and energy cycles and by preventing soil degradation and erosion. They provide much-needed recreation and their continued existence contributes to stabilizing rural communities.

Forests are managed for timber production and species, habitat and process conservation. A subtle shift from *multiple-use management* to *ecosystems management* is being observed and the new ecological perspective of *multi-functional forest management* is based on the principles of ecosystem diversity, stability and elasticity, and the dynamic equilibrium of primary and secondary production.

Making full use of new technology is one of the challenges facing forest management today. Resource information must be obtained with a limited budget. This requires better timing of resource assessment activities and improved use of multiple data sources. Sound ecosystems management, like any other management activity, relies on effective forecasting and operational control.

The aim of the book series **Managing Forest Ecosystems** is to present state-of-the-art research results relating to the practice of forest management. Contributions are solicited from prominent authors. Each reference book, monograph or proceedings volume will be focused to deal with a specific context. Typical issues of the series are: resource assessment techniques, evaluating sustainability for even-aged and uneven-aged forests, multi-objective management, predicting forest development, optimizing forest management, biodiversity management and monitoring, risk assessment and economic analysis.

The titles published in this series are listed at the end of this volume.

Landscapes, Genomics and Transgenic Conifers

Edited by

Claire G. Williams
Duke University,
Durham, NC, U.S.A.

A C.I.P. Catalogue record for this book is available from the Library of Congress.

ISBN-10 1-4020-3868-2 (HB)
ISBN-13 978-1-4020-3868-6 (HB)
ISBN-10 1-4020-3869-0 (e-book)
ISBN-13 978-1-4020-3869-3 (e-book)

Published by Springer,
P.O. Box 17, 3300 AA Dordrecht, The Netherlands.

www.springer.com

Printed on acid-free paper

All Rights Reserved
© 2006 Springer
No part of this work may be reproduced, stored in a retrieval system, or transmitted
in any form or by any means, electronic, mechanical, photocopying, microfilming, recording
or otherwise, without written permission from the Publisher, with the exception
of any material supplied specifically for the purpose of being entered
and executed on a computer system, for exclusive use by the purchaser of the work.

Printed in the Netherlands.

Contents

Contributors   vii

Introduction – Claire G. Williams   1

Section I. Pros And Cons For Transgenic Conifer Plantations

Chapter 1.   13
Foresters and DNA
    Jesse H. Ausubel
    Paul E. Waggoner
    Iddo K. Wernick

Chapter 2.
The Question of Commercializing Transgenic Conifers   31
    Claire G. Williams

Chapter 3.
It's Just A Crop?: Public Perception and Transgenic Trees   45
    Steven Anderson

Section II. Genomics Methods, Resources And Alternative Applications

Chapter 4.
Genomics Resources for Conifers   55
    Jeffrey F.D. Dean

Chapter 5.
A New Direction for Conifer Genomics
    Kermit Ritland, Steven Ralph, Dustin Lippert, Dainis Rungis, Jörg Bohlmann   75

Chapter 6.
Using Genomics to Study Evolutionary Origins of Seeds   85
    Eric D. Brenner, Dennis Stevenson

Chapter 7.
Metabolic Profiling for Transgenic Forest Trees   107
    Hely Häggman, Riitta Julkunen-Tiitto

Section III. Viewing Transgenic Conifer Plantations On A Landscape Scale

Chapter 8.
Dispersal of Transgenic Conifer Pollen   121
    Gabriel Katul, Claire G. Williams, Mario Siqueira, Davide Poggi, Amilcare
    Porporato, Heather McCarthy, Ram Oren

Chapter 9.
Gene Flow in Conifers   147
    Jeffry B. Mitton
    Claire G. Williams

CHAPTER 10.
Pines as Invasive Aliens: Outlook on Transgenic Pine Plantations in the Southern Hemisphere — 169
    David M. Richardson
    Remy J. Petit

SECTION IV. ECONOMICS OF TRANSGENIC TECHNOLOGY ADOPTION

CHAPTER 11.
Economic Prospects and Policy Framework of Forest Biotechnology for the Southern U.S.A. and South America — 191
    Frederick W. Cubbage
    David N. Wear
    Zohra Bennadji

CHAPTER 12.
Private Forests and Transgenic Forest Trees — 209
    Mark A. Megalos

SECTION V. GOVERNMENT REGULATIONS AND BIOSAFETY

CHAPTER 13.
Canada's Regulatory Approach — 229
    Anne-Christine Bonfils

CHAPTER 14.
Biosafety of Transgenic Trees in the United States — 245
    Ruth Irwin
    Phillip B.C. Jones

Contributors

ANDERSON, Steven
Forest History Society, 701 Vickers Avenue, Durham NC 27701 USA

AUSUBEL Jesse
Rockefeller University, Program for the Human Environment, 1230 York Ave, Box 234, New York, NY, 10021 New York NY USA

BENNADJI Zohra
Instituto Nacional de Investigación Agropecuaria, Forestry Research Department, Tacuarembó, Uruguay

BOHLMANN Jörg
University of British Columbia, Department of Botany, Vancouver B.C. V6T 1Z4 Canada

BONFILS, Anne-Christine
Canadian Forest Service, Natural Resources Canada
580 Booth Street Ottawa Ontario K1A 0E4 Canada

BRENNER, Eric
New York Botanical Garden, Bronx, NY 10458-5126 USA

CUBBAGE, Frederick
North Carolina State University, Department of Forestry and Environmental Resources, Raleigh North Carolina 27695 USA

DEAN, Jeffrey F. D.
University of Georgia, Warnell School of Forest Resources, Athens Georgia 30602 USA

HÄGGMAN, Hely
University of Oulu, Department of Biology, PO Box 3000, FIN-90014 Oulu Finland

IRWIN, Ruth, Information Systems for Biotechnology, Virginia Tech University, Blacksburg VA 24061 USA

JONES, Phillip
Freelance Biotechnology Writer, Spokane Washington USA

JULKUNEN-TIITTO Riitta
University of Joensuu, Department of Biology, PO Box 111, FIN-80101 Joensuu Finland

KATUL, Gabriel
Duke University, Nicholas School of the Environment and Earth Sciences, Box 90328, Durham NC 27708 USA

LIPPERT Dustin
University of British Columbia, Department of Forest Sciences
2424 Main Mall, Vancouver BC V6T1Z4 Canada

MCCARTHY Heather
Duke University, Nicholas School of the Environment and Earth Sciences, Box 90328, Durham NC 27708 USA

MEGALOS, Mark
Forest Stewardship / Legacy ProgramNorth Carolina Forest Service, 1616 Mail Service Center Raleigh, NC 27699-1616 USA

MITTON, JEFFREY
University of Colorado, Department of Ecology and Evolutionary Biology
Campus Box 334, Boulder, CO 80309 USA

OREN, Ram
Duke University, Nicholas School of the Environment and Earth Sciences, Box 90328, Durham NC 27708 USA

PETIT, Remy
INRA Laboratoire de Génétique et Amélioration des Arbres Forestiers, F-33611 Gazinet Cedex, France

POGGI Davide
Dipartimento di Idraulica, Trasporti ed Infrastrutture Civili Politecnico di Torino, Torino, Italy

PORPORATO, Amilcare
Duke University, Department of Civil and Environmental Engineering, Box 90287, Durham NC 27708 USA

RALPH, Steven
University of British Columbia, Department of Forest Sciences
2424 Main Mall, Vancouver BC V6T1Z4 Canada

RICHARDSON David
University of Stellenbosch, Private Bag X1, Matieland 7602, South Africa

RITLAND Kermit
University of British Columbia, Department of Forest Sciences,
2424 Main Mall, Vancouver BC V6T1Z4 Canada

RUNGIS Dainis
University of British Columbia, Department of Forest Sciences, 2424 Main Mall, Vancouver BC V6T1Z4 Canada

SIQUEIRA Mario
Duke University, Nicholas School of the Environment and Earth Sciences, Box 90328, Durham NC 27708 USA

STEVENSON Dennis
New York Botanical Garden, Bronx, NY 10458-5126 USA

WAGGONER Paul
Connecticut Experiment Station, P.O. Box 1106, New Haven, CT 06504-1106 USA

WEAR DAVID
USDA-Forest Service, P.O. Box 12254, Research Triangle Park, NC 27709-2254 USA

WERNICK Iddo
Environmental Knowledge Management Consulting Group, 116 Meade Avenue Passaic, NJ 07055 USA

WILLIAMS Claire
Duke University, Department of Biology, Box 90338, Durham NC 27708 USA

INTRODUCTION

What is the state of knowledge about transgenic conifer plantations in Northern and Southern Hemispheres? The content of this edited volume addresses this question directly – and indirectly – using language drawn from policy, forest history, genomics, metabolism, pollen dispersal and gene flow, landscape ecology, evolution, economics, technology transfer and regulatory oversight. Although the book takes its title from a Nicholas School Leadership forum held November 17-19 2004 at Duke University, its *de novo* contents move past the forum's deliberations. The result is a trans-disciplinary book composed of 14 chapters written by a total of 31 authors working in North America, South America, Europe and Africa.

The book is designed for policy experts, life scientists, government and business leaders, biotechnology writers and social activists – hence the lexicon at the end. Few decision-makers realize the unprecedented degree to which transgenic technology is now possible on a commercial scale – and that for a research area so underfunded in the public sector, this breakthrough marks a real achievement over a short period of time. But its emergence still so recent that dialogue about the pros and cons are confined to the scientific community. We must open dialogue. Without an outlet or content for public dialogue on the subject of transgenic forests, we can expect alienation of other interest groups. Alienation will lead to heightened clashes in the public policy arena or even radical environmental action. But how to move public deliberation forward? First, we must re-frame the question of transgenic forest commercialization more narrowly than ever before – down to the level of one technology and one taxon.

By re-framing the question, fresh options can surface. Dialogue has been stymied in part by a tendency to frame the issues for all transgenic plants too broadly. Transgenic forest trees and food crops have more differences than similarities. To see this misfit, compare life history, production systems and domestication history.[1] Does this mean that we should frame the question for forest biotechnology as a whole? No. This is still too broad. Forest biotechnology refers to a smorgasbord of technology choices ranging widely in benefits and risks. Benefits and risks for cloning urban forest trees or transgenics for heritage tree restoration are distinct from those associated with transgenic forest plantations grown for wood products. Choosing transgenics technology for wood production frames the technology side of the question. But what about choice of taxon?

Conifers are the right choice for framing issues related to transgenic technology. Lumping transgenic hardwoods and conifers together is still too broad because it obscures a major evolutionary divide between angiosperm and gymnosperms. As gymnosperms, conifers are separated from hardwood angiosperm trees by 100+ million years of evolutionary progression. The evolutionary divide between conifers and hardwoods have present-day consequences, affect the type of wood, its product value and most profoundly, the reproductive system and subsequent gene flow. Conifers are the more logical choice because they have a profound economic impact on the world's wood supply in developed countries. Of the 550 living species of gymnosperms, only a few conifers are used for plantation forestry and only a handful of pine species have being developed as transgenic plantation species. Field trials started within the last seven years but no transgenic pine plantations exist at this time. Pines are mostly indigenous to the Northern Hemisphere, so the potential for introducing transgenic pines into the Southern Hemisphere adds an interesting duality into the dialogue.

Having now framed the issue as transgenic conifers, the goal of our book is to provide content for science-based public deliberations on the subject of genetic composition of our future forests. The book does not represent a compendium of all facets of ongoing research; instead its goal is trans-disciplinary balance. Of the five sections in this book, Section I provides an overview of pros and cons of transgenic conifer plantations followed by Sections II and III on conifer genomics and ecology research followed by Section IV technology adoption scenarios and Section V on the status of regulatory oversight of transgenic forest trees in Canada and the United States.

Section I. Pros and Cons for Transgenic Conifer Plantations

Section I is a presentation of pros and cons on whether to commercialize transgenic conifers. Its chapters draw from public policy, natural resources management, forest biology and history. In the opening chapter, Jesse Ausubel, Paul Waggoner and Iddo Wernick address the provocative question of whether editing a few bytes of the genetic message for a tree to fit human desires will do harm or good. They argue that providing more wood via transgenic forest trees is part of a social bargain arising from burgeoning human population growth rates and concomitant demands for a higher quality of life. In Chapter 2, Claire Williams takes the opposing position, that transgenic forest trees have not been subjected to either rigorous risk or benefits analysis and thus could compromise adaptation of our indigenous conifer forests in North America. She discusses several future scenarios, ending with a proposal for a technology trust to spur completion of benefits-risks analyses. In Chapter 3, Steve Anderson draws on rich examples from forest history to conclude that it is only a matter of time before transgenic plantations become accepted as part of the technology portfolio of the high-yield forest.

Section II. Genomics Methods, Resources and Alternative Applications

Section II describes the immensity of new knowledge flowing from conifer genomics. Its primary message is that the potential of genomics is becoming far broader than simply finding new genes for transgenic conifers. Alternative applications are emerging in several countries and this trend is consistent with forestry's historical tendency to diverge from agricultural innovation. As such, the position of conifers within each new wave of genomics policy deserves a closer look because justification seems to rise with each new wave of public-sector funding.

The first wave, sequencing the entire human genome, was justified by a greater knowledge of human diseases. Here, conifers received spare attention due to taxol and other medicinal derivatives from the Taxaceae. The second wave, directed at plant genomics, was justified as a means for increasing the world's food supply. This effort included improving wood quality, as shown by the conifer genomics resources reviewed in Chapter 4 by Jeff Dean. Garnering these resources has been no trivial feat because conifers have colossal genomes four to seven times larger than the human genome. The next step is to sequence the entire *Pinus taeda* genome, roughly 21,000 megabases of DNA sequence.

In Chapter 5, Kermit Ritland and his co-authors are concerned with problems facing western Canadian forests. Here, forest genomics is coming into its own. The vast forests of British Columbia are characterized by many indigenous conifer species, long rotation ages and a prevalence of public forest ownership. Elevated exposure to exotic pests and pathogens introduced by global trade is mandate for better forest protection. Consistent with this national mandate, their research is directed at protecting forest health using genomics of plant defense genes. Their approach raises the interesting possibility of triggering defense gene expression in non-transgenic trees under siege from pest or pathogen.

The third wave of genomics is directed at ecology and evolution. Here, conifers and other gymnosperms become central to the question of early seed plant evolution. As the most ancient of extant seed plants, conifers and other gymnosperms provide inferences about the progymnosperms, extinct ancestors for angiosperm and gymnosperm alike. In Chapter 6, Eric Brenner and Dennis Stevenson present a set of hypotheses under test by the Gymnosperm Genome Project, a joint effort shared among the New York Botanical Garden, New York University and Cold Spring Harbor Laboratory. Coupled with high throughput genomics, studying conifers and their gymnosperm relatives is possible by comparing hundreds or even thousands of expressed genes during the transition from vegetative to reproductive structures. A rich knowledge base about seed plant evolution is emerging from the Gymnosperm Genome Project.

Metabolic profiling brings a new genomics tool to the study of transgenic conifers as discussed in Chapter 7 by Hely Hägmann and Riitta Julkunen-Tiitto. Metabolic profiling measures the small-molecule inventory expressed in a specific tissue at a specific point in time. Such metabolites can be compared between transgenic and non-transgenic plants with identical genetic backgrounds. This is important in testing secondary or unintended effects. These occur if a transgenic plant shows a new expression pattern outside its target metabolic pathway compared to its non-transgenic control. Presence of secondary effects raises the question of altered fitness in the transgenic organism and can pinpoint the need for additional data collection at the whole-organism level. Otherwise, measurements at the whole-organism are too laborious because results can vary with each genetic transformation event, even if it is the same DNA construct being inserted into the same genetic background. This can be further complicated by different methods of genetic modification, use of different genetic backgrounds using different DNA constructs or even novel combinations of DNA constructs. The authors of Chapter 7 emphasize that metabolic profiling, as a rapid, high-resolution alternative, holds potential value for other forestry applications in addition to comparing transgenic versus non-transgenic conifers.

Section III. Viewing Transgenic Conifer Plantations on a Landscape Scale

Section III addresses the potential interface between transgenic conifer plantations and less managed ecosystems. Northern Hemisphere countries, as the indigenous source of many commodity conifer species, are concerned with the scale of gene flow from transgenic forests to surrounding conspecifics and close relatives. In Chapter 8, Gabriel Katul and co-authors show windborne pine pollen disperses on a scale of kilometres and that this distance can be predicted with a high degree of certainty using fluid dynamics principles developed for environmental engineering applications. The distances for pine exceeds the distances for annual crops by one or more orders of magnitude. The open question then becomes how long does the pollen remain viable and if viable, how competitive is long-distance pollen compared to local sources. In any event, biocontainment zones used for transgenic food crops are not effective for transgenic pines.

As a part of the gene flow process, pine seeds and pollen move by way of two separate processes. The first process is local neighborhood dispersal (LND) which accounts for 99% of the seeds and pollen falling near source. The second process, long-distance dispersal (LDD), accounts for a tiny fraction (1%) of escaped seeds or pollen yet poses the greatest ecological concern. This latter process causes the most concern as LDD seeds and pollen are vertically uplifted above the forest canopy by air currents then moved by wind on the order of kilometers from source.

The relationship between LDD seeds or pollen and actual gene flow is hard to measure. Gene flow distances may be shorter than dispersal distances, as suggested in Chapter 9 by Jeff Mitton and Claire Williams but the question is whether the scale is still sufficient that corporate landowners adopting transgenic technology must consider neighboring public and private landowners. Pollen dispersal is first of many steps in the gene flow process. Conifers, as outcrossing organisms, have long-distance gene flow rates which can be monitored using polymorphisms in nuclear and preferably by organellar DNA but this has not be applied to the question of transgenic conifers yet. Unlike angiosperms, conifers have a long interval between pollination and fertilization during which pollen growth can halt altogether. These and other unusual reproductive features mean that model angiosperm plants provide little insight into gene flow in conifers. Movement of transgenic pollen and seeds remains the most serious deterrent to commercial-scale use of transgenic conifer plantations in the North Hemisphere.

Some believe that limiting transgenic conifer plantations to Southern Hemisphere countries is the best solution because here pines are exotic introductions and gene flow to indigenous forests cannot be a problem. This idea is challenged in Chapter 10 by David Richardson and Remy Petit who point out seed dispersal and subsequent colonization from non-transgenic pines in parts of the Southern Hemisphere are already a problem and that molecular domestication could compound the problem further. They point out that pines transgenic for pest- and herbicide-resistance in particular could create new invasion problems for treeless ecosystems in the Southern Hemisphere. They draw on the southern African experience with invasive pines. Use of transgenic pine plantations is consistent with a 100-year history of exotic pine introductions into the Southern Hemisphere but ecological impact may not be nil.

Section IV. Economics of Transgenic Technology Adoption

Section IV addresses potential scenarios for future transgenic conifer technology adoption. In Chapter 11, the question of transgenic conifers in Southern Hemisphere continues. Authors Fred Cubbage, David Wear and Zohra Bennadji compare economics of biotechnology adoption between the American South and South America. Drawing on the Uruguay experience, they note some resistance to use of transgenic conifers. They also predict that the first transgenic pine plantations will eventually be planted as exotics in South American countries by private-sector landowners. Even with strong economic incentives in place, these authors identify some emerging cross-trends which could slow or hasten transgenic forest tree technology adoption in South America. Multinational corporations have acquired land or strategic partnerships in the forest-rich South American countries but land prices are rising. At present, international forest certification programs deter transgenic plantations but this is likely to change as new certification agencies organized at the national level emerge. Flux in forest certification standards is a pivotal determinant for adoption of transgenic forests because at present, certification tends to block transgenic wood products from entering world markets. In contrast to regulation, certification is an elective process driven by the profit motive.

In Chapter 12, Mark Megalos addresses the same question for a different segment of forest landowners in the American South: individuals, not corporations. This region supplies 16% of the world's wood supply and individuals constitute the largest group of landowners and the question is whether they will adopt or forego molecular domestication technology. Megalos draws on innovation theory as a means of predicting that the landowners will accept specialty clonal forest trees for non-timber use but that probability is low that even wealthy individuals will be early-adopters of transgenic conifers. Whatever the outcome, molecular domestication of the family forest will mark a turning point in public dialogue.

Section V. Government Regulations and Biosafety

Section V addresses status of regulatory oversight in Canada and in the United States. Anne-Christine Bonfils, author of Chapter 13, shows that Canada, as a nation that relies more heavily on its forest products than its U.S. neighbor, is proceeding slowly with caution. Only one transgenic forest tree field trial is planted at this time. Two western provinces have enacted voluntary planting moratoria or restrictions against transgenic forest trees on Crown lands. In contrast, Ruth Irwin and Phillip Jones show biosafety of transgenic forests in the U.S. remains closely tied to agricultural policy and regulatory oversight. This means biocontainment zones and reproductive sterility are favored as deterrents for all transgenic plants but neither method fits for conifers. Regulatory reform for forest trees in the U.S. is underway and the costs are likely to be high.

In summary, one must ask if managing transgenic conifers will exceed human limits to biotechnology governance. Once released, movement of transgenic conifer pollen or seeds cannot be halted or even truly regulated. Look at the pine pollen cloud that settles on your windshield each spring. It is a rebuke to technology's promise, a murky reminder that we may have reached "the ragged fringes of human understanding – the unknown, the uncertain, the ambiguous and the uncontrollable."[2]

Claire Williams, Editor
April 22, 2005

---

[1] Williams C. G. 2005. Framing the issues on transgenic pine forests. Nature Biotechnology 23:531-532.

[2] Quote from Jasanoff, S. 2003. Technologies of humility: citizen participation in governing science. Minerva 41: 223-244.

The editor gratefully acknowledges awards from the Canada-U.S. Fulbright Program, USDA-Forest Service's Southern Research Station and the National Science Foundation (DEB #0454650).

INTRODUCTION

A lexicon of life science terms frequently used in reference to transgenic conifers.

ANGIOSPERM: Also known as flowering plants, these were the last of the seed plant groups to evolve, appearing over 140 million years ago during the late Cretaceous. All angiosperms produce flowers and fruits. The angiosperms (angios = hidden) produce modified leaves grouped into flowers that in turn develop fruits and seeds. Seeds include a zygote nourished by a triploid endosperm. Today there are ~ 235,000 known living species.

CHROMOSOME: A chromosome consists of a long, continuous strand of DNA and associated proteins. It resides within the nucleus of a cell. Each parent contributes one chromosome to each offspring, so an individual receives half of its chromosomes from its mother and half from its father.

DNA CONSTRUCT: An artificially constructed segment of nucleic acid that is going to be injected into resident DNA of the host organism's tissues. Transgenic constructs are new combinations of genes.

DOMESTICATION: A process by which behavior, life cycle, or physiology of animals, plants, and other organisms has been altered as a result of being under human control for multiple generations. A domesticated plant, strictly defined, is one whose reproductive success depends on human intervention.

FORESTRY: The profession of creating, managing, using, and conserving forests for human benefit.

FUNCTIONAL GENOMICS: Functional genomics links gene expression to cell or tissue function (or lack of function), thus aims to find the biological function of gene sets and determine how their gene products work together. Functional genomics expands the study of individual genes and proteins to a more systematic approach of studying all expressed genes or their proteins at once.

GENES: A gene is a region of DNA on a chromosome which codes for biological information. A gene refers to a functional unit composed of coding DNA sequences, non-coding introns and its regulatory DNA sequences. A gene corresponds to a sequence used in the production of a specific protein or another nucleic acid, RNA.

GENE FLOW: Exchange of genes between species, usually taking place by reproduction, such as cross-pollination, or directly through horizontal gene transfer.

GENE POOLS: Refers to genetic information segregating within a species. The genetic information is coded by genes which reside on chromosomes. Variants of genes, defined as alleles, range from common to rare.

GENETIC MODIFICATION: Eliminating or adding copies of specific genes often from the same organisms or other organisms through using genetic transformation and other molecular biology techniques. Refers to a series of techniques used to transfer the genes from one organism to another or to alter the expression of an organism's genes. For example, a genetic modified plant may produce a new protein or be prevented from producing its own proteins. Also known as gene splicing, recombinant DNA (or rDNA) technology or genetic engineering.

GENETIC TRANSFORMATION: see Genetic modification.

GENOME: The entire complement of genetic material in a cell. This definition includes nuclear chromosomes in addition to organellar DNA in chloroplasts and mitochondria.

GYMNOSPERM: Gymnosperms are a class of plants which bear seeds but have neither fruits nor flowers. Seeds have a zygote nourished by a haploid female gametophyte (i.e. no triploid endosperm). Gymnosperms first occurred in the fossil record during the Paleozoic Era and became the dominant vegetation by the early Mesozoic Era. Dominance was subsequently lost with the evolution and spread of angiosperm plants. Living gymnosperm species includes conifers as the major group in addition to cycads, ginkgo, and gnetales.

HETEROZYGOSITY: The state of having two different alleles at a single gene locus residing on a chromosome.

HORIZONTAL TRANSFER: Transfer of genes between organisms without reproduction. Horizontal or lateral transfer of genes has occurred over millions of years without sexual reproduction or genetic engineering techniques.

METABOLIC PROFILING: Refers to the cumulative effects of all the expressed and modified proteins within a cell or tissue type. It results from the expression of the genome and proteome in response to the cellular environment.

SILVICULTURE: The science and practice of tending forests for human use.

SOMATIC EMBRYOGENESIS: Clonal propagation of a single individual or genotype by culturing undifferentiated cells from immature embryos.

STRUCTURAL GENOMICS: Describing physical aspects of the genome through the construction and comparison of sequences and genetic maps in addition to gene discovery, localization and characterization. Can also include describing large numbers of proteins (see METABOLIC PROFILING).

TRANSGENE: A foreign gene(s) or DNA construct transferred into another organism using genetic modification methods.

Section I

Pros and cons for transgenic conifer plantations

CHAPTER 1

FORESTERS AND DNA

JESSE H. AUSUBEL

PAUL E. WAGGONER
Connecticut Agricultural Experiment Station, New Haven CT, USA

IDDO K. WERNICK
The Rockefeller University, New York NY, USA

Abstract. Would editing a few bytes of the genetic message for a tree to fit human desires do harm or good? To meet demands of larger populations and changing diets, farmers have used a series of innovations to lift yields and thus reduce the area of land needed to support a person. Since 1950 rising yields have stabilized land for agriculture and now promise a Great Restoration of nature on land spared. Foresters have also lifted yields and could lift them much higher, thus sparing natural forests while meeting demand for wood products, whose growth is anyway slowing. While weak demand, numerous worries, and vague promises will slow penetration of genetically modified trees, any technology that improves spatial efficiency has appeal, and editing DNA could lift yields. Both farmers and foresters must work precisely, using fewer hectares and more bits. Fortunately, foresters have several decades in which to test and monitor their practices before genetically modified trees will diffuse widely.

1. INTRODUCTION

The decoding of DNA messages produces magnificent structures, perhaps none more magnificent than a tree. Our question is, would editing a few bytes of the message for a tree to fit human desires do harm or good?

Forests do an admirable job of collecting solar energy and storing it in stable chemical form. The problem is spatial. The collected energy of trees is spatially dilute and in forms awkward to handle. Harvesting requires lots of manpower and sophisticated machinery. Further, harvesting is only the beginning. The bulky, round, solid biomass harvested from trees is unsuitable for convenient transport techniques developed for oil and gas, or for molding techniques matching plastics.

Consequently, Americans and most of the rest of the world have been abandoning wood in favor of the so-called fossil fuels and the plastics they become. Since 1800, wood and hay have plunged from a 90 percent market share of primary energy to less than 10 percent for the world and less than 2 percent for the United States in the year 2000. During the twentieth century, plastics replaced much timber, too (Figure 1). Even paper now struggles to hold its share of a dollar spent, as ubiquitous flat-screen monitors and e-books threaten to replace it. Paradoxically, in large tropical regions, people hack down forests resplendent with life for little gain of useful material or income.

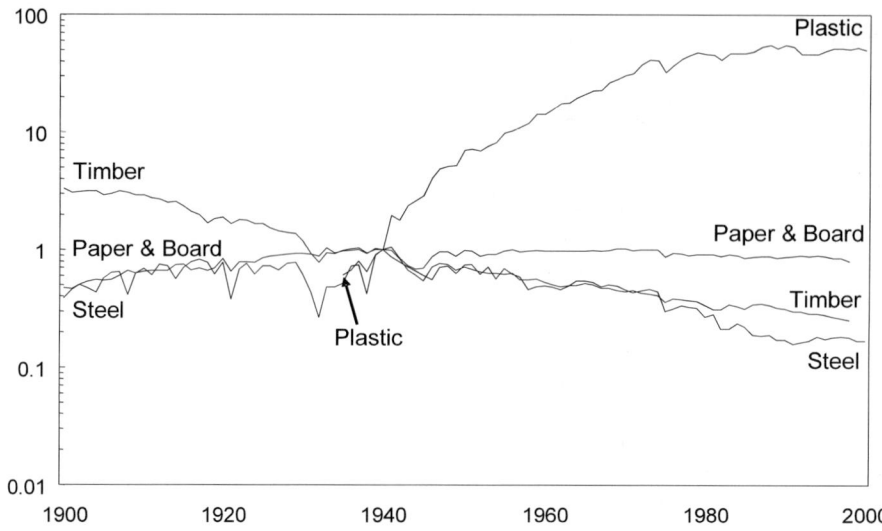

Figure 1. U.S. intensity of materials use, 1900-2000. The y-axis shows kilograms per constant dollar indexed to 1940 on a semi-log scale. For example, consumption data for timber are divided by GDP in constant 1996 dollars and indexed to their value for 1940. Data sources: Ince, 2000; U.S. Geological Survey, 2004; Chemical & Engineering News, 2001; Wernick and Ausubel, 1995.

The idea of a tree remains supremely elegant: living hardware automatically produced and maintained by coding of genetic messages whose raw materials are collected mainly from the atmosphere (MARCHETTI 1979a). Could science help forestry survive while at the same time sparing the forests? Our answer is yes, through improving spatial efficiency, a path that leads foresters ultimately to DNA.

2. THE FARMER'S PLOT

Let us begin to grasp spatial efficiency by considering first the much more numerous cousins of foresters, farmers, who have edited DNA messages longer. Analysis of farming shows a coherent pattern of evolution from Neolithic times up to our new millennium (MARCHETTI 1979b). Farmers have exploited technical advances fundamentally for intensification, to increase the specific productivity of land, to earn more from a plot. Yields per hectare measure the productivity of land and the efficiency of land use. Low yields squander land, and high yields spare land.

Beginning as hunter-gatherers, humans differed little from other animals. We met the pressure to grow by extending our geographical habitat. But we also extended our range of digestible foods, achieving great breakthroughs with energy. Plants defend themselves against predators with a panoply of armor and weapons. The most important are chemical and tend to make the plant indigestible and occasionally poisonous. Animals developed other defenses. Human genius was to apply thermal treatment to upset or destroy the delicate organic chemistry of defense. Boiling softens flinty rice and maize, and ovens convert pasty wheat into bread. Seven minutes of boiling soybeans denatures the trypsin inhibitor that would otherwise render tofu useless to us. Fire revolutionized food, permitting digestion of much plant material and seeds in particular, and in most cases improving taste as well. Fire extracted more nourishment from the same acres as well as nourishment from acreage formerly yielding none.

Farming in turn amplifies the production of biological material that we can assimilate directly or by thermal treatment. Humans ally with certain plants by collaborating in their reproductive cycle and by fighting their natural enemies. We put ourselves first among selective forces, picking the plants most profitable from our point of view.

What then has driven the laborious development of agriculture? After filling available geographical niches, the only way to expand is intensification. Like fire, agriculture essentially reduces the amount of land needed to support a person. The fruits of agriculture consequently support the human drives to multiply and to increase consumption.

Draft animals were the first big advance. Draft animals did not reduce human toil. Peasants with animals sweat as much as those without, as Paul Bunyan and his Blue Ox, Babe, would attest. Nor did oxen, buffalo, and horses drastically lift the productivity per worker, though an Iowan with a team could till far more than an Incan with a spade. Draft animals did increase the specific productivity of the land. Ruminants are the most successful symbiotic draft animals, consuming little human food but digesting roughage and poor pasture, as they extract energy from cellulose and manage nitrogen in the rumen's flora. Still, draft animals take land. In some farming systems, oxen, buffalo, and horses eat the yield of one-quarter of the land.

After World War II, the automobile industry produced solid, cheap, dependable tractors that pulled as powerfully as ten teams of oxen. Tractors proportionately increased the productivity of labor, but without substantially intensifying production. By draining land, they extended farming, and by freeing land that had grown timothy and oats for draft animals, they shrank it. Tractors released workers from the farms, but alone they grew little more corn per hectare. The story of forest machinery is the same.

Chinese agriculture represents an important counterpoint, an improvement of yield per hectare that saves land rather than labor. By 1900, without machines but using a thousand bioinformatic tricks, Chinese farmers reduced to 100 square meters the amount of land needed to support a person. Compare this space, about equal to a one-bedroom American apartment, to a few square kilometers for a hunter-gatherer. The difference is a factor of 10^4, or 10,000 times in intensification.

The ecological systems farmers create, although often visually appealing, bear no resemblance to any natural ecosystem, if only because of great structural simplification. Equilibrium and resilience tend to be lost, and the spatially efficient system becomes unstable and challenging to manage. No farm reproduces itself year after year without a farmer. The wits and toil of almost half the Chinese population are still employed to keep their farms going.

Because few societies could approach the summit the Chinese reached by labor, for most of the world farm evolution could continue only with a qualitative breakthrough. It came, like cooking, with the introduction of external energy, in this case fossil fuels. Starting around 1900, we not only tamed machines for the same purposes as draft animals but also started to synthesize as well as mine chemicals that hugely increased yields.

The effect of chemicals fits the master trend of intensification perfectly. Fertilizers, most obviously, are intensifiers. In the form of dead fish and animal droppings such as guano and manure, they have always been used to increase yield per plot. The external energy of fossil fuels permitted massive, economical, and convenient nitrogen synthesis beginning about 1950. Geneticists called crop breeders also began to deliver plants that yielded more product per hectare.

The diffusion of the innovations made average U.S. grain yields, rising very slowly for two centuries until about 1940, leap fourfold by 2000. In fact, on all continents during the past half century, ratios of crops to land for the world's major grains—corn, rice, soybean, and wheat—have soared (WAGGONER 1996). The breakthroughs in harnessing external energy and editing DNA allowed farmers to spare land, as production grew by intensification much faster than population. By tripling yield since the mid-1960s, India's wheat farmers, for example, have spared about 50 million hectares, about 80 percent of the present area of India's woodlands (Figure 2).

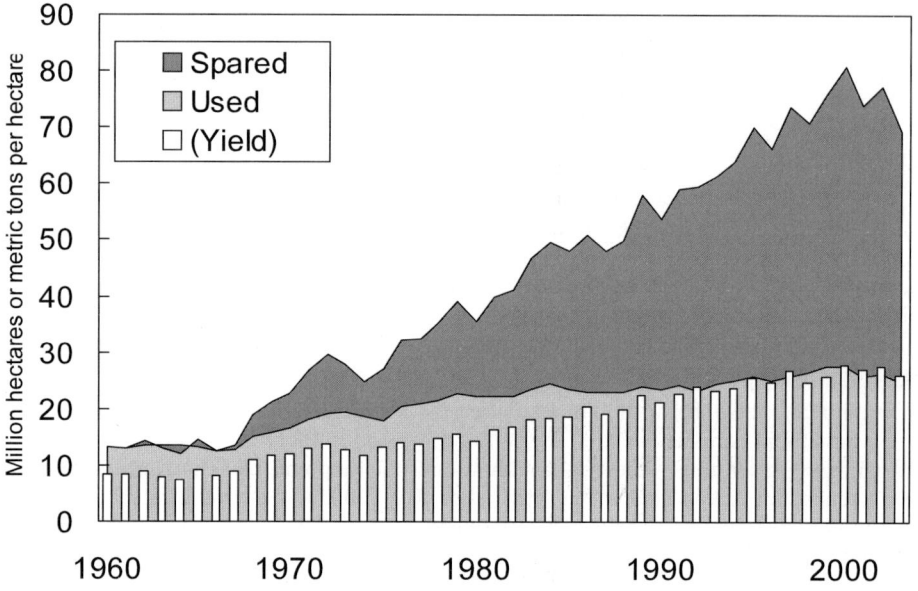

Figure 2. Land spared by Indian farmers by raising wheat yields. The white bars show annual yields in metric tons per hectare. The light gray area shows land actually farmed in million hectares, while the dark gray area shows the additional hectares that would have been needed to produce actual output had yields remained at the level of the early 1960s. Source: Waggoner, 1996 (updated).

3. THE GREAT REVERSAL

For centuries, globally, cropped land expanded and cropland per person rose, as more people sought more protein and calories. But about 1950, by rapidly lifting the specific productivity of land, the world's farmers stopped plowing up nature, and the worldwide area of cropland per person began dropping steeply. While diet improved, the land used to feed a person halved from almost half a hectare in mid-century to about one-quarter in 2000, signalling a great reversal of human extension into nature (AUSUBEL 2001). Per hectare, the global Food Index of the Food and Agriculture Organization (FAO) of the UN, which reflects both quantity and quality of food, rose 2.3 percent annually between

1960 and 2000 (Food and Agriculture Organization, various years). In the United States in 1900, the protein or calories raised on one Iowa hectare fed four people for the year. By the year 2000, a hectare on the Iowa farm of master grower Francis Childs could feed 80 people for the year, comparable to the most intensive Chinese agriculture (National Corn Growers Association). The Chinese, of course, kept lifting the comparison as they lifted cereal yields 3.3 percent per year between 1972 and 1995. A cluster of innovations including not only tractors, chemicals, and seeds but also irrigation, joined through timely information flows and better organized markets, raised the yields to feed billions more without clearing new fields.

High-yield agriculture need not tarnish the land. The key is precision agriculture. This approach to farming relies on technology and information to help the grower use precise amounts of inputs—seeds, fertilizers, pesticides, and water—exactly where they are needed. We have mentioned two revolutions in agriculture in the twentieth century. First, the tractors of mechanical engineers saved not only the oats that horses ate but also labor. Then chemical engineers and plant breeders made more productive plants. The present agricultural revolution comes from information engineers, some of whose code is DNA. What do the past and future agricultural revolutions mean for land?

The agricultural production frontier remains open. If during the next 60 to 70 years, the world farmer reaches the average yield of today's U.S. corn grower, the 10 billion people then likely to live on Earth will need only half of today's cropland. This will happen if farmers maintain on average the yearly 2 percent worldwide growth per hectare of the FAO Food Index, slightly less than the record achieved since 1960. Even if the rate slows to half, an area the size of India, more than 300 million hectares, could revert from agriculture to woodland or other uses (WAGGONER AND AUSUBEL 2001).

Meanwhile, the unnecessarily high cost in energy of modern agriculture should be reduced. The cost can be split between machines and chemicals. In energy terms, they represent about equal inputs. Most of the work of the machines goes into tillage, whose main objective is to kill weeds. Low-tillage techniques are, however, improving and spreading. Low-tillage farming uses herbicides to control weeds after seeds are planted by injection in the soil.

Herbicides and pesticides that now operate on the principle of carpet bombing are moving progressively to the hormonal and genetic level and require less and less energy as the amounts of product needed are reduced. The big slice of energy taken for fertilizers, nitrogen in particular, could be produced by grains capable directly, or through symbiosis with bacteria, of fixing nitrogen from the atmosphere. Improved tractors, minimum tillage, targeted herbicides and pesticides, and an extended capacity for nitrogen fixation might reduce farmers' energy consumption by an order of magnitude.

Lifting yields while minimizing environmental fallout, farmers can offer hundreds of millions of hectares for a Great Restoration of nature. The strategy is precision agriculture. MARCHETTI (1979b) describes it as more bits and fewer kilowatts.

4. LAND NEEDED FOR WOOD

Farmers may no longer pose much threat to nature. What about lumberjacks? As for food, the area of land needed for wood begins with a multiple of population and income, and then continues with the ratio of the wood products to the economy measured as gross domestic product (GDP). Let us focus on industrial wood—cut for lumber, plywood and veneer, pulp for paper, and fuel—and on the United States, always a pioneer and exemplar in resource use, good and bad.

Between 1900 and 2000, the national use of timber products grew about 70 percent. Meanwhile, at the end of the century, Americans numbered more than three and a half times as many as at the beginning, and an American's average share of GDP had grown nearly fivefold. Had timber consumption risen in constant proportion to population and income, Americans would have consumed 16 times as much timber in the 1990s as in 1900, not a mere 70 percent more (WERNICK et al. 1998).

Industrial ecologists call a ratio of material to GDP its intensity of use. Because the annual percentage change of GDP is the sum of the changes in population and an individual's share of GDP, a constant intensity of use means consumption is rising in step with the combined rise of population and personal GDP or income. A constant intensity of timber use would mean timber played the same role in the economy in 2000 as in 1990 or 1900.

Practically, what lowered the intensity of timber use, or the ratio of timber products to GDP? For lumber, its replacement during the century by steel and concrete in applications from furniture and barrels to railroad ties and lath lowered the intensity of use. Living in the stock of existing houses and prolonging the life of timber products by protecting them from decay and fire lowered it. For pulp, more widespread literacy and the shift to a service economy raised the intensity of use in the early twentieth century, and then television and the Internet replaced newspapers, lowering the intensity of use.

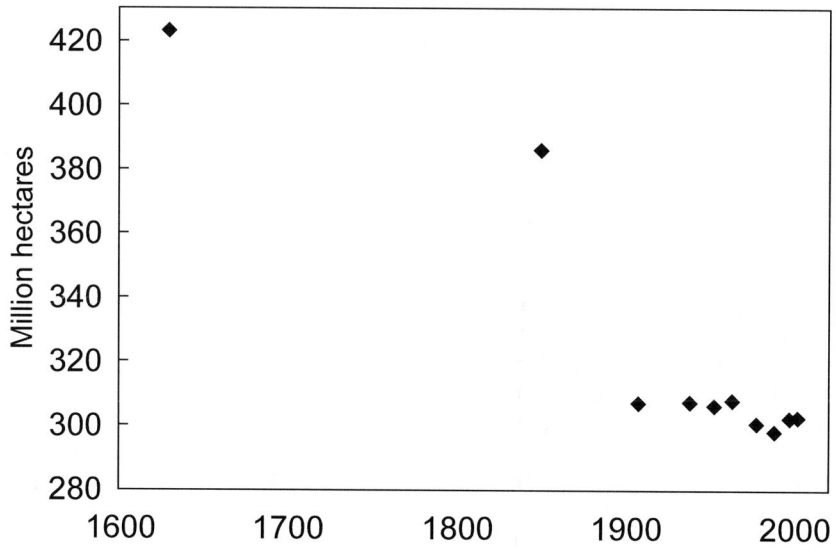

Figure 3. U.S. forest land area, 1630-2002. Data sources: Fedkiw, 1989; USDA Forest Service, 1997 and 2002.

Large areas formerly cleared have regenerated in New England and the upper Great Lakes states. Overall, in the United States, timberland plus forest area reserved for wilderness increased 9 percent during the period from 1992 to 2002, about 1 percent per year. Furthermore, reversing hundreds of years of depletion, the volume of wood on American timberland has risen about 40 percent since 1952 (Figure 4). Analysts have observed such a transition from deforestation to reforestation and afforestation in scores of countries (MATHER et al. 1999; MYNENI et al. 2001; UN ECE/FAO 2000).

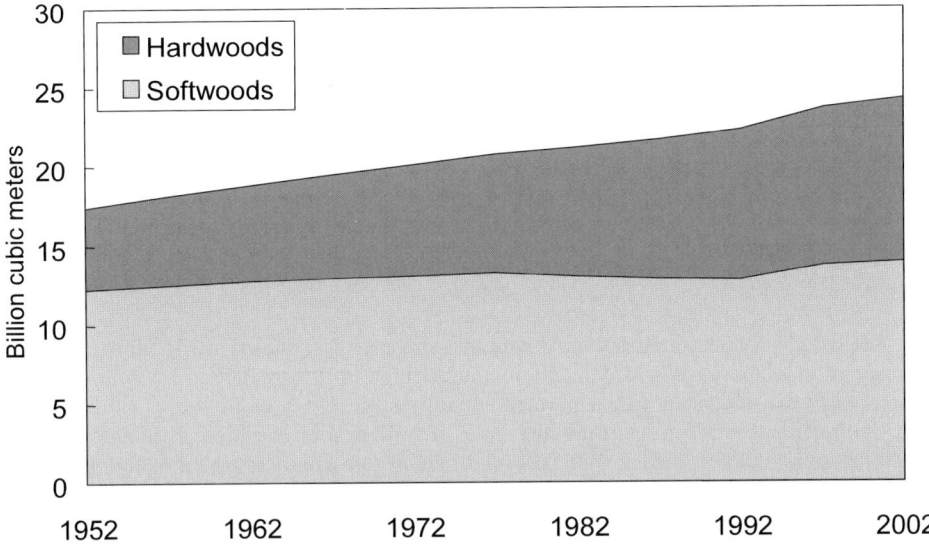

Figure 4. U.S. forest volume, 1952-2002. Data sources: USDA Forest Service 1992, 1997, 2002.

Because the average contemporary American annually consumes only half the timber for all uses as a counterpart in 1900, U.S. forests expanded rather than shrank. Amidst a housing boom, demand for lumber has become sluggish, and world consumption of boards and plywood has actually declined in the last decade. Even the appetite for pulpwood that ends as sheets of paper has levelled.

Meanwhile, more efficient mills carve more value from the trees people cut (Figure 5). Because waste is costly, the best mills, which operate under tight environmental regulations and the gaze of demanding shareholders, already use everything but the whine of the saw as meatpackers once used everything but the squeal of the pig. In the United States, for example, leftovers from lumber mills account for more than a third of the wood chips that are turned into pulp and paper, and what is still left is burned for power. In 1970, American consumers recycled less than one-fifth of their paper, while today the world average is double that. Recycling closes leaks in the paper cycle.

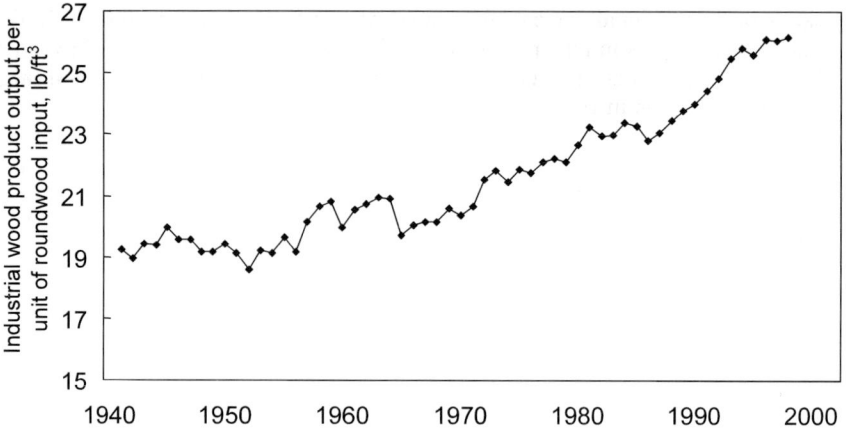

Figure 5. Industrial wood productivity in the United States, 1941-1998. Date source: Ince, 2000.

The wood products industry has learned to increase its revenue while moderating its consumption of trees. World demand for industrial wood, now about 1.5 billion cubic meters per year, has risen only 1 percent annually since 1960, while the world economy has multiplied at nearly four times that rate. If millers improve their efficiency, and if manufacturers deliver higher value through the better engineering of wood products while consumers recycle and replace more, demand for timber in 2050 could be only about 2 billion cubic meters per year and thus permit reduction in the area of forests cut for lumber and paper.

5. SKINHEAD EARTH OR GREAT RESTORATION?

The permit, as with farming, comes largely from lifting yield. The cubic meters of wood each hectare grows each year provide large leverage for change. Like fishers and hunters, foresters for centuries hunted and fished out local resources and then moved on, returning only if trees regenerated on their own. The effect was to shave Earth's forests from about 6 billion hectares 8,000 years ago to about 2.3 billion hectares of nonindustrial forests and 0.9 billion hectares of industrial forests now. Most of the world's forests still deliver wood the old-fashioned way, with an average annual yield of perhaps 2 cubic meters of wood per hectare.

Fortunately, industrial foresters are rising to the challenge of spatial efficiency. Forest yields have grown steadily during the past 50 years, as a series of innovations, mechanical, chemical, and informational, have diffused through the forestry sector (Figure 6). Yields have multiplied six times in pine plantations in the southern United States. Logically, the area of pine plantations has climbed steeply, in the South from about 800,000 hectares in 1950 to about 14 million in 2000.

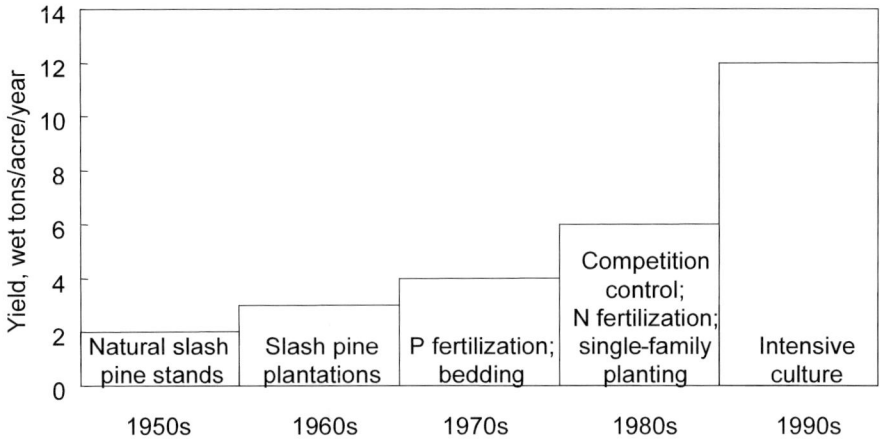

Figure 6. Sources of rising yield in pine plantations in the southern United States. Source: Wann and Rakestraw, 1998.

At likely planting rates, at least one billion cubic meters of wood—half the world's supply—could come from plantations by the year 2050 (SEDJO2001). Seminatural forests that regenerate naturally but are thinned for higher yield could supply most of the rest. Small-scale traditional "community forestry" could also deliver a small fraction of industrial wood. Such arrangements, in which forest dwellers, often indigenous peoples, earn revenue from commercial timber, can provide essential protection to woodlands and their inhabitants.

More than a fifth of the world's virgin wood is already produced at yields above 7 cubic meters per hectare. In Brazil, Chile, and New Zealand, plantations of selected species and varieties sustain yearly growth of more than 20 cubic meters per hectare. In Brazil hardwood eucalyptus good for some kinds of paper delivers more than 40 cubic meters per hectare, and the Aracruz Cellulose Company has recorded yields greater than 70 cubic meters per hectare per year. In the rainy Pacific Northwest and British Columbia, hybrid poplars deliver 50 cubic meters per hectare. The requisite informational innovations are increasingly biological.

If yield remains at the "natural" rate of 2 cubic meters per hectare, by 2050 lumberjacks will regularly saw nearly half the world's forests (Figure 7), a dismal vision of a chainsaw every other hectare, "Skinhead Earth" (VICTOR AND AUSUBEL 2001). The spatial efficiency of higher yields, however, will spare forests. Raising average yields 2 percent per year would lift growth over 5 cubic meters per hectare by 2050 and shrink production forests to just about 12 percent of all woodlands—a Great Restoration. Today's 2.4 billion hectares used for crops and industrial forests expand to 2.9 billion hectares on Skinhead Earth, while in the Great Restoration they contract to 1.5 billion.

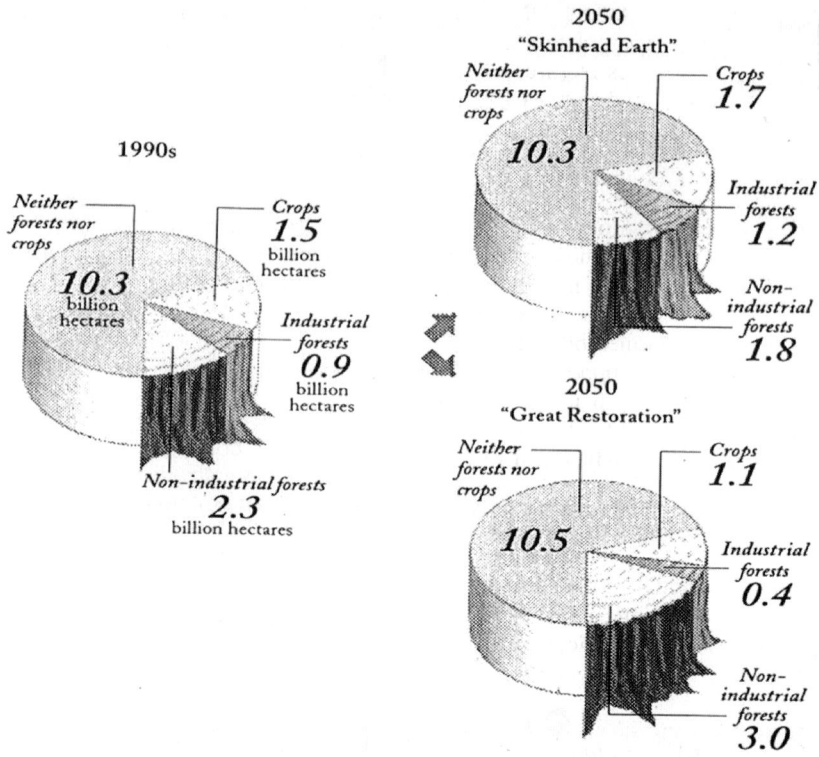

Figure 7. Present and projected global land use and land cover. Source: Ausubel, 2002.

6. DNA WORRIES

Facing paths toward Skinhead Earth or Great Restoration, we must now address the controversy at the root of forest genomics. Would editing a few bytes of the message for a tree in the same way the messages for food crops have already been edited do harm or good? We have clearly declared ourselves for lifting yields and intensification. As for means, anything that works with minimal fallout interests us. The foresters in Brazil and British Columbia, like Chinese rice farmers, have achieved high yields with a bunch of clever tricks, many cellular and genetic, but few would be labelled "genetic engineering" in the formal sense of gene-splicing. In fact, the breeding and management that transformed crop production in the twentieth century have scarcely been tapped in forestry and could lift it for several decades. Still, DNA is the sanctum sanctorum of biology, and now that humans have unlocked it, humans should consider the consequences. Evaluation of genetically modified trees requires examining the promises that would tempt people to plant genetically modified trees as well as the risks that should worry the planters. First, the worries.

6.1 Genetically modified plants encourage plantations

Because genetically modified trees are likely to be set in plantations, they raise the issue of unnatural forest plantations. Worriers should clarify whether they worry about plantations or genetically modified trees and not confuse the two issues. Some environmentalists worry that industrial plantations will deplete nutrients and water in the soil and produce a vulnerable monoculture of trees where a rich diversity of species should prevail. Would opponents rule out all genetically modified trees for fixing greenhouse gases or only plantations of single-species forests like eucalyptus, acacia, or pines? What if new genetically modified tree plantations are established on abandoned croplands, which are already abundant and accessible?

6.2 Genetically modified plants encourage concentration on a few species

Although foresters now use about a thousand tree varieties for industrial wood production, geneticists are likely to edit the DNA of only a few. Genomics might further concentrate forestry on a handful of species and thus increase vulnerability to catastrophic failure. In agriculture, nine plants and three animal species provide 70 percent of the world's food, and research has tended to concentrate on only those rather than diversify food sources.

6.3 Genetically modified plants violate nature

A purist might deplore any meddling or violation by humanity of nature's ancient genetic pool, especially in forests. With new mixtures of genes, genetically modified plants increase global biodiversity in a sense, as mutation does. Nevertheless, because humanity made genetically modified plants, some will worry philosophically that, whether genetically modified trees expand or shrink biodiversity, they violate nature. A proper philosopher would go on to examine and define the meanings of violate and natural. Baking bread is hardly natural, either. Bread does not grow on trees. At the same time, ethicists can easily imagine irresponsible or frivolous applications such as leaves that glow in the dark with a product endorsement.

6.4 Genetically modified plants endanger

Beyond *philosophical* worries lie tangible, *practical* dangers to worry planters and onlookers. For genetically modified *food crops,* health or dietary worries might be expected to head the list. Little if any evidence so far shows genetically modified foods cause damage to health. Admittedly, viewing genetically modified trees as a risk to health is a stretch, but research could bring relief to worried onlookers.

Economic and political worries will surely arise. Opponents of genetically modified crops assert the debate is in part about who controls the food chain from the seed to production and even distribution. Genetically modified trees are likely to elicit analogous worries about economic domination associated with highly capitalized, hierarchical enterprises that can afford to influence governments. Advocates for indigenous peoples, who have witnessed the harm caused by crude industrial logging of natural forests, warn that big corporations concentrated on global profit will dislocate forest dwellers and upset local economies with plantations.

Ecological worries certainly exist. Genetically modified transplants, seed, and especially pollen will inevitably escape. To survive, the escapee must be evolutionarily fit. The question that follows, and a hard one to answer, asks what practical danger an evolutionarily fit escapee poses. Although kudzu and multiflora rose are not genetically modified escapees, they give the perspective of historical experience to worries about practical dangers. A more speculative worry is that a transgene might accidentally switch on "sleeper" genes or silence currently active genes. Although pests might overcome genetically modified host resistance, the selection of successful pests is not peculiar to genetically modified pest control. Worries like escapees and resistant pests are less about the trees than about the forest, that is, that parts of the landscape in some way may become less appealing.

7. DNA PROMISES

What then are promises that might drive the planting of genetically modified trees?

7.1 Faster growth and land sparing

Faster growth drives the planting of genetically modified trees. For entrepreneurs faster growth appeals because it means profits sooner. For us, the appeal of faster growth is spatial efficiency. Low yields squander land, while high yields spare it. Appropriately, the foremost promise of genetically modified trees is for the landscape, to hasten the Great Restoration.

7.2 Improved product

Investors seeking more practical benefits than the Great Restoration of nature will care about improved product as well as faster delivery. Conceivably, modified genes could tune trees more closely to users' needs, for example, for tree form or uniform wood quality, than simply selecting species or varieties. Tested over time, genetically modified trees may grow shapes with fewer branches and more trunks that can be formed into products.

Practical screening of genetically modified cells among millions of candidates demands a well-defined, easily assayed characteristic. In crops, abilities to emit a toxin that clears the bacterium in the culture medium or survive a herbicide in the medium have made ideal characteristics to speed screening. In trees, lignin content is well defined, and a clever assay for cells rich or poor in lignin may be designed. On the one hand, lignin makes lumber strong. Builders lay hardwood floors rich in lignin. On the other hand, paper mills prefer softwood pulp already low in lignin because it otherwise must be removed with costly, environmentally hazardous chemicals. Thus, the analogue in trees of the genetically modified, golden rice with richer vitamin A could be trees rich in lignin for lumber or poor for pulp.

7.3 Pest resistance

Because corn and cotton transformed to produce *Bacillus thuringiensis* have spread across millions of hectares and saved tons of insecticide, one looks for analogues of this success in forestry. Forests have pests, too. After World War II, some proposed spraying DDT from used bombers onto millions of acres of forest. Although trees' long life cycle could give pests more chance to develop resistance to the toxins, transforming trees to produce pesticide and prevent defoliation holds promise.

7.4. Herbicide resistance

The relatively straightforward selection of transformed cells that survive the herbicide Roundup has made Roundup Ready crops a pre-eminent success in agronomy. Although herbicide resistance would not likely help to maintain a plantation, it might help establish it.

7.5. Growth in harsh environments

Editing DNA so that trees could reclaim land damaged by erosion or salt would be beneficial. Other analogues of crops transformed to grow in harsh environments could be imagined.

7.6 Carbon fixation

Transforming trees into faster producers of wood and fixers of greenhouse gas could benefit those who fight climate change.

7.7 New varieties quickly and cheaply

Producing a single new variety of a well-studied species such as the apple costs about half a million dollars and 15 to 20 years even with marker-assisted breeding and tissue culture. Although forest genomics may cost a lot at the outset as genomes are sequenced, subsequent innovation could be quick and thus cheap.

7.8 Wood at a competitive price

As the wood products industry struggles to compete, large parts of it will suffer unless the price of stumpage compares favorably with plastics and other alternatives. Jobs for foresters, millers, and carpenters may finally depend on keeping the commodity price low. The time when genetic engineering could be a top contributor to cost control seems distant, but the several ways genetic modification could eventually boost wood growth and quality finally might sum high.

8. HOW FAST MIGHT MODIFIED TREES PENETRATE THE FOREST?

Many worries about genetically modified trees may be ill defined, but the promises are, too. Until something more effective than a United Nations recommendation to fix carbon dioxide in forests, genetically modified trees are likely to remain scarce and little reason to worry. Still, how fast might genetically modified trees penetrate the forest?

The economy and effectiveness of pest control have propelled the rapid diffusion of genetically modified *crops* (Figure 8). As a reference for speed, we can compare the current national speed of adoption of genetically modified soy, corn, and cotton with the classic adoption of hybrid corn in Iowa, Kentucky, and Alabama. While hybrid corn took 6 to 10 years to diffuse within American states, the new crops are taking 10 to 20 years. Globally, genetically modified crops appear likely to diffuse over about 20 years. Recent decisions from Brazil and Paraguay to the European Union suggest the global pace will be maintained.

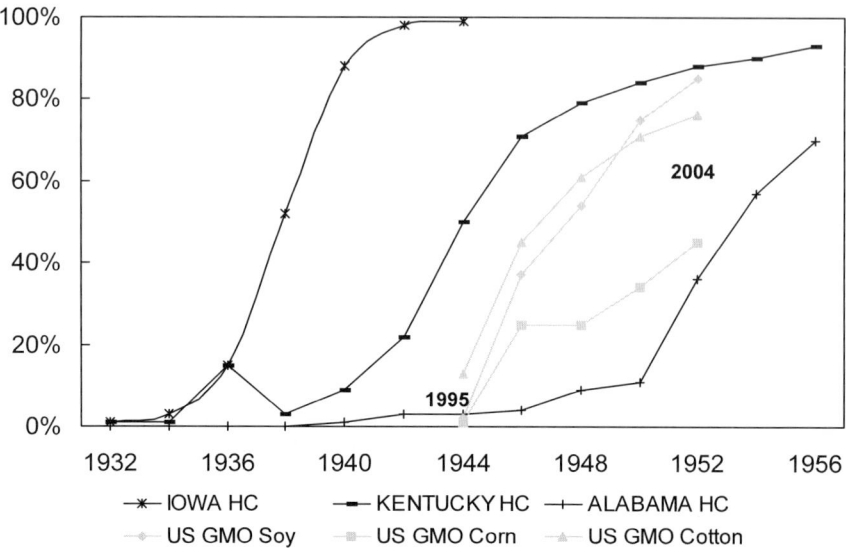

Figure 8. Comparative diffusion of hybrid corn and genetically engineered crops. Adoption of hybrid corn is shown as a percent of the total corn crop for three U.S. states from 1932 to 1956; in gray is adoption as a percent of total crops of genetically engineered soy, corn, and cotton crops in the entire United States from 1995 to 2004. Sources: Griliches, 1988; International Service for the Acquisition of Agri-Biotech Applications, 2004.

The slow life cycle of 25 to 100 years from planting to harvest will inevitably retard the diffusion of new kinds of trees. But adopting genetically modified trees as fast as genetically modified crops is most likely for plantations. Even in plantations where the adoption of modified trees is most likely, the cropping cycle of, say, 30 years for pine versus 1 year for corn must be remembered. Should modified trees achieve a cycle of, say, 15 years to harvest, it would speed planting on fresh land. Existing plantations would be little affected until the present growing stock could be economically replaced. If trees planted in 2010 or 2020 matured as fast as hemp, plantations would change much sooner.

Although forest owners might be propelled to plant genetically modified trees, how much land do they own? Owners in the U.S. forest industry, who are most likely to plant modified trees, own only about a tenth of the 300 million hectares of U.S. forest. Penetration of modified trees requires planting on land they already own or on abandoned land they may buy. As a guide to what might be planted, consider present planting rates. In 2000 the United States had about one-tenth of the world's 187 million hectares of tree plantations. In relative terms, the world plantations comprised 13 Iowas. U.S. plantations had an extent slightly broader than one Iowa. Worldwide plantations cover 5 percent as much as forests, and in the United States they cover 7 percent as much as total forest. Planting per annum in 2000 was only a small 1 percent to 2 percent of plantation area. In the United States, planting in 2000 proceeded at only 121,000 hectares, or one-eighth the area of Yellowstone National Park, curiously slow.

Another standard for the rate of penetration of genetically modified trees is the planting of genetically modified crops. Because genetically modified crops are annuals, the rate of planting genetically modified crops should be compared to the planting of plantations. Worldwide the annual planting of trees in plantations proceeded only 6.6 percent as fast as the planting of genetically modified crops. In the United States, where two-thirds of genetically modified crops are planted, the planting of all trees in plantations in 2000 proceeded only 0.3 percent as fast as the planting of genetically modified crops, a difference caused at least by economic fluctuations, social anxiety, and technical prowess.

9. MONITORING

Although the expected penetration of genetically modified trees is slow, the worry that "Something terrible could happen" may slow it even more. As with most innovations, achieving the promise of genetically modified trees, or high-yield forestry generally, will require feedback from a watchful public, and we should start watching now. Although experience and research may reduce some unknowns underlying worries, some things that happen are fundamentally unknowable. No level of foresight will anticipate every contingency or confluence of factors. There is no map of the future because no one who has been there ever comes back. So, like good scouts, people should be prepared.

The overwhelming volume of trade dims hope that quarantines will catch all invaders, which will continue mutating and sneaking in. The southern corn leaf blight of 1970 that invaded the United States from the Philippines during wet weather is probably the nearest analogue to genetic engineering gone astray, as it attacked male-sterile corn on which the U.S. corn crop had become heavily dependent. Because quarantines leak, people must monitor vigilantly and innovate in vigilance. Drone planes with spore traps now catch spores of tobacco blue mold and potato leaf blight aloft. Monitoring requires the sequels of reporting and communicating, which the Internet aids, and eventually action. Preparedness to act, whether through chemical control or other means, minimizes danger. U.S. farmers impressively replaced the vulnerable male-sterile corn in one season.

10. CONCLUSION

Human numbers, now 6 billion and heading for 8 billion to 10 billion in this new century, mean we already have a Faustian bargain with technology. Having come this far with technology, we have no road back. If wheat farmers in India allow yields to fall back to the level of 1960, to sustain the present harvest they would need to clear nearly 50 million hectares, about the area of Spain.

Through further intensification, farmers can be the best friends of the forest. Alternatively, they can plow through it. Technology can double and redouble farm yields and spare wide hectares of land for nature. We have confidence that farmers and their partners in the scientific community and elsewhere will meet the challenge of lifting yields per hectare close to 2 percent per year through the new century.

Freed and encouraged by the sparing of farmland, humanity can set a global goal of a 10 percent spread in forest area, about 300 million hectares, by 2050. Furthermore, foresters should concentrate logging on about 10 percent of forestland. Behavior can moderate demand for wood products, and foresters can make trees that speedily meet that demand, minimizing the forest that people disturb. The main benefit of the new approach to forests will reside in the trees spared by more efficient forestry. An industry that draws from planted forests rather than cutting from the wild will disturb one-fifth or less of the area for the same volume of wood. Instead of logging half the world's forests, humanity can leave almost 90 percent of them minimally disturbed. Social acceptance of the vision of the Great Restoration is key, both for farmers and for foresters.

The essence of the strategy for foresters to achieve the Great Restoration is the same as that for farmers, more bits and fewer hectares. Call it precision forestry. Working precisely, people can spare farmland and spread forests. Precise bits of information called DNA are finally the forester's inevitable and most powerful tool. Would editing a few bytes of the message for a tree to fit human desires do harm or good? Fortunately, foresters have several decades during which to answer by testing and monitoring wisely.

Acknowledgment: We thank Cesare Marchetti and David G. Victor.

11. REFERENCES

AUSUBEL, J. H., 2001 The great reversal: Nature's chance to restore land and sea. Technol. Soc. **22**: 289-301

USUBEL, J. H., 2002 Maglevs and the Vision of St. Hubert – Or the Great Restoration of Nature: Why and How, Pp.175-182 in *Challenges of a Changing Earth*, W. Steffen, J. Jaeger, D. J. Carson, and C. Bradshaw (eds.), Springer, Heidelberg.

CHEMICAL & ENGINEERING NEWS, 2001 Data for plastics for 1992-2000. June 25, 2001, pp. 44-51.

FEDKIW, J., 1989 The evolving use and management of the nation's forests, grasslands, croplands, and related resources. Gen. Tech. Rep. RM-175. USDA-Forest Service, Fort Collins, Colorado.

FOOD AND AGRICULTURE ORGANIZATION OF THE UNITED NATIONS, various Yearbooks. http://faostat.fao.org/faostat/collections?subset=agriculture.

GRILICHES, Z., 1988 Hybrid Corn: An Explanation in the Economics of Technological Change, pp. 27-52 in *Technology, Education, and Productivity: Early Papers with Notes to Subsequent Literature*. Basil Blackwell, New York.

INCE, P. J., 2000 *Industrial Wood Productivity in the United States, 1900-1998.* Res. Note FPL-RN-0272. U.S. Department of Agriculture, Forest Service, Forest Products Laboratory, Madison, Wisconsin.

INTERNATIONAL SERVICE FOR THE ACQUISITION OF AGRI-BIOTECH APPLICATIONS, 2004 *Global Status of Commercialized Biotech/GM Crops: 2004*, C. James, chair, ISAAA Board of Directors, http://www.isaaa.org.

MARCHETTI, C., 1979a Genetic engineering and the energy system: How to make ends meet. Technol. Forecast. Soc., **15**: 79-86.

MARCHETTI, C., 1979b On energy and agriculture: From hunting-gathering to landless farming. RR-79-10. International Institute for Applied Systems Analysis, Laxenburg, Austria.

MATHER, A. S., J. FAIRBAIRN, and C. L. NEEDLE, 1999 The course and drivers of the forest transition: The case of France. J. Rural Stud. **15**: 650-690.

MYNENI, R. B., J. R. DONG, C. J. TUCKER, R. K. KAUFMANN, P. E. KAUPPI, J. LISKI, L. ZHOU, V. ALEXEYEV, AND M. K. HUGHES, 2001 A large carbon sink in the woody biomass of Northern Forests. Proc. Natl. Acad. Sci. USA **98**(26): 14784-14789.

NATIONAL CORN GROWERS ASSOCIATION, Production, Corn Yield Contest, AA Non-Irrigated Class, http://www.ncga.com/02profits/CYC/profiles/aanonir.htm (accessed March 25, 2005).

SEDJO, R. A., 2001 The Role of Forest Plantations in the World's Future Timber Supply. Forest Chron. **77**(2): 221-226.

UN ECE/FAO, 2000 *Forest resources of Europe, CIS, North America, Australia, Japan and New Zealand (industrialized temperate/boreal countries), contribution to the Global Forest Resources Assessment 2000.* United Nations, New York.

U.S. DEPARTMENT OF AGRICULTURE FOREST SERVICE, 1992 Forest Statistics of the United States, 1992, Washington DC.
U.S. DEPARTMENT OF AGRICULTURE FOREST SERVICE, 1997 Forest Resources of the United States, 1997, Washington DC.
U.S. DEPARTMENT OF AGRICULTURE FOREST SERVICE, 2002 Forest Resources of the United States, Draft RPA 2002 Forest Resource Tables, Washington DC.
U.S. GEOLOGICAL SURVEY, 2004 *Historical Statistics for Mineral Commodities in the United States.* Open-File Report 01-006 http://minerals.usgs.gov/minerals/pubs/of01-006/
VICTOR, D. G., and J. H. AUSUBEL, 2000 Restoring Forests. Foreign Aff. **79**(6): 127-144.
WAGGONER, P. E., 1996 How Much Land Can Ten Billion People Spare for Nature? Daedalus **125**(3): 73-93.
WAGGONER, P. E., and J. H. AUSUBEL. 2001 How much will feeding more and wealthier people encroach on forests? Popul. Dev. Rev. **27**(2): 239-257.
WANN, S. R., and J. L. RAKESTRAW, 1998 Maximizing hardwood plantation productivity in the Southeastern United States—Lessons learned from loblolly pine. In Conference Proceedings, Improving Forest Productivity for Timber, A Key to Sustainability, 1-3 December 1998, Duluth, Minnesota. http://www.cnr.umn.edu/FR/publications/proceedings/papers/Wann-Duluth.pdf.
WERNICK, I. K., and J. H. AUSUBEL, 1995 National material metrics for industrial ecology, Resour. Policy **21**(3): 189-198.
WERNICK, I., P. E. WAGGONER, and J. H. AUSUBEL, 1998 Searching for leverage to conserve forests: The industrial ecology of wood products in the United States. J. Ind. Ecol. **1**(3): 125-145.

CHAPTER 2

THE QUESTION OF COMMERCIALIZING TRANSGENIC CONIFERS

CLAIRE G. WILLIAMS
Duke University, Department of Biology

Abstract. Genomics is creating new knowledge in forestry and making novel transgenic forest trees is one of its many emerging economic deliverables. What is the decision matrix for using transgenic conifers on a commercial scale? Three major determinants will shape decisions to use plant transgenic forest trees on a commercial scale: reproductive biology of the forest species, production systems and the type of transgene. One example is given as a case study of *Pinus taeda* in the southeastern United States. Exploring likely scenarios for commercial use suggested the need for a public-private technology partnership or technology trust. A technology trust proposed for transgenic *Pinus taeda* has three components for gathering sound information about transgenic conifer use: 1) a gene conservation program as a hedge against molecular domestication which would be subsidized by 2) a technology tax on transgenic field testing which includes a hold-harmless provision for the payee and 3) a public-private partnership for evaluating benefits-risks analysis which would also serve as an outlet for public deliberation.

1. INTRODUCTION

High throughput genomics is providing a wealth of new DNA sequences available for gene discovery and genetic modification in plants. Forest trees are no exception. Genomics and gene discovery, when combined with clonal technology, offers the opportunity for capital accumulation through production of transgenic trees as one of many applications. And pursuit of transgenic technology research is principally corporate, shaped by the imperatives of private investment, market forces and government regulatory institutions. In simplest terms, novel forest tree phenotypes can now be created on a commercial scale as a means to increase shareholder value of investor companies so biotechnology firms are preparing for commercial use in the near future. From the vantage point of societal good, transgenic forest trees are appealing as a new way to meet the world's need for higher-yielding forests while reducing the pressure on more fragile ecosystems.

Concerns about these justifications for transgenic forest trees surface in the absence of relevant ecological risk information. First, use of transgenic trees on a commercial scale means potential benefits will accrue to shareholders yet its ecological risks, if any, will be shouldered by all citizens. Second, using transgenic trees as part of the high-yield forest management technology portfolio without gauging its ecological impact opens the prospect of doing harm to the very ecosystems it purports to protect. Better information about ecological risks on a case-by-case basis is needed although public awareness tends to low.

Ecological risk analysis has not paralleled recent breakthroughs in transgenic technology for conifers. Long-distance dispersal for conifer pollen has been observed empirically for over a hundred years (BESSEY 1884) compared to the recent advent of genetic transformation for a conifer less than 20 years ago (SEDEROFF et al., 1988). A couple of reasons are offered: 1) the distinctions between transgenic conifers and transgenic food crops are so poorly understood to the extent that gene flow questions for conifers do not command research priority in agricultural funding programs and 2) the concentration of private investment in transgenic conifer research and development means that no formal outlet exists for public deliberation or policy debates which in turn would raise public awareness. This is consistent with the observation that national forests, not private forests, have been the playing field for debates about genetic composition (FRIEDMAN AND FOSTER 1997).

1.1. Distinctions between transgenic conifers and transgenic food crops

If transgenic conifers are traditionally viewed as another type of transgenic agricultural crops then we must consider the possibility that the question of ecological risk has become passé even before transgenic conifers were feasible on a commercial scale. Extensive gene flow studies from transgenic crops have been conducted; one outcome has been deregulation for many crop species. Another outcome is that those transgenic crops still under regulation are surrounded by biocontainment zones to prevent transgenic pollen escape. But gene flow is more complex for conifers than annual plants and biocontainment zones are ineffective (WILLIAMS 2005; WILLIAMS et al., 2006). Few public policy decision makers have delved into these distinctions. Unless the question of ecological risks for transgenic conifers is treated separately from those associated with food crops, it is reasonable to expect that tensions will escalate between private and public interests. If so, the question of commercialization, pro or con, will remain mired for lack of relevant information.

1.2. Determinants for ecological risks associated with transgenic conifers

Gene flow for forest trees is complex, occurring on micro-, meso and macro-transport scales. Seeds and pollen can move kilometers from source with 100% certainty (WILLIAMS et al., 2006). In the most recalcitrant cases, each transgenic conifer can produces abundant seeds and pollen several years before it reaches harvest age. The taller the tree, the farther the dispersal distances. The older the tree, the greater seed and pollen production becomes. The scale of gene flow is too great to be deterred by biocontainment zones. Placing a 1-km biocontainment zones around the perimeter of every transgenic plantation requires excessive use land and even so, the risk of transgenic escape would still be high.

Certainty of escape leads us to examine the consequences of escaped transgenic pollen and seed. Here lies another distinction for commodity conifers which tend to be weedy colonizers (HOLM 1979). In southern Africa, transgenic escapes will thrive in less managed ecosystems without human intervention as a result of deliberate genetic diversity retention and recent domestication. Domesticated conifers have no breed structure or inbred lines. They are one to three generations removed from their wild conspecifics (WILLIAMS *et al.*, 1994; MCKEAND *et al.*, 2003). In the case of indigenous conifers, one must also consider the high potential for interspecific hybridization. Compare this to U.S. food crops such as maize, cotton and soybeans which have no wild or weedy relatives in the vicinity, making gene flow from transgenic varieties less likely.

Life history is only one source of relevant distinctions between transgenic conifers and food crops. Production or silvicultural systems as well as landownership objectives also play into the decision to commercialize transgenic conifers. High-yield plantations can be adjacent to less managed or even unmanaged forest lands. Neighboring ownerships may prove problematic to those planting transgenic conifers if technology adoption rates vary widely among landowners.

Another risk determinant is the type of transgene. DNA constructs can come from genes discovered in organisms other than conifers (heterologous constructs) or from genes discovered in the same taxon (homologous constructs). DNA constructs are also classified based on perceived risk (STRAUSS 2003). The higher-risk groups include toxin genes, allergenic pharmaceutical genes or even pesticidal genes. The class with the least direct ecological impact is hypothesized to be domestication transgenes (STRAUSS 2003).

This framework shown by contrasting of two extreme examples (Table 1). Example 1 is an exotic species characterized by short rotations, sparse or delayed reproduction and perhaps facultative or obligate self-pollination. If this set of life history characters is coupled with a prevalence of high-yield private forests and long-term commitment to forest ownership then Example 1 becomes a case where transgenics are chosen as a part of the technology portfolio. Contrast this with the more recalcitrant Example 2 in Table 1. Indigenous conifers with close relatives, early reproduction, windborne seed and pollen, long rotations coupled with a prevalence of public forest ownership raise probability of ecological risk. A milder variant on Example 2 can be examined more closely using *Pinus taeda* in the southeastern United States.

Table 1. Factors affecting decision to commercialize transgenic trees shown with two opposing examples.

Factors	Example 1	Example 2
Species source	Exotic introduction	Indigenous
Sympatric relatives	None	Many
Pollination system	Selfing	Outcrossed wind-pollination
Seed dispersal system	Close-range animal dispersal	Windborne
Onset of reproduction	Late	Early
Type of propagules	Seeds	Vegetative and seeds
Abundance of propagules	Sparse, masting species	Abundant annual production
Rotation age	Short	Long
Land ownership type	Private	Mixed or public
Ownership commitment	Long-term	Shorter than full rotation
Neighboring ownership	Private, intensively managed plantations	Public forest owners and small woodlot owners
Type of transgenes	Domestication genes; homologous constructs	Toxin genes; Heterologous constructs

2. *PINUS TAEDA* AS A CASE STUDY

Pinus taeda is selected here as a less extreme variant of Example 2 (Table 1) and it is less extreme because domestication transgenes, not toxin transgenes, are the norm. It is a wind-pollinated indigenous species characterized by abundant reproduction via seed starting after the age of 10 to 15 years, well before its rotation age of 25 to 35 years. Its range is widespread, extending throughout the southeastern United States from Maryland to Florida and to Texas. Several sympatric relatives within its *Australes* subsection have ranges which overlap with *Pinus taeda*. Its share of the world's market in timber production has now reached 15.8%; the southeastern United States produces more timber than any other country in the world (WEAR AND GREIS 2002). Forest ownership in southeastern U.S. is a mosaic of private and public landowners. Over 89% of the timberlands are privately held and small woodlot owners are the predominant owners (WEAR AND GREIS 2002). Production increases have been dramatic over the last few decades and the addition of genetic improvement to the *Pinus taeda* technology portfolio has added substantially to increased yields (MCKEAND et al., 2003). No transgenic plantations exist although field testing for transgenic *Pinus taeda* is extensive. The most technology-intensive portfolio for plantation forestry is still a rarity even among timber companies so commercial use of transgenic plantations can be envisioned as a set of small foci amidst a mosaic of private and public *Pinus taeda* forests.

2.1. Technology breakthroughs enabling molecular domestication

Molecular domestication for *Pinus taeda* has become feasible due to breakthroughs in somatic embryogenesis and genetic modification. Somatic embryogenesis is a process by which a single genotype can be cloned into millions of trees. In theory, genetic modification of a single genotype which is then cloned by somatic embryogenesis makes large-scale use of transgenic pines possible in the near future. Biotechnology firms such as Arborgen and CellFor are offering clonal *Pinus taeda* seedlings to timber companies. No transgenic plantations exist yet but the approach of commercialization can be measured in the increasing numbers of transgenic field trials (MANN AND PLUMMER 2002). Molecular domestication of *Pinus taeda* plantations is technically feasible on a large scale. Testing has focused on wood quality, a source of domestication genes for transgenic *Pinus taeda*.

Molecular domestication with somatic embryogenesis and genetic transformation represents a departure from conventional domestication. Widespread use of clonal *Pinus taeda* forests with or without transgenes could narrow the numbers of genotypes, opening the question of how to protect genetic diversity of undomesticated *Pinus taeda* and its close relatives. This emerging issue deserves closer scrutiny because the federal government has contributed so little to collective genetic improvement. As a result, domesticated *Pinus taeda* germplasm is mostly privately owned by timber companies and more recently, institutional investors. A few state agencies share germplasm with these corporate owners.

Public repositories or gene conservation programs formalized by the federal government are needed as a counterbalance for molecular domestication. Gene conservation is an insurance policy against transgenic mistakes or genetic bottlenecks caused by cloning too few genotypes. In the case of *Pinus taeda*, starting a federally funded gene conservation program protects against any chances that highly domesticated or transgenic conifers prove poorly adapted outside intensively managed plantation systems.

2.2. Transgenic pollen and seed move long distances

Transgenic *Pinus taeda* is outcrossing wind-pollinated, producing an abundance of unwanted pollen and seeds. Perennial production of seeds and pollen, occurring years before timber harvest, adds up to complex gene flow dynamics. Transgenic seeds and pollen move by way of two separate processes (HENGEVELD 1989). The first process is local neighborhood dispersal (LND) which accounts for 99% of the seeds and pollen falling near source. The second process, long-distance dispersal (LDD), accounts for a tiny fraction (1%) of escaped seeds or pollen yet poses the greatest ecological concern. LDD seeds and pollen are vertically uplifted above the forest canopy by air currents then move on the order of kilometers from source.

Consider that a fraction of seeds uplifted above the forest canopy can move as far as 11.9 to 33.7 kilometers from source (NATHAN *et al.*, 2002). Out of 10^5 seeds produced per hectare per year in a 16-year old plantation, roughly of those 70 seeds will reach distances in excess of 1 km from the source, a distance too great to serve as an biocontainment zone (WILLIAMS *et al.,* 2006). The value of biocontainment zones are apparent for hardwood species (LINACRE AND ADES 2004) but negated by the greater dispersal scale for conifers. Although 99% of *P. taeda* seeds and pollen fall near the source tree, it is the remaining 1.0% which poses the greatest ecological concerns. Transgenic *P. taeda* seeds and pollen moving in less managed areas, public or privately owned, can establish remote satellite colonies.

If the transgene is favored by selection, then only a small fraction of dispersed seeds will cause its spread (MEAGHER *et al.,* 2003). To be harmful, the transgenic phenotype must exhibit increased invasiveness properties compared to its wild-type. Genes conferring increased invasiveness can result in displacement of local endemic species or even maladaptation (LEE 2003). Ecological risks are critical if invasiveness is increased in transgenic conifers.

2.3. Hybridization between transgenic Pinus taeda and its close relatives

So far, discussion has centered on gene flow between domesticated and undomesticated *Pinus taeda*. Natural hybridization does occur between *P. taeda* and its numerous relatives in the southeastern US although phenological conditions rarely overlap among species from year to year. For example, close relative *Pinus serotina* hybridizes with *Pinus taeda* on rare occasion (SAYLOR AND KANG 1973). Hybrids can freely cross with the two parental species or even a third species so the transgene may remain segregating in the gene pool indefinitely. Unless transgenic *P. taeda* trees have altered phenology, genetic swamping of related species by transgenic *P. taeda* has low probability but potentially high impact if the transgene confers higher fitness to the hybrid seedlings.

2.4. Exploring regulatory reform

Federal regulation of transgenic *Pinus taeda* as field tests or commercial plantations will continue given the unknown nature of ecological risks. Regulation of transgenic organisms has matured since inception of a tripartite government regulatory entity among three agencies: USDA, EPA and FDA. Primary responsibility resides with USDA-APHIS's Biotechnology Regulatory Services who recognizes that one set of regulations no longer fits all transgenic plants so U.S. regulatory reform is underway for transgenic conifers.

To aid this effort, several focus groups during our 2004 Nicholas School Leadership Forum *Landscapes, Genomics and Transgenic Conifers* were asked to evaluate four hypothetical regulatory options and results shown here were first summarized in our conference report. Continued regulatory oversight was a common response (Table 2). One individual voiced a need for a fourth option, a permanent moratorium for transgenic forest trees.

Table 2. Pros and cons of four hypothetical options for regulating transgenic or genetically modified (GM) forest trees.

Option	1) Moratorium	2) Research Agenda	3) Relaxed Regulations	5) Free Market
Description	Halt outdoor planting of GM confers for ten years on public and private lands. Permit laboratory and greenhouse research.	Continue to test GM conifers with domesticated neogenes in field tests under current APHIS regulations. Shift priority for competitive funding to exploratory research on other genomic applications suited to a wide range of silvicultural applications.	Allow certain GM field trials to reach timber harvest age in order to assess full benefits. These selected trials will test trees with DNA constructs only from functional genes discovered in conifers.	Remove government regulation. GM seedlings can be sold for-profit to any customer who wants to buy them.
Pros	No risk of seed or pollen escape	Stimulates new funding for forest tree genomics	Benefits from current investment in R&D pipeline realized	Benefits from current investment in R&D pipeline realized
	Technology develops during moratorium. For example, can test for unintended effects	Allows time to shift investment into new areas without losing materials in current pipeline	Timber and seedling industry can stay competitive in global markets	Timber and seedling industry stay competitive in global markets
	Provide time for public input	Acquire real data for benefits analysis	Acquire better data on gene flow data from GM trees for a full rotation	Industry forced to self-regulate
	Potential harm suspended	Increased funding for marker-assisted breeding and for functional genomics	Test methods for reproductive sterility	Reduced cost of final product, decrease time to market
	More time to identify societal benefits	Increased research funding		
	Basic research focus will be deepened with less pressure on immediate application			

	Consider release and field testing in Southern Hemisphere where there are no indigenous pine forests			
Cons	All transgenes do not have the same adverse (or beneficial) effects	Stifles GM research	Potential benefits suspended yet potential harms not avoided	Early adopters could face unforeseen problems without data collection, tracking or regulatory oversight
	Cost borne by private sector	Current regulations are inadequate for preventing escape of GM pollen and seeds	Increased gene flow from GM conifers	Effect on indigenous forests questionable due to high rate of transgenic escapes from commercial plantations
	Current investment in R&D pipeline wasted	Current regulations would have to be changed to allow reproduction	International embargo against GM timber from US	GM timber might meet with embargo on world market
	No field data would be available at the end of the moratorium for calibrating regulations	Too little competitive funding even at this time so not likely to have enough to research or develop genomics alternatives		Setting bad precedent for any other technology or methods for producing novel organisms
	Potential benefits suspended	No emphasis on forest genomics funding as is. Research funding occurs in the for-profit sector.		Buyer obligations and liability are unclear so harm will not borne by those who benefit financially
	Subversion of investor commitment	Decrease private investment		Untested risks from gene flow
	Other countries have less regulatory restriction and will capture market	Delay benefits from forest biotechnology by starting other avenues of inquiry		Increase in legal action and litigation
	Decreased funding for research			Patchwork of local and state regulations would restrict GM trees and this is the wrong scale for

				biotechnology governance
	Create backlog in R&D pipeline which would release an huge amount of regulatory works after moratorium is over			
	Permanent ban is needed for GM forest trees because they exceed our limits of biotechnology governance; no regulatory oversight would be a cost savings			

2.5. Exploring future scenarios for commercial use of transgenic Pinus taeda

Regulatory reform carries several implicit assumptions about use of transgenic conifer technology. Exploring future scenarios offers some insight about the validity of regulation. Two likely scenarios fit: 1) transgenic technology is not commercialized for *Pinus taeda* and 2) transgenic technology deregulation. Evaluating these served as a means of introducing a third intermediate scenario: starting a technology trust to ensure sound use of transgenic *Pinus taeda*.

The first scenario predicts commercial use of transgenic technology could be reduced or even abandoned for *Pinus taeda* if the trend towards timber company land divestiture continues. Recent sales of U.S. timberlands to institutional investors points to less intensively managed forest plantations. In 2005 alone, Boise Cascade sold 891,000 hectares of U.S. timberland to Forest Capital Partners LLC; the sale includes a long-term wood supply agreement for Boise's mills. Likewise, MeadWestvaco Corp. divested 360, 000 hectares of its timberlands to a private investment firm, Cerberus Capital Management. Less intensively managed timberlands might be expected in the future because the high costs of genetically enhanced seedlings or emblings incurred at plantation establishment must be carried through rotation. If the new institutional buyer wants to maximize profit from supplying wood to the seller, then less costly regeneration practices seem likely. If the standing timber is sold by the new buyer before harvest then genetic enhancements will also represent a loss because the timberland value is the same with or without genetic enhancements.

To illustrate this point, genetically enhanced emblings cost more than the five-cent *Pinus taeda* seedling sold today. Biotechnology firms are offering clonal seedlings or somatic emblings to timber companies and the price of a somatic embling is thought to be at least eight to ten times higher than a seedling. Price of a genetically modification to somatic emblings will be even higher in order to obtain a return on research, development and regulatory costs unless costs are offset by increased production efficiency or volume sales. Volume sales for *P. taeda* planting stock are low at 1 billion trees annually and seem unlikely to expand with offshore planting. Success of *Pinus taeda* as an exotic has been limited (BRIDGWATER *et al.*, 1997) and even curtailed by successful introduction of several Mexican and Central American pine species (DVORAK *et al.*, 1996). One might conclude that adding transgenic trees to the technology portfolio for *Pinus taeda* will be curbed by trends towards U.S. timberlands divestiture. Will timberland divestiture diminish chances of transgenic technology adoption for *Pinus taeda* in the southeastern United States?

The second likely scenario predicts deregulation. One can argue stringent U.S. regulations could provide incentive for multinational timber companies to plant transgenic forest plantations in South America rather than in North America. Given a profitable outcome for these Southern Hemisphere transgenic plantings, political pressure will mount to introduce transgenic conifer technology in North America and with this pressure will come demands for transgenic deregulation. Otherwise, cost of regulation will be too high to extract profits from local seedling markets and global wood products consumers alike. Will pressure to deregulate come from other countries with no regulations or relaxed forest certification flooding the world's markets with cheaper wood?

A second argument for deregulation is based on low public awareness in the United States. The issue of transgenic *Pinus taeda* falls outside of the issue-attention cycle for the following reasons: 1) too few citizens suffer enough on daily basis to stay riveted to the problem, 2) profits are at stake for a powerful minority of multinational corporations, 3) the timescale for adverse impact of transgene escape is beyond comprehension and 4) the problem lacks dramatic qualities to make it newsworthy. Low public awareness coupled with increasing scientific illiteracy adds up to no political interest in the outcome. Without political interest, research on ecological risks will not receive priority in competitive federal programs and without ecological risks data, one is left with intuitive arguments dismissing potential harm rather than the objective outcome from experimental data. In short, low public awareness seems a likely outcome because forests are rural, remote and scenic to urban dwellers. Who among us will know or even care if transgenic forests are even planted?

A third argument for deregulation can be made on the basis of pragmatism: that regulation of transgenic conifers in the absence of reproductive sterility will prove costly but futile. Tracking transgenic plantations bought and sold before harvest age will be prone to regulatory failure. Compliance can fail due to attrition, poor records, mergers and acquisitions so that lost transgenic plantings are abandoned. If their location is proprietary then location may not be disclosed as part of the sale. Even if a transgenic plantation is tracked, regulation will be costly but only be nominal due the sheer volume of escaped seeds and pollen. Is transgenic *Pinus taeda* a resource which cannot be truly contained, tracked or regulated?

3. A TECHNOLOGY TRUST OPTION FOR *PINUS TAEDA*

A public-private partnership or a technology trust is proposed as an alternative to the deregulation scenario. Focus group results shown in Table 2 suggests a public-private partnership might be favored as a vehicle for evaluating transgenic *Pinus taeda* for commercial use. A technology trust has three components: 1) a federal gene conservation program as a hedge against molecular domestication, 2) a technology tax on transgenic field testing which carries a hold-harmless provision to the payee or protection against future liability claims and 3) a designated subset of transgenic field tests would become long-term study sites for collecting relevant data for sound benefits and risk analyses. Relevant data on benefits and risk analysis would be made publicly available, thus providing an outlet for public dialogue. Such a technology trust could be designed and run with scientific oversight from government, university and private-sector research organizations.

A possible drawback to a technology trust is that its public-private scientific partnership could slow or even halt private investment in North America and Europe. Timber companies may not welcome public participation in its research agenda. Its commitment to research tends to rise and fall so sustaining a technology trust could a problem in this market sector. Without corporate interest, a technology trust will fail. Although it offers public relations value in the short-term, its true scientific contribution depends on sustained corporate commitment to landownership within a given geographic region and to high-yield forestry practices.

Direct benefits are that studies conducted under the technology trust would be re-evaluated at specific intervals and they would require a formal outlet for public dialogue. The public forum would bring a sharper focus to relevant goals, strategic experimental designs and efficient data collection. The added transparency to technology portfolio decision-making would provide a needed conduit for public deliberation within a time-sensitive framework. Public deliberation would raise needs for specialized knowledge for landowners, business managers, foresters and environmental activists. Genetic composition is a matter of good stewardship yet this topic has received less attention in classroom and in continuing-education venues than it deserves. It deserves more professional education than merely updating curricula for students enrolled in college and university classrooms.

In summary, a technology trust for *Pinus taeda* in the southeastern United States is proposed as a model program for providing sound benefits-risk analysis, initiating a gene conservation program, addressing long-term liability claims and sharpening relevant public dialogue. The price for adding transgenics to its technology portfolio for increasing plantation productivity without this timely information has yet to be determined.

4. CONCLUSIONS

Private investment can fund the creation of novel transgenic trees at a rate that will outpace scientific assessment of environmental concerns. Public-sector funding of ecological research has not kept up for several reasons, perhaps because transgenic conifers incorrectly viewed as a type of transgenic food crop. Transgenic conifers do in fact differ in a number of important ways, including 1) a wider radius for pollen and seed movement, 2) recurring and abundant seed and pollen production from each plant which begins years before its harvest and 3) diverse objectives and production systems in adjacent forest ownerships.

Major factors driving the decision to use transgenic conifers on a commercial scale are 1) reproductive systems of the forest species and 2) landownership profiles and 3) types of transgenes. To illustrate this point, a case study for *Pinus taeda* is presented. Two future scenarios of technology abandonment and technology deregulation were explored then followed by proposal of a technology trust, which was defined as a public-private partnership for evaluating transgenic *Pinus taeda* commercialization. The Technology Trust would provide a formal gene conservation program, address liability claims, serve as a formal outlet for benefits-risk analysis and open an outlet for better public dialogue. Public dialogue in turn would prompt better education about genetic composition for landowners, environmental activists and foresters alike. Under certain conditions, a technology trust could lead to maintaining healthy, well-adapted indigenous forests.

5. REFERENCES

BESSEY, C. 1883. Remarkable fall of pine pollen. *American Naturalist* **17**:658.

BRIDGWATER, F.E., R.D. BARNES AND T. WHITE. 1997. Loblolly and slash pines as exotics. pp. 18-32. In Proceedings of the 24th Southern Forest Tree Improvement Conference, Orlando Florida June 9-12, 1997.

FRIEDMAN S.T. AND G.S. FOSTER. 1997. Forest genetics on federal lands in the United States: public concerns and policy responses. *Canadian Journal of Forest Research* 27(3): 401-408.

DVORAK W.S. AND J.K. DONAHUE. 1992. CAMCORE Cooperative Research Review 1980-1992. North Carolina State University, Raleigh North Carolina.

HENGEVELD, R. 1989. *Dynamics of biological invasions*. Chapman and Hall, London. 160 p.

HOLM L., J.V. PANCHO, J.P. HERBERGER AND D.L. PLUCKNETT. 1979. *A geographical atlas of world weeds*. John Wiley and Sons, New York. 391 p.

LEE, C.E. 2002. Evolutionary genetics of invasive species. *Trends in Ecology and Evolution* **17**: 386-391.

LINACRE N.A. AND P.K. ADES. 2004. Estimating isolation distances for genetically modified trees in plantation forestry. *Ecological Modelling* **179**: 247-257.

MANN, C.C. AND M.L. PLUMMER. 2002. Biotechnology – forest biotechnology edges out of the lab. *Science* **295**: 1626-1629.

MCKEAND S.E., T. MULLIN, T. BYRAM AND T. WHITE. 2003. Deployment of genetically improved loblolly and slash pines in the South. *Journal of Forestry* **101 (4/7)**: 32-37.

MEAGHER, T.R., F.C. BELANGER AND P.R. DAY 2003. Using empirical data to model transgene dispersal. *Philos. Trans. R. Soc. Lond. B* **358**: 1157-1162.

NATHAN, R, G.G. KATUL, H.S. HORN, S.M. THOMAS, R. OREN, R. AVISSAR, S.W. PACALA AND S.A. LEVIN. 2002. Mechanisms of long-distance dispersal of seeds by wind. *Nature* **418**: 409-413.

SAYLOR L.C. AND K.W. KANG. 1973. A study of sympatric populations of *Pinus taeda* L. and *Pinus serotina* Michx. in North Carolina. *Journal of Elisha Mitchell Society* **89**: 101-110.

SEDEROFF, R., A. STOMP, B. GWYNN, E. FORD, C. LOOPSTRA, P. HODGKISS AND W.S. CHILTON. 1987. Application of recombinant DNA techniques to pines: a molecular approach to genetic engineering in forestry. Chapter 19. In Cell and Tissue Culture in Forestry, vol. I (J.M. Bonga and D.J. Durzan, eds). Nijhoff Publishing. Leiden, The Netherlands.

STRAUSS, S.H. 2003. Genomics, genetic engineering and domestication of crops. *Science* **300**: 61-62.

WEAR D.N. AND J.G. GREIS. 2002. Southern Forest Resource Assessment: summary of findings. *Journal of Forestry* **100**: 6-14.

WILLIAMS, C.G., J.L. HAMRICK AND P.O. LEWIS. 1994. Genetic diversity levels in a multiple population breeding strategy: a case study using *Pinus taeda* L. *Theoretical and Applied Genetics* **90**:384-394.
WILLIAMS, C.G. 2005. Framing the issues on transgenic pine forests. *Nature Biotechnology* **23**: 1-3.
WILLIAMS, C.G., S.L. LaDEAU, R. OREN AND G.G. KATUL. 2006. Modeling seed dispersal distances: implications for transgenic *Pinus taeda*. *Ecological Applications* (in press).

CHAPTER 3

IT'S JUST A CROP?

Public Perception and Transgenic Trees

STEVEN ANDERSON

Forest History Society
Durham, NC

Abstract. Genetic modification of forest trees has followed similar efforts in agriculture. Over 100 years ago, those conducting purposeful forest management in the United States have engaged terminology from Europe that was naturally adapted from agricultural pursuits. Early conservationists used such phrases as "trees are a crop" to help the American public understand that forests deserved their investment. Later, the same language was used and continues to be used to justify active management in the face of environmental pressures. Public perception of genetically modified trees is connected in this basic way to transgenics in agriculture and this increases the challenge to understanding the driving forces of public opinion. This paper traces this historic connection and suggests that public discourse will be a vital part of the process that decides how quickly and under what conditions transgenic forest trees are used.

1. INTRODUCTION

Until very recently, trees were genetically improved through conventional breeding techniques, in much the same way as livestock are bred to produce certain desirable characteristics in their offspring. Molecular domestication has arrived with advances in science and technology and the issue of genetic engineering is front and center in forest management. Part of the issue is that the general public has little knowledge about genetic engineering. However, the result of one study suggests that providing more public education alone about genetic modification will not necessarily lead to acceptance. The relationship between levels of support for genetically modified food, for instance, and objectively measured knowledge is weak at best and there may be little difference in knowledge between supporters and opponents (HALLMAN 1995). It may be unlikely that one can simply "educate people into acceptance" (HALLMAN AND ALQUINO 1993). Potential consumers must be viewed as taxpayers and citizens who will deliberate about use of this technology and if "perception is reality" then producers and policymakers must also concern themselves with public perception.

2. PUBLIC PERCEPTION AND TRANSGENIC TREES

Public perception of the science and social aspects of landscapes, genomics, and transgenics in forests seems to be inextricably linked with our perceptions of transgenics in agriculture; from the killer tomato, to Dolly the sheep, to finding genetically manipulated corn intended for animal feed in corn supplies for human consumption. On the surface, we can assume that fears in one arena can influence and lead to fears in another. But how might our dialogue in the past about forests and agriculture affect our debates today? While scientists face the challenge of discussing how transgenics in forests may differ from agriculture, our past language suggests that differentiation will be complex.

It is not hard at all to see that there is a natural and undeniable connection between the concepts of forestry and agriculture. Trees grow from seed, trees can be planted, they can be tended, we employ site preparation and weed control, it is periodically harvested, and then the process can begin again. If someone wanted to put forth a justification for treating forests as something to be managed, as something that is best rigorously controlled, then what better analogy than an agricultural field. As well, the historical labors to move plants (and their genomes) across large distances and fledging tree improvement approaches were born out of early agricultural and horticultural efforts. The beginnings of plant husbandry in North America began in earnest in the Late Archaic period from 3,000 to 1,000 B.C. (YARNELL 1998). In an 1882 book, *Elements of Sylviculture*, G. Bagneris, who was inspector of forests and professor at the forest school of Nancy, France, wrote about reproduction of the forest. He writes "…recourse must be had to artificial means in order to restore the good condition of the forest, or a satisfactory composition of the crops." He said "To exploit a forest or crop means to fell it in accordance with the principles of sylviculture." He defines annual yield as "…the quantity of produce that can be taken out of a forest annually on the condition that this quantity can be maintained at a constant figure" (BAGNERIS 1882).

Less than ten years later, Gifford Pinchot, attended classes at this same school at Nancy France, for less than a year of study that represented his formal training in forestry. The intellectual ancestor of forestry in the United States was exposed in early training to a system of highly regulated forestry. This early use of the term "crop" as it relates to trees continued and it would eventually be used in public relations by the forest industry that previously had abandoned their land instead of paying the taxes. At the American Forest Congress in 1905, F.E. Weyerhaeuser put forth time, fire and taxes as primary obstacles "that must be reckoned with in the profitable production of timber…," claiming that it certainly was not "...just that land which can produce but one crop in forty years should be taxed on the same scale as land which produces an annual crop" (WEYERHAEUSER 1905).

The American Forest Congress of 1905 was, in part, conducted to secure political support to transfer the forest reserves from the U.S. Department of the Interior to the U.S. Department of Agriculture. This would bring the foresters and the land they would manage together and it seems probable that this reference to crops assisted Pinchot and President Roosevelt to orchestrate the transfer. In the House of Representatives debate about the transfer, Pinchot wrote that Congressman John F. Lacey of Iowa defined Forestry as "... a great system of tree farming" and that it "...is of vital importance to the farmers of the United States (PINCHOT 1987).

In 1909, Carl Alwin Schenk, founder of the Biltmore forest school in North Carolina, published his book entitled *Forest Protection*. He uses phrases such as "Work towards immediate reforestation after making a clean sweep of the old crop", a reference to clearcutting with obvious agricultural imagery. The language became the norm. In 1929, two professors from Cornell, in their book on *Forestry; A Study of its Origin, Application and Significance* in the United States, wrote that "...like any other crop, timber becomes a partial or total loss if not utilized when mature...". In a 1949 text entitled "*Farm Wood Crops*," the author notes that "...forestry is not a forestry job on farmland; it is a farm job on forest land." A special point is made that "...wood products are interchangeable with clean-tilled crops." Comparisons of trees to crops are provided such as "...rayon from wood is the equivalent of that made from cotton" and that "Coppice, in hardwoods, is produced by clean-cutting; the next crop, like asparagus, originates from sprouts on the stump..." (PRESTON 1949).

It is no wonder that industry would eventually adopt the concept that timber is crop. Early in the 1900s, George Long, Weyerhaeuser's manager for west coast operations was reported to have used the term consistently. In 1938 the Weyerhaeuser Timber Company made a decision to pay the taxes on some of their cutover lands and allocated $.50 per acre to plant trees among other things. They subsequently took out an ad in the Seattle Post-Intelligencer – "Timber is a Crop." In 1941, the Tree Farm program was launched marking industry's formal commitment to purposefully managing their lands. By 1948, "Timber is a Crop" appeared on the front cover of Weyerhaeuser's annual report. Only 18 years later, Weyerhaeuser would continue on this track by announcing their commitment to "high yield forestry," a reference to assisting nature through planting, site preparation, weed control and nutrient amendments. For decades previously, researchers had explored high yield environments for hybrid corn.

Interestingly, we see more about trees growing like crops rather than <u>forests</u> growing like crops. The difference is subtle but not insignificant. In the February 1950 Forest Farmer magazine, the West Virginia Pulp and Paper Company, which later became Westvaco and more recently MeadWestvaco, placed an ad that was entitled "Timber is a Crop." It stated that "Trees don't grow as rapidly as corn or wheat–and they are not harvested as often. But timber is a crop – a crop that will turn idle acres into productive acres."

In this sense most could agree that, yes, timber is a crop. But the reality is that it is not <u>only</u> a crop. Inherently we recognize the difference between forests and crops. We recognize that corn may grow as high as an elephant's eye but we do not visit a corn field to rejuvenate ourselves, we do not visit a wheat field to reconnect with a world not so dominated by humans, and we do not stand in awe of the majestic height of a soybean field. Why? Because forests can represent a place that is beyond human dominance; because trees live past our lifetimes, both past and future; because forests offer varied opportunities for recreation; and because, while agricultural fields also function as ecosystems, we inherently understand that the complexity of the agricultural field ecosystem does not nearly approach the complexity or aesthetic appeal of a forest.

What is true is that both enterprises, farming and forestry, are dominated by economics. Aldo Leopold, perhaps the most abused of all writers in American literature, recognized this. Often quoted and taken out of context more times than not, his writings are used by pundits who wish to substantiate their own personal philosophy. Take for instance, one of his most popular quotes "A thing is right when it tends to preserve the integrity, stability, and beauty of the biotic community. It is wrong when it tends otherwise." Surely this is an antithesis to the "Timber is a Crop" moniker. But, rarely is this quotation followed by his next sentences…"It of course goes without saying that economic feasibility limits the tether of what can or cannot be done for the land. It always has and always will" (LEOPOLD 1968 pp. 224-225). It is this economic, ecological and social paradox that Leopold consistently identified but yet fell short of presenting a clear framework for reconciliation of competing land uses. Leopold talked of biotic rights and he was a proponent of wilderness. He was also a devote student of history. Above all, though, he counseled managers to pay attention to their effects on the land; what can we learn from our management decisions, both good and bad.

Leopold took exception to the simplicity of the "trees are a crop" approach. In an example of extremes, he generalized about two groups of thought in specialized fields. Group (A), he remarked "…regards the land as soil, and its function as commodity-production;" while Group (B) "regards the land as biota and its function as something broader." He chastised his own field of forestry as being predominately Group (A),

"…quite content to grow trees like cabbages, with cellulose as the basic forest commodity." He thought that Group (B) professionals would see "forestry as fundamentally different than agronomy" and would manage for both biotic and economic concerns. Leopold asserted that the farmer "must modify the biota more radically than the forester or the wildlife manager."

LEOPOLD, along with fellow forester BOB MARSHALL, helped establish the Wilderness Society in 1935. Initially it was considered a group made up of rich easterners and they were partially right. But 20 years later, as the forestry community was concentrating on explaining that timber was a crop, the Wilderness Society had surpassed 5,000 members across the country and the Sierra Club was starting a grass roots set of chapters. Part of the public's perception was no longer focused on how to get industry to invest in long-term management of the land; it was beginning to focus on how forests were managed on both private and public land. A confluence of factors including RACHEL CARSON's *Silent Spring* highlighted the unintended effects of pesticides, increased harvesting of the national forests to meet the public demand for building products, and a growing population having increased time for recreation, among others, contributed to produce an intensified public discernment of environmental consequences of land management and business decisions. This discerning public eye, whether knowledgeable or not, continues today and will certainly cast its stare to biotechnology and genetic engineering. And this is a good thing. When we are arguing about an issue we know the system is working. Scientific authentication is part of the process as well as social acceptability. Grappling with how to build a sustainable human society is an excellent sign of a vibrant democracy.

While industry and foresters may not have had the foresight in the mid twentieth century to keep pace with public opinion, the general trend of management improvements in the United States during the last hundred years has been on a path toward sustainability (FEDKIW 2004). From conservative lumbering, to sustained yield, multiple-use, ecosystem management and sustainable forestry, the progress has been incremental and adaptive. As well, the scientific knowledge gained during the last past century about ecosystems, genetics, biological diversity, global nutrient cycles, climate, and energy is genuinely astounding (FLOYD 2001). Molecular domestication is part of the current wave of new technology that will be integrated into our quest for sustainability.

Some basic arguments for the use of somatic embryogenesis and genetic transformation in forest ecosystems include that (1) it has the potential to provide solutions to providing adequate wood supplies to meet the needs and desires of tomorrow's human population, (2) by providing these wood needs, we eliminate the need to use substitutes such as various metals, brick, cement, and plastics that require more energy, emit more greenhouse gases, and introduce more toxic materials into the environment over their life cycle, from extraction and manufacture to eventual disposal, and (3) purposeful genetic changes that are intended to increase the productivity and value of forest trees that are planted, grown, and harvested for wood products can keep more land in reserves and wilderness, as well as open space that increases recreational opportunities. But the general public cannot be viewed as merely consumers and the world is more than economics and the free market system. The actual route to sustainability will be political and a result of public deliberation. Merely helping the public to understand that genetically modified trees can improve their standard of living will not educate them into acceptance. The concerns about unintended consequences will require that safeguards and risk management be integrated into the process, and arguably should have already started. If ample evidence doesn't already exist that indicates genetically modified plants used in agriculture can be hard to contain, the current dialogue in the media surely points this out (CLAYTON 2005). With the ability of pollen from trees to travel hundreds if not thousands of miles (KATUL *et al.,* 2006), it is not likely that any safeguards will assuage the segment of the public with greatest angst. As well, the use of genetically engineered trees most likely demands an intensive management approach. This relationship invokes the agriculture-forestry scenarios discussed above.

Ironically, it might be the use of genetic engineering in a horticultural sense that might ease public acceptance to transgenic trees. Already, a genetically modified elm resistant to the Dutch Elm disease, is being outplanted and transgenic plum and papaya are undergoing USDA review. Perhaps the most interesting opportunity is restoration of the American Chestnut to the forests of the eastern United States. The transgenic chestnuts will not be in plantations. The concept is to install them individually or in small plots so that they can regenerate on their own. Those trees will receive intensive management, but it will be far different than for industrial plantations. If successful, this example will help frame, not circumvent, public discourse about molecular domestication in forest trees.

During the last century we have had parallel discourse about forests and trees as they relate to farms and crops. One thread purports that trees are a crop and can be managed intensively to produce public goods. This was a productive analogy while informing public opinion about a possible timber famine and gaining industrial commitment to purposeful forest management. The other thread, that forests are more than crops, that aesthetics as well as economics will provide the greatest public good, lurked in the shadows and emerged with the second conservation movement in the mid 1900s. Both are relevant today as we continue the public debate about transgenic trees because they both still pervade the public consciousness. We can be sure that the political decisions necessary to use the technology under defined constraints will be enabled when the public perceives that the benefits of genetic engineering in forest trees outweigh the risks. Public perceptions, mutual understanding, debate and discourse will be necessary parts of the process.

2. REFERENCES

BAGNERIS, G. 1882. *Elements of Sylviculture.* London: William Rider and Son. 283 pp.

CLAYTON, M. 2005. Now, bioengineered trees are taking root. Christian Science Monitor. March 10, 2005 edition. http://www.csmonitor.com/2005/0310/p14s02-sten.html.

FEDKIW, J, D.W. MACCLEERY, AND V.A. SAMPLE. 2004. Pathway to Sustainability: Defining the Bounds on Forest Management. Durham: Forest History Society, Inc. 64 pp.

FLOYD, D.W. 2001. Forest Sustainability: The History, the Challenge, the Promise. Durham: Forest History Society, Inc. 81 pp.

HALLMAN, W. K. 1995. Public perceptions of agri-biotechnology. *Genetic Engineering News*, 15(13): 4-5.

HALLMAN, W.K. AND H. ALQUINO. 2003. Public Perceptions of Genetically Modified Food: An Intenational comparison. Presented at the 7th ICABR International Conference on Public Goods and Public Policy for Agricultural Biotechnology. Ravello, Italy June 19-July 3, 2003. http://www.economia.uniroma2.it/conferenze/icabr2003/papers/index.htm

KATUL G.G., C.G. WILLIAMS, M. SIQUEIRA, D. POGGI, A. PORPORATO, H. MCCARTHY AND R. OREN. 2006. Spatial modeling of transgenic conifer pollen. Chapter 9 in *Landscapes, Genomics and Transgenic Conifers* (C.G. Williams, editor). Springer, New York.

LEOPOLD, A. 1968. *A Sand County Almanac*: and sketches here and there. New York: Oxford University Press. 226 pp.

PINCHOT, G. 1987. *Breaking New Ground.* Washington D.C.: Island Press. 522 pp.

PRESTON, J.F. 1949. Farm Wood Crops. New York: McGraw Hill Book Company. 302 pp.

RECKNAGEL, A.B. AND S.N. SPRING. 1929. *Forestry: A Study of its Origin, Application and Significance in the United States.* New York: Alfred Knopff.

SCHENCK, C.A. 1909. *Forest Protection.* Asheville: The Inland Press. 159 pp.

WEYERHAEUSER, F.E. 1905. Interest of lumbermen in conservative forestry. In: Proceedings of the American Forest Congress: Held at Washington, D.C. January 2 to 6 1905 under the auspices of the American Forestry Association: 137-141. H.M. Suter, Washington D.C.

YARNELL, S.L. 1998. The Southern Appalachians: A History of the Landscape. USDA Forest Service. Southern Research Station General Technical Report SRS-18. 45 pp.

SECTION II

GENOMICS METHODS, RESOURCES AND ALTERNATIVE APPLICATIONS

CHAPTER 4

GENOMICS RESOURCES FOR CONIFERS

JEFFREY F.D. DEAN
*Daniel B. Warnell School of Forest Resources,
University of Georgia, Athens, Georgia, USA*

Abstract. Genomic techniques and technologies are revolutionizing the biological sciences, and the impact that this new approach to biology will have on forestry will be nothing short of profound. A hallmark of the genomic sciences is rapid release of large datasets to the public via the Internet with the expectation that important and useful information will be gleaned more efficiently when multiple groups of researchers are allowed to apply a wide variety of bioinformatics tools independently to the analysis of these datasets. Although the development of genomic datasets for conifers has lagged the torrent of information being made available for various model systems and organisms of biomedical interest, the resources are growing rapidly and with them come great opportunities to mine the data for new research hypotheses. This chapter summarizes the content of conifer information available at the major public repositories of genomic data, as well as conifer data housed at smaller online sites. Available bioinformatics tools are also discussed.

1. INTRODUCTION

In 1980, complete nucleotide sequences were known for a mere 80 genes. A decade has passed since publication in 1995 of the first complete genome sequence for a free-living organism (*Haemophilus influenzae*) (FLEISCHMANN *et al.,* 1995), an event that provides a convenient marker for the start of the genomics revolution in biology. From those early days trends toward automation and miniaturization in molecular biology and biochemical analysis have continued to accelerate, ever increasing the speed at which data from all levels of organismal biology is acquired. This, in turn, has driven demand for even faster and more efficient techniques for the electronic winnowing of that data, which has spawned the field of bioinformatics. It is a thrilling time for biologists as raw data floods public repositories, and every online search holds the potential for revelation as disparate pieces of metabolic pathways and biological responses are assembled from the ether to provide glimpses of the organismal fabric from which they were drawn. Perhaps nowhere is the potential for this new mode of biological science any more appreciated than among those who specialize in the biology of trees, whose characteristics of size, growth rate, retarded puberty, and disinclination to inbreeding so limit research by experimentation. Now, relying on the innately conservative nature of evolution, tree biologists can use bioinformatics to compare genomes, gene sequences, and gene expression patterns across phylogenetic space and infer from more tractable model organisms likely functions and responses of tree genes, thereby improving the efficiency of subsequent experimental tests with those trees.

All enthusiasm aside, though, the genomics revolution has greatly challenged biologists who must constantly retool their research skills in order to keep pace with the information age. This chapter was written to provide conifer biologists, in particular, with an overview of those information resources and bioinformatics tools available online that may be used to obtain molecular data related to conifers and other forest trees. However, much of the following discussion focuses on "entry

Table 1. Web addresses for resources discussed in this chapter.

ORGANIZATION / Database	URL
NCBI	http://www.ncbi.nlm.nih.gov/
Taxonomy Browser	http://www.ncbi.nlm.nih.gov/Taxonomy/taxonomyhome.html/
Entrez	http://www.ncbi.nlm.nih.gov/Entrez/
Site Map	http://www.ncbi.nlm.nih.gov/Sitemap/
Organelle Genomes	http://www.ncbi.nlm.nih.gov/genomes/static/euk_o.html
UniGene DDD	http://www.ncbi.nlm.nih.gov/UniGene/info_ddd.html
SAGEmap	http://www.ncbi.nlm.nih.gov/SAGE/
PIR	http://pir.georgetown.edu/home.shtml
EMBL-EBI	http://www.ebi.ac.uk/
UniProt	http://www.ebi.ac.uk/uniprot/index.html
Site Map	http://www.ebi.ac.uk/services/index.html
EMBL-Heidelberg	http://www.embl-heidelberg.de/
Bioinformatics Tools	http://www-db.embl.de/jss/servlet/de.embl.bk.emblGroups.EmblGroupsOrg/serv_0?t=1&p=1
ExPASy	http://au.expasy.org/
DDBJ	http://www.ddbj.nig.ac.jp/
Top 1000 Organism	http://www.ddbj.nig.ac.jp/ddbjnew/org1000/top100-e.html
Database Collection	http://www.ddbj.nig.ac.jp/infobio/linksddbj-e.html
TIGR	http://www.tigr.org/
Plant Gene Indices	http://www.tigr.org/tdb/tgi/plant.shtml
Loblolly Pine Project	http://www.tigr.org/tdb/e2k1/pine/index.shtml
PlantGDB	http://www.plantgdb.org/
openSputnik	http://sputnik.btk.fi/
DENDROME	http://dendrome.ucdavis.edu/
Treegenes	http://dendrome.ucdavis.edu/treegenes.html
Tree Projects	http://dendrome.ucdavis.edu/forest_tree_genome.htm
INRA-PIERROTON	
Pine Markers	http://www.pierroton.inra.fr/genetics/labo/index.html#V
Conifer Review	http://www.pierroton.inra.fr/genetics/labo/mapreview.html
Proteomics	http://cbi.labri.fr/outils/protic/
CSIRO	
Radiata Pine SSRs	http://www.ffp.csiro.au/tigr/molecular/microsatellites.html
MGEL	
Loblolly Pine BAC	http://www.msstate.edu/research/mgel/awards/nsf0421717.htm

CCBG Loblolly Pine	http://www.ccgb.umn.edu/ http://pine.ccgb.umn.edu/
FUNGEN Loblolly Pine	http://fungen.org/ http://fungen.org/Projects/Pine/Pine.htm
ARBOREA Spruce ESTs	http://www.arborea.ulaval.ca/en/ http://www.arborea.ulaval.ca/en/results/est-sequencing.php
TREENOMIX	http://www.treenomix.com/
FFPRI *Cryptomeria*	http://ss.ffpri.affrc.go.jp/ http://ss.ffpri.affrc.go.jp/labs/cjgenome/database/cjdatae.html
EXPRESSO Conifer Microarrays	http://bioinformatics.cs.vt.edu/~expresso/ https://bioinformatics.cs.vt.edu/~vsinghal/
PICME	http://www.arcs.ac.at/U/UB/picme/PICME
CELL WALL NAVIGATOR	http://bioinfo.ucr.edu/projects/Cellwall/index.pl

points" from which a researcher may gain access to a broad swath of information available on any given species before burrowing down to specific items of interest. Consequently, these examples should also prove equally helpful to researchers interested in virtually any other species. Because online resources change so rapidly, something as static book chapter stands a good chance of being outdated by the time it is published and read. As a consequence, much of the following presentation focuses on larger, better-funded repositories and resources with the presumption that these are most likely to remain accessible in the future. No doubt some of the smaller resources mentioned in this chapter will disappear by the time readers attempt to use them, but it increasingly seems that it is not the data that disappears, since it tends to get mirrored and copied at multiple online repositories. Instead, it is the tools developed to explore specific datasets that get discarded as better ones become available. Table 1 lists the web addresses (URLs) for the online resources described in this chapter.

Knowledgeable genomics researchers make a habit of repetitively examining data at multiple sites in order to stay abreast of newest tools developed for datamining. As a final piece of advice to biologists wishing to become more skilled in the mining of online data, take time to review the help and resources pages provided with these online tools as the time invested will always return dividends in faster and more efficient searches.

2. GLOBAL DATABASES AND RESOURCES

The principal worldwide repositories for nucleotide sequence information are GenBank, maintained by the National Center for Biotechnology Information (NCBI, Bethesda, MD, USA), the Nucleotide Sequence Database (EMBL-Bank) maintained by the European Molecular Biology Laboratory (EMBL/EBI, Hinxton, UK), and the DNA Data Bank of Japan (DDBJ, Mishima, Japan). These three institutions, which collectively form the International Nucleotide Sequence Database Collaboration, receive more than 20,000 individual and 200,000 automated sequence submissions per month, with submitted sequences being shared amongst the three centers on a

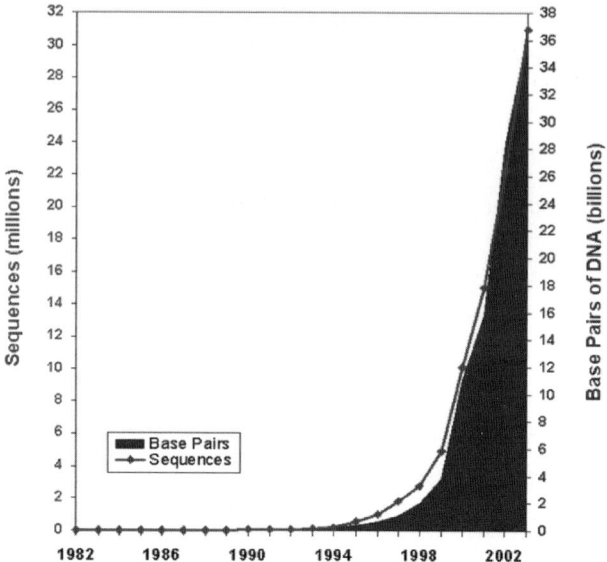

Figure 1. Growth statistics for the International Nucleotide Sequence Database Collaboration through the end of 2003 (http://www.ncbi.nlm.nih.gov/Genbank/genbankstats.html).

daily basis. As shown in Figure 1, through the end of 2003, the combined resource contained more than 31 million sequence accessions representing more than 36 billion base pairs (Gbp) of sequence. The exponential growth rate for the resource has resulted in an average doubling in the total number of nucleotide bases every 14 months since 1982 (when GenBank started recording statistics), although recent accelerations in submission volume have led to a current doubling rate closer to nine months. At the end of 2004 some 40 million accessions containing more than 44 Gbp of sequence were available. Although conifer sequences make up only a small proportion of this data collection, it nonetheless remains the richest source of conifer genomic data in existence.

Although the three collaborating centers mirror the same complete dataset, each provides a unique set of datamining tools that can be used to probe the database. A selection of primary search tools from each center is profiled here with examples of relevance to conifer biology.

2.1. National Center for Biotechnological Information

In addition to maintaining GenBank, the National Center for Biotechnology Information (NCBI), a division of the National Library of Medicine at the National Institutes of Health (NIH), curates several other molecular databases, most related to human medical issues, as well as databases covering scientific literature (BENSON et al., 2000; WHEELER et al., 2000). NCBI also has a mandate to develop software and bioinformatics tools for biological discovery. Two of the primary search tools at NCBI will be discussed here in more detail, but other NCBI tools and resources can be easily accessed using the Site Map URL provided in Table 1.

2.1.1. Taxonomy Browser

More than 130,000 species are represented by at least one entry in GenBank. For researchers wishing to focus on a specific species or phylogenetic division, the NCBI Taxonomy Browser provides a means to quickly collect GenBank entries of interest. From the NCBI home page, selection of the "TaxBrowser" tab opens the Taxonomy Browser, from which a researcher may either go directly to the taxonomic division of choice using the search box, or burrow down through a taxonomic lineage starting with the Kingdom links on the left side of the page.

Starting with approximately 48,000 species of plants (*Viridiplantae*), burrowing allows researchers to generate a list of the 486 species, subspecies and hybrids of conifers (*Coniferales*) represented in the database. Check boxes at the top of the page enable the display of quantitative information about the number of database entries of various types that are available for each species. Thus, from the *Coniferales* summary it can be determined that 97% of all conifer nucleotide sequences in GenBank come from the *Pinaceae*, with most of the remaining entries (2%) coming from *Cryptomeria japonica*, a member of the *Cupressaceae*. Further breakdown of the *Pinaceae* numbers shows that 68% of the entries come from pines (*Pinus*), mostly loblolly pine (*Pinus taeda*) (61%), while 26% are from spruces (*Picea*), primarily the North American complex of *Picea glauca, P. sitchensis,* and *P. engelmannii x P. sitchensis*, while about 2% of the entries are for Douglas-fir (*Pseudotsuga menziesii*). This, of course, serves to highlight the huge amount of work that remains for the conifer research community to generate sufficiently balanced datasets for fundamental genomic analyses across the entire *Coniferales* division. In reviewing these numbers it is also important to keep in mind that they represent all nucleotide sequence entries, including genomic sequences, like those for genetic markers (SSRs) and organellular genome sequences, and expressed sequences, such as cDNAs and ESTs.

As shown in Figure 2, species-level pages provide summaries of the NCBI databases that contain data for the target organism, including a self-referential Taxonomy database entry. Selecting any these links takes the researcher to that database where searches of the collected data may be further refined by adding Boolean operatives and search terms to the taxonomic identifier that automatically appears in the search box. Thus, the UniGene link for *P. taeda* generates a list of the 13,038 genes that can be predicted so far from clustering of expressed *P. taeda* sequences with all other sequences in GenBank. Similarly, selecting the GEO Expressions link, which leads to a gene expression-profiling tool that queries and summarizes data from the Gene Expression Omnibus (GEO) database, and refining the search query to show only the information related to "laccases", yields 34

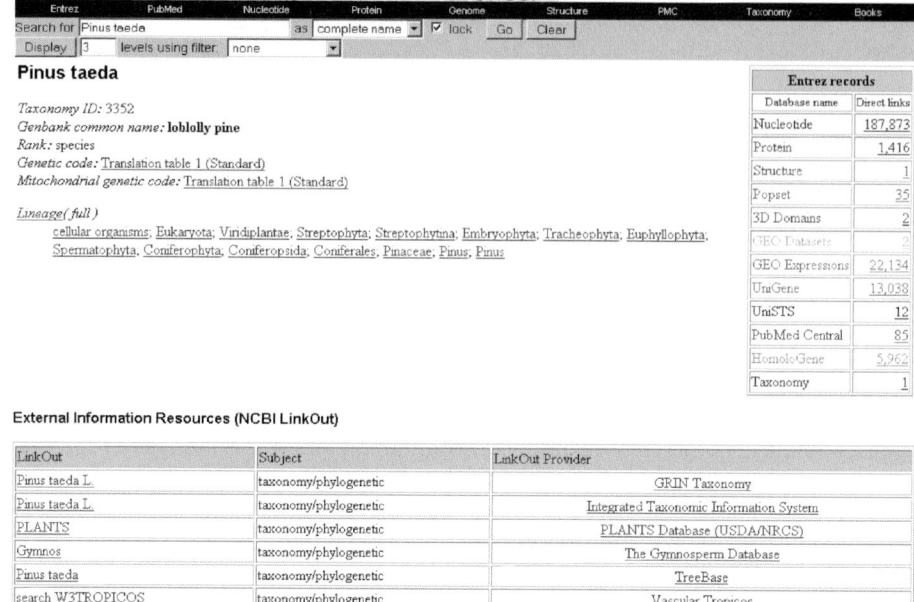

Figure 2. NCBI Taxonomy Browser search results for Pinus taeda (January 2005).

separate comparisons of expression data for the 10 different loblolly pine laccase genes characterized to date. While GenBank, as the best-known NCBI resource, is routinely used by molecular biologists as the point of first deposit for sequence data, the newer, gene expression databases, like GEO, are less familiar and not so well populated with data. However, as more labs become versed in genome-scale, gene expression analyses, such as SAGE and DNA microarrays, these public repositories will also grow rapidly. The conifer research community stands to gain enormous benefits from efforts to encourage rapid submission of functional genomic datasets to this repository.

As with all database search tools, a researcher's ability to gather all relevant information from a database will be limited by how well each database entry has been categorized and cross-referenced. Several good examples of problematic entries that would be of interest to conifer biologists, yet difficult to find using the Taxonomy Browser, are to be found by selecting "Eukaryota" from the opening page and scrolling down to "-unclassified eukaryotes", "-mixed EST libraries". Four of the 19 entries under this heading are for EST collections representing various combinations of three different species of pine infected with three different fungal pathogens. (Note that two other forest trees, *Betula pendula* and *Populus tremula* x *P. tremuloides*, are also among the 19 entries in this unusual category.) GenBank

Figure 3. NCBI Entrez search results for Pinus taeda (January 2005).

entries are structured such that only a single organism may be listed per accession, but because these EST collections might contain either pine or fungal sequences the only appropriate organismal identifier was neither. Obviously, when gathering data for comprehensive genomic analyses it is a good idea to use multiple search tools to query the database.

2.1.2. Entrez

As the better known and most recommended search tool for comprehensive datamining across all NCBI databases, Entrez provides an excellent alternative to the Taxonomy Browser for collecting conifer-specific information. The Entrez search engine performs test searches in each database and returns lists of all accessions containing any instance of the entered text. Figure 3 shows the summary of returns for a search using "*Pinus taeda*" as the search term performed on the same day as the Taxonomy search shown in Figure 2. Although the number of records returned from each database by the two searches tends to be roughly equivalent, there are typically discrepancies, which reflect the fundamental differences in the search tools. Thus, the number of GenBank accessions reported by Entrez is increased over the Taxonomy Browser report by the number of ESTs submitted for the mixed organism libraries that were noted above as being missed by the Taxonomy Browser.

However, the fine mesh of the Entrez net also collects a great deal of non-relevant information. For example, the six "Genome" entries correspond to the five *Arabidopsis thaliana* full-length chromosome sequences and a nearly complete sequence for *Plasmodium falciparum*, each of which were identified by Entrez because some of the gene annotations in these entries noted similarities to *P. taeda* sequences. Similarly, none of the 26 "Gene" entries were for loblolly pine genes; 21 were for *A. thaliana* genes, four were for rice (*Oryza sativa*) genes, and one was for a gene from *P. falciparum*. Filtering the search using Boolean operatives, such as "butnot Arabidopsis" will eliminate some non-relevant items, but restrictive search terms must be chosen carefully to avoid loss of pertinent information. At some point it always becomes necessary to actually review the specific database entries returned by the search parameters, and in this manner learn that three of the hits from the PubChem Compound database were for herbicides in a formulation used to kill pine trees, while the other three hits were for variations on a compound extracted from *Pinus parvifloria* [sic] cones that has been tested for anti-HIV activity. (The misspelled species name is shown as returned by the search, and highlights how human error can complicate searches – the compounds in question did not appear on the Taxonomy Browser page for this species, Txid71644. In this case, the author contacted the PubChem database custodian to recommend adherence to established taxonomic database terms for all natural product sources. Readers are encouraged to help improve our online resources by taking similar actions when even small errors are noted in these public resources.)

2.2. European Bioinformatics Institute

In addition to partnering the EMBL-Bank nucleotide sequence database, the European Bioinformatics Institute (EBI) supports partnered databases for protein sequence (UniProt) and protein structure (EMSD) information, as well as stand-alone databases for vertebrate genome comparisons (Ensembl) and microarray data (ArrayExpress). While both the EBI and the EMBL bioinformatics group at Heidelberg have developed a wide array of computational tools for sequence and structure analysis, most of which are readily accessible using the URLs provided in Table 1, it is the protein sequence database that contains the largest amount of information specific to conifers. There is also an excellent collection of protein sequence analysis tools available from the ExPASy (Expert Protein Analysis System) proteomics server at the Swiss Institute of Bioinformatics (SIB).

2.2.1 The Universal Protein Resource

The Universal Protein Resource (UniProt) is a combined protein sequence and information database that was formed from the consolidation of Swiss-Prot, a protein sequence database started in 1986 at the University of Geneva (now at the

Figure 4. UniProt search results for Pinus taeda proteins (January 2005).

SIB), TrEMBL, a computer-annotated supplement to Swiss-Prot that contains nucleotide translations from EMBL-Bank not already in Swiss-Prot, and the Protein Information Resource (PIR) supported by NIH and housed at Georgetown University. The UniProt database, is primarily queried through either textual or protein sequence (BLAST) searches. Figure 4 shows the results returned from a quick search of the term "*Pinus taeda*." Advanced search tools (Power Search) provide refinements similar to those described for Entrez and the Taxonomy browser, such as restriction of the search term to taxonomy fields (in which case only 138 records were returned in response to "*Pinus taeda*"). Although the content of conifer-specific data in UniProt is as yet fairly modest – even a text search using "*Coniferales*" returned only 3222 entries – the non-redundant protein sequence database at the heart of UniProt (UniRef) is the most comprehensive resource of its kind, and will no doubt remain a primary reference source for genome annotation for some time to come (CAMON *et al.*, 2004). Consequently, all genomics researchers should be familiar with this resource.

2.3. Other Major Centers

The DDBJ develops and maintains a variety of tools for analyzing the nucleotide and protein sequence information, as well as specialty databases, most of which are directed to biomedical research. One source of general information that might be of interest to conifer biologists is the TOP 1000 Organisms list, which provides a running summary of the total number of database entries and sequenced nucleotides available for the 1000 species with the most available data. As of the end of December 2004, *Pinus taeda*, the highest ranked forest tree on the list was #38, with almost 175,000 entries and 98 Mbp of available sequence. Other trees making the top 100 were *Populus balsamifera* subsp. *trichocarpa* (#83), *Citrus sinensis* (#84), *Picea glauca* (#85), and *Populus tremula* x *Populus tremuloides* (#89). However, it should be noted that the imminent inclusion of the completed, 500 Mbp *Populus trichocarpa* genome sequence into the database will likely vault this species into the top 10 on this list.

There are numerous other large genome centers housed at institutions worldwide, and many of which develop and maintain databases and bioinformatics tools that would be useful to conifer biologists interested in genomics. However, The Institute for Genomics Research (TIGR) is singled out here because of the specific tools and databases it maintains for pine and spruce gene sequences. The TIGR Gene Indices Project creates organism-specific databases to organize EST and gene sequence data in such a way as to identify the distinct and individual genes expressed by that organism (QUACKENBUSH *et al.,* 2001). Among the plant indices created by TIGR are genus-level compilations for pine (*Pinus*) and spruce (*Picea*), which most recently identified 35,053 and 27,194 unique sequences for pine (Release 5.0) and spruce (Release 1.0), respectively.

Another resource similar to the TIGR Gene Indices is the collection of compiled EST assemblies made available by the Plant Genome Database (PlantGDB) at Iowa State University (DONG *et al.,* 2004). One of the scientific goals behind the establishment of these assemblies is an improved understanding of the extent and conservation of alternative splicing in plants. With a tighter focus at the species-level, PlantGDB provides assemblies for *P. taeda* and *P. pinaster*, as well as *Gingko biloba*, from the gymnosperms. From among the angiosperm trees and woody allies, PlantGDB has assemblies for *Citrus sinensis, Coffea arabica, Hevea brasiliensis, Manihot esculenta, P. tremula* x *P. tremuloides, Prunus persica, Theobroma cacao,* and *Vitis vinifera*.

The openSputnik comparative genomics platform, recently established at the Centre for Biotechnology in Turku, Finland, provides another set of tools for exploring EST assemblies from a wide variety of plant species, including the conifers, *C. japonica* and *Cycas rumphii*, in addition to *P. taeda* and *P. pinaster*.

3. SPECIFIC RESOURCES FOR CONIFER GENOMICS

One of the more useful online clearinghouses for links and information specific to forest tree species is Dendrome, set up by Dr. David Neale and jointly supported by the US Forest Service, Institute of Forest Genetics and the Department of Plant Sciences at the University of California, Davis. Dendrome maintains curated lists of forest tree genome websites, directories of contact information for researchers studying forest tree genetics and genomics, and information about meetings and jobs in the field. Researchers interested in conifer genomics are strongly encouraged to contribute and update the public information available at Dendrome.

3.1. Physical Genomic Resources

3.1.1. Markers and Maps

The potential that molecular markers have for improving the efficiency of plant breeding programs was recognized early on by the forest tree community, and marker-assisted breeding (MAS) strategies are an important part of most forest tree improvement programs (AHUJA 2001; WANG and SZMIDT 2001). Although difficult to work with in many aspects of their biology, the haploid nature of conifer megagametophyte tissue makes these trees particularly well-suited for certain mapping approaches (TULSIERAM et al., 1992). Consequently, a large number of molecular markers and maps are available for conifers of commercial interest, although it is important to note that given the large size of the typical conifer genome, marker densities are still not at the level necessary for screening progeny at high resolution. While most of the breeding programs employing MAS are using microsatellites, markers based on simple-sequence repeats (SSRs), development efforts are underway to identify markers based on single-nucleotide polymorphisms (SNPs), which should provide the marker density that will be required for high-resolution screening.

In recent years, Neale and co-workers have comprised one of the groups most actively involved in developing public-domain molecular markers for several conifer species (BROWN et al., 2001; CHAGNE et al., 2003; KOMULAINEN et al., 2003; KRUTOVSKY et al., 2004; SEWELL et al., 1999; TEMESGEN et al., 2001). As a consequence, TreeGenes, a component database of Dendrome, is the most data-rich public repository for conifer markers and genetic maps. Work is currently underway to implement the CMAP viewer (FANG et al., 2003) in TreeGenes to facilitate comparative mapping of molecular markers across multiple conifer species. Figure 5 shows a view of one marker collection for loblolly pine as displayed by the CMAP viewer. Although only a few of the marker catalogs housed at Dendrome are currently viewable via the CMAP viewer, the selection will improve as the new database is populated. Under Genome Resources, Dendrome also hosts lists of SSR and EST markers for *P. taeda*, *P. strobus*, and *C. japonica*.

There are other online repositories containing information on specific sets of conifer molecular markers, such as RFLP restriction enzymes and fragment mobilities, SSR primer sets, and candidate gene SNPs. Among the more easily accessed, data-rich databases are ones housed at The Laboratory of Forest Genetics and Tree Improvement of INRA Cestas, which contains SSR and some SNP marker collections for *P. pinaster* and *Picea abies* (CHAGNE et al., 2004; LE DANTEC et al., 2004), as well as a nice collection of marker analysis protocols and literature reviews. (Note that as this chapter was being written, this research group became affiliated with BIOGECO, and the marker datasets were being migrated to a new

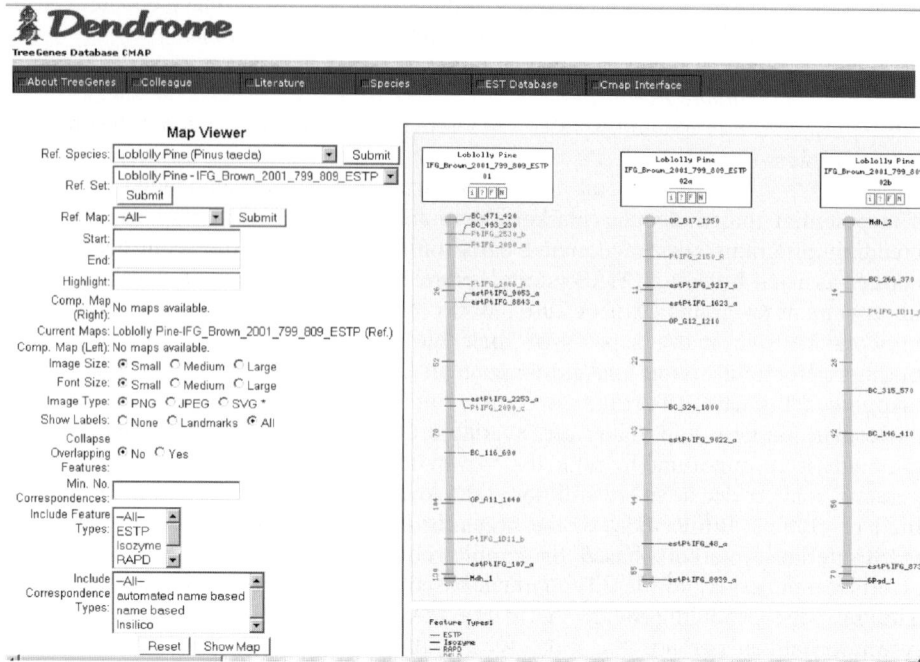

Figure 5. Part of CMAP Browser display of a Pinus taeda molecular marker map showing the relative locations for a collection of isozyme, EST,RAPD, and AFLP markers in TreeGenes.

server. An alternative URL from the one noted in Table 1 is http://cbi.labri.fr/outils/SAM/COMPLETE/index.php.) Information on *Pinus radiata* SSR primers is available from the CSIRO Forestry and Forest Products division.

There is a great deal of information on conifer molecular marker sets for conifers that remains in private databases that are not public or at least widely accessible. It is hoped that reticence on the part of both public and private researchers to the release of such datasets will wain as the conifer research community realizes how much more quickly progress is achieved for all when these genomic resources are made available for general use and testing.

3.1.2. Complete Genome Sequences

Genome sequencing of *Populus trichocarpa* was completed in 2004 and a final draft of the assembled sequence is scheduled for public release in March 2005, making it the first forest tree, and only the third higher plant, to have its genome completely sequenced. Unfortunately, conifer genomes are uniformly large; for example, the average pine genome is roughly 50 times larger than the *Populus* genome and more than 6 times larger than the human genome. Although technical improvements will no doubt continue to increase the speed and reduce the cost of automated DNA sequencing, it is difficult to envision complete sequencing of a conifer genome in the near future, although plans for such a project targeting loblolly pine have been drafted (http://dendrome.ucdavis.edu/lpgp/pdf/prospectus.pdf).

Bacterial Artificial Chromosome (BAC) Libraries.

BAC libraries are a fundamental resource for numerous genomic techniques, including complete genome sequencing (ZHANG and WU 2001). However, because the large size of conifer genomes, relatively few efforts have been made to construct such resources. One report described production of a BAC library for *P. taeda* containing enough inserts to represent approximately 0.1x coverage of the genome (ISLAM-FARIDI et al., 1998), but there were no further reports of work with the library. Recently, a project to create a 10x coverage BAC library for *P. taeda* was initiated by researchers at Mississippi State University, the US Forest Service Southern Research Station, and the Clemson University Genomics Institute. When completed, this BAC library or libraries will comprise approximately 1.7 million clones, making it the largest BAC library ever produced. Ancillary work to be performed in conjunction with BAC library construction will include characterization of repetitive and moderately repetitive DNA sequences through construction of differential $C_o t$ libraries (PETERSON et al., 2002), fluorescence in situ hybridization (FISH) analysis of repetitive DNA location on pine chromosomes, and mapping of repetitive and non-repetitive DNA to anchor the BAC library to the pine chromosomes. The genotype providing the DNA for this BAC resource is clone 7-56, a *P. taeda* selection that is perhaps the most widely distributed genotype in pine breeding programs in the southeastern US (MCKEAND et al., 2003). Some interesting background information on 7-56 and other particularly successful pine breeding stock can be found at http://www.genfys.slu.se/staff/dagl/Successful_trees/ successful_trees.htm.

There is also a report of a BAC library for *P. abies* (CATTONARO et al., 2001), but insufficient information was available from the report to determine the coverage achieved by these researchers or whether the library would be used to develop a public resource.

Organellular Genomes.

Complete chloroplast genome sequences are available from NCBI for two pine species, *P. koraiensis* and *P. thunbergii*; however, no complete conifer mitochondrial genomes have been reported.

3.2. Transcriptomics Resources

Given the technical challenges to characterization of a complete conifer genome any time in the near future, many researchers have focused on characterizing the active gene space of these trees and using functional genomic approaches to assess changes in gene expression under different development and environmental conditions. So far, most of this work has focused on EST discovery and DNA microarrays.

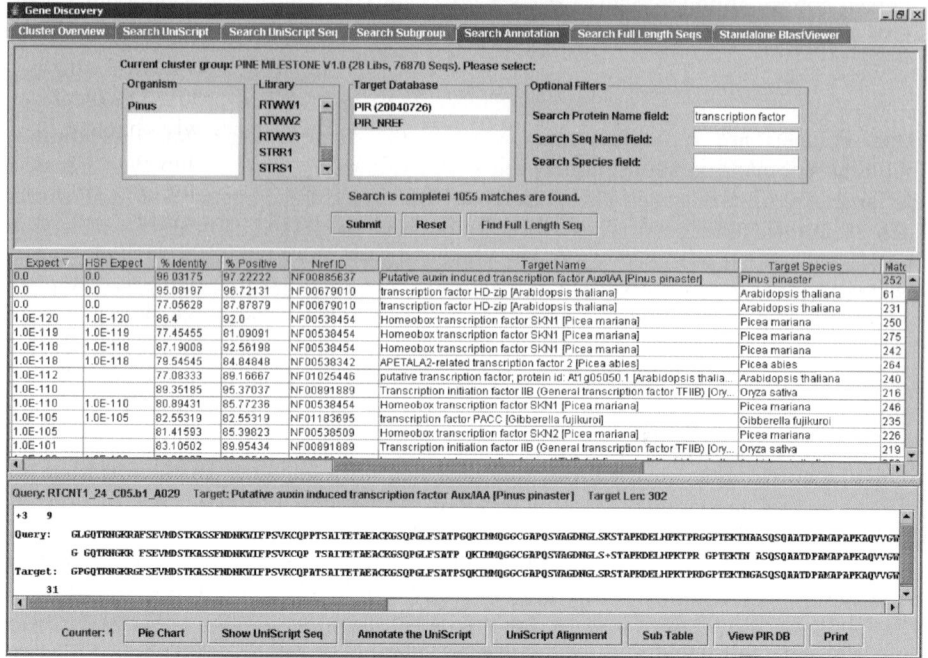

Figure 6. The MAGIC Gene Discovery tool used to display an amino acid sequence alignment between a loblolly pine EST in the Fungen dataset with an homologous AUX/IAA transcription factor identified in Pinus pinaster.

3.2.1. Expressed Sequence Tag (EST) Collections

The first large-scale conifer EST discovery project was a private enterprise, undertaken in the early mid-1990s by Genesis Research and Development Corporation Limited with funding provided through a joint venture of Fletcher Challenge Forests (now Rubicon), International Paper, Westvaco (now MeadWestvaco), and Monsanto, which subsequently pulled out of the venture. The several hundred thousand ESTs that were identified in the project, primarily derived from *P. radiata*, remain in private hands as the underpinning for functional genomics work at Arborgen (KODRZYCKI *et al.*, 2004).

The first public EST discovery effort in a conifer species was initiated at North Carolina State University and focused on the genes in *P. taeda* involved in wood formation (ALLONA *et al.*, 1998). The >70,000 ESTs characterized in that project were deposited in GenBank, but contig assemblies and tentative annotations for the sequences are available from the Center for Computational Genomics and Bioinformatics (CCGB) at the URL listed in Table 1.

Pinus taeda was the species of choice for two more recent EST projects directed by the author at the University of Georgia that have been focused on gene expression patterns in pine roots exposed to a variety of environmental stresses (LORENZ et al., 2005). Approximately 90,000 of an eventual 175,000 ESTs resulting from these projects have so far been deposited in GenBank, but newly characterized ESTs are available from the project's public web server (Fungen, Table 1) immediately upon completion of DNA sequencing. Contig assemblies and tentative annotations for the ESTs are also available. The MAGIC DB Gene Discovery Viewer provides a number of powerful, interactive Java-based tools that enable users to mine the available EST datasets on the local server using such parameters as differential expression, full-length sequences, annotation, or BLAST (PRATT et al., 2005).

In addition to the developing the gene indices mentioned previously, TIGR has collaborated with Georgia Tech researchers on a project to characterize ESTs related to somatic embryogenesis in *P. taeda*. As is the case for the preceding projects, the ESTs generated at TIGR are deposited in GenBank (≈25,000 to date), but are also available in FASTA format from the TIGR website (Table 1). ESTs for *P. pinaster* are being developed and deposited in GenBank by Plomion and co-workers (DUBOS et al., 2003). Two Canadian researchers projects, Treenomix and Arborea, headquartered at the University of British Columbia and Laval University, respectively, are characterizing spruce ESTs. Sequences generated by the Arborea team are available from their web server (Table 1), but both projects are placing ESTs in GenBank (≈80,000 to date). The only other conifer project releasing EST data to public databases is located at Japanese Forestry and Forest Products Research Institute and focuses on *C. japonica*.

With respect to datamining of conifer ESTs, the UniGene Digital Differential Display (DDD) tool available at NCBI enables users to compare UniGene clusters for a species across EST datasets in GenBank to identify differentially expressed genes. Results from DDD comparisons are provided with a measure of statistical significance based on the Fisher Exact Test. So far, UniGene clusters have only been generated for *P. taeda*, but NCBI UniGenes will be generated for other conifers as sufficient numbers of ESTs are deposited in GenBank.

3.2.2. Serial Analysis of Gene Expression (SAGE) Tag Collections

The UniGene DDD tool performs transcriptional profiling by direct quantitation of DNA sequence (EST) information associated with UniGene clusters. Two other functional genomic techniques, Serial Analysis of Gene Expression (SAGE) (VELCULESCU et al., 1995) and Massively-Parallel Signature Sequencing (MPSS) (BRENNER et al., 2000), provide quantitative gene expression data based on counts of short, but unique, DNA sequences derived directly from mRNA transcripts isolated from tested tissues. Although MPSS data has been described for conifers, a SAGE study of gene expression differences between wood-forming tissues in the crown and base of a single *P. taeda* tree has been published (LORENZ and DEAN 2002). The complete SAGE dataset is available from a couple of sources, including the GEO database at NCBI. Because the SAGE technique has been heavily utilized for work in molecular oncology, NCBI has developed a set of bioinformatics tools, known as SAGEmap, specifically for the analysis of the SAGE datasets deposited with NCBI. Conifer biologists can, thus, take advantage of these tools to probe the SAGE datasets for the identity of genes whose expression may be related to differences in wood quality.

3.2.3. DNA Microarrays

Increasingly, DNA microarrays are the technique of choice for researchers looking to perform transcriptional profiling studies. However, the front-end costs for developing these tools are steep, and access to microarrays designed for conifer studies remains limited. The NCSU *P. taeda* EST program developed a cDNA microarray that was used to study wood formation (EGERTSDOTTER *et al.*, 2004; WHETTEN *et al.*, 2001) and drought stress responses (HEATH *et al.*, 2002; WATKINSON *et al.*, 2003) in *P. taeda*, as well as adventitious root formation in *Pinus contorta* (BRINKER *et al.*, 2004) and somatic embryo development in *P. abies* (STASOLLA *et al.*, 2004; VAN ZYL *et al.*, 2002). However, the NCSU microarray is not publicly available. The University of Georgia group developing *P. taeda* ESTs recently announced the availability of a 13,500 element cDNA array for public research projects. Information on access to the UGA array is available from the Fungen webserver. A second *P. taeda* cDNA microarray has been proposed by the TIGR-Georgia Tech collaboration, but details are not yet available. Plans are also in the works for development of a spruce cDNA microarray based on the ESTs being released from the Arborea and Treenomix projects, but details are likewise unavailable. Only one commercial venture – PICME (Table 1) – supplies cDNA microarrays based on conifer (*P. pinaster*) gene sequences.

With respect to public access to microarray data, the drought stress datasets generated using the NCSU microarrays are available from the Expresso microarray data management system at Virginia Tech (Table 1), while the *P. contorta* adventitious rooting dataset has been deposited in the GEO database where it may be queried using the NCBI toolset. (Hopefully, all microarray datasets from publicly funded conifer research projects will eventually be deposited in the GEO database.) As part of the MTA required for access to the UGA microarray, users will be required to deposit their raw data scans to the MAGIC Microarray database housed at Fungen. The agreement specifies that deposited data will be held in confidence for up to 180 days, but after that period will be made available to the public.

3.3. Proteomics Resources

The next link in the genomics chain running from genes to phenotype is at the level of proteins. The complexity of protein structure and function makes proteomics studies much more difficult than the analysis of nucleic acids in physical or functional genomics studies, even in model systems for which the complete genome sequence is available. Nonetheless, a few groups are pursuing proteomics studies in conifers (COSTA et al., 1999; LIPPERT et al., 2005; POULIS and ADERKAS 2003; THIELLEMENT et al., 2002), although publicly available datasets and tools remain scarce. The INRA group at Pierroton has developed the Protic database (Table 1), which provides public access to protein information that they have collected from a variety of *P. pinaster* tissues (PLOMION et al., 2000). According to their website the Treenomix group is using a proteomics approach to study the response of spruces to insect herbivory, in addition to the somatic embryogenesis studies (LIPPERT et al. 2005); however, the site does not currently provide public access to the data.

Just as there are online resources for the independent collection and analysis of transcriptional information from all manner of plant species, there are specialty databases for protein information. One that should be of interest to all manner of forest tree researchers is one at the Center for Plant Cell Biology (UC Riverside) focusing on plant cell wall proteins (COSTA et al., 1999). ExPASy maintains a database of 2-dimensional polyacrylamide gels (2D-PAGE) images, and although no conifer data has yet been deposited in that repository there is good reason to expect conifer proteomics data to enter this database in the future. An excellent collection of reviews on the current state of plant proteomic research is to be found in a 2-volume special issue of Phytochemistry from June 2004 (BOLWELL et al., 2004a; BOLWELL et al., 2004b).

3.4. Metabolomics Resources

Metabolomics has been described as the missing piece of the plant genomics puzzle (HALL et al., 2002; SUMNER et al., 2003). There are large public and private centers being developed to house projects, instrumentation and databases focused on plant metabolomics (e.g. http://www.metabolomics.bbsrc.ac.uk/ and http://www.plant.wur.nl/), but very little data is yet available, particularly for conifers. That said, community standards for plant metabolomic data release along the lines of the MIAME standards for microarray experiments have been recommended (JENKINS et al., 2004), so it is only a matter of time before metabolomic datasets are more widely released and subjected to comparative analyses in a manner similar to what is being done with other functional genomic datasets. No doubt the field of plant metabolomics will bear watching by conifer researchers.

4. CONCLUSIONS

Genomic approaches are changing everything about the manner in which biological research is pursued. The information and resources available for genomics studies across all kingdoms of living organisms are increasing at rates that challenge researchers to keep up. Although modest in comparison to the materials available for biomedical or model system research, there is already a great deal of information available for conifers and other forest tree species. This chapter provides a starting point from which interested researchers are encouraged to explore and contribute to the online resources available for conifer biology.

5. REFERENCES

AHUJA, M. R., 2001 Recent advances in molecular genetics of forest trees. Euphytica **121:** 173-195.

ALLONA, I., M. QUINN, E. SHOOP, K. SWOPE, S. ST CYR et al., 1998 Analysis of xylem formation in pine by cDNA sequencing. Proceedings of the National Academy of Sciences of the United States of America **95:** 9693-9698.

BENSON, D. A., I. KARSCH-MIZRACHI, D. J. LIPMAN, J. OSTELL, B. A. RAPP et al., 2000 GenBank. Nucleic Acids Res **28:** 15-18.

BOLWELL, G. P., A. R. SLABAS and J. P. WHITELEGGE, 2004a Proteomics: empowering systems biology in plants. Phytochemistry **65:** 1443-1447.

BOLWELL, G. P., A. R. SLABAS and J. P. WHITELEGGE, 2004b Proteomics: empowering systems biology in plants. Phytochemistry **65:** 1665-1669.

BRENNER, S., M. JOHNSON, J. BRIDGHAM, G. GOLDA, D. H. LLOYD et al., 2000 Gene expression analysis by massively parallel signature sequencing (MPSS) on microbead arrays. Nature Biotechnology **18:** 630-634.

BRINKER, M., L. VAN ZYL, W. B. LIU, D. CRAIG, R. R. SEDEROFF et al., 2004 Microarray analyses of gene expression during adventitious root development in Pinus contorta (1[w]). Plant Physiology **135:** 1526-1539.

BROWN, G. R., E. E. KADEL, D. L. BASSONI, K. L. KIEHNE, B. TEMESGEN et al., 2001 Anchored reference loci in loblolly pine (Pinus taeda L.) for integrating pine genomics. Genetics **159:** 799-809.

CAMON, E., M. MAGRANE, D. BARRELL, V. LEE, E. DIMMER et al., 2004 The Gene Ontology Annotation (GOA) Database: sharing knowledge in Uniprot with Gene Ontology. Nucleic Acids Research **32:** D262-D266.

CATTONARO, F., I. JURMAN and A. ZUCCOLO, 2001 Microscale organization of repetitive sequences in the Norway spruce genome (Picea abies L., Karst.), pp. in *Proceedings of the XLV Italian Society of Agricultural Genetics - SIGA Annual Congress*, Salsomaggiore Terme, Italy.

CHAGNE, D., G. BROWN, C. LALANNE, D. MADUR, D. POT et al., 2003 Comparative genome and QTL mapping between maritime and loblolly pines. Molecular Breeding **12:** 185-195.

CHAGNE, D., P. CHAUMEIL, A. RAMBOER, C. COLLADA, A. GUEVARA et al., 2004 Cross-species transferability and mapping of genomic and cDNA SSRs in pines. Theoretical and Applied Genetics **109:** 1204-1214.

COSTA, P., C. PIONNEAU, G. BAUW, C. DUBOS, N. BAHRMANN et al., 1999 Separation and characterization of needle and xylem maritime pine proteins. Electrophoresis **20:** 1098-1108.

DONG, Q. F., S. D. SCHLUETER and V. BRENDEL, 2004 PlantGDB, plant genome database and analysis tools. Nucleic Acids Research **32:** D354-D359.

DUBOS, C., G. LE PROVOST, D. POT, F. SALIN, C. LALANE et al., 2003 Identification and characterization of water-stress-responsive genes in hydroponically grown maritime pine (Pinus pinaster) seedlings. Tree Physiology **23:** 169-179.

EGERTSDOTTER, U., L. M. VAN ZYL, J. MACKAY, G. PETER, M. KIRST et al., 2004 Gene expression during formation of earlywood and latewood in loblolly pine: Expression profiles of 350 genes. Plant Biology **6:** 654-663.

FANG, Z., M. POLACCO, S. CHEN, S. SCHROEDER, D. HANCOCK et al., 2003 cMap: the comparative genetic map viewer. Bioinformatics **19:** 416-417.

FLEISCHMANN, R. D., M. D. ADAMS, O. WHITE, R. A. CLAYTON, E. F. KIRKNESS et al., 1995 Whole-Genome Random Sequencing and Assembly of Haemophilus-Influenzae Rd. Science **269:** 496-512.

HALL, R., M. BEALE, O. FIEHN, N. HARDY, L. SUMNER et al., 2002 Plant Metabolomics: The Missing Link in Functional Genomics Strategies. Plant Cell **14:** 1437-1440.

HEATH, L. S., N. RAMAKRISHNAN, R. R. SEDEROFF, R. W. WHETTEN, B. I. CHEVONE et al., 2002 Studying the functional genomics of stress responses in loblolly pine with the Expresso microarray experiment management system. Comparative and Functional Genomics **3:** 226-243.

ISLAM-FARIDI, N., Y.-L. CHANG, H. ZHANG, C. KINLAW, R. L. DOUDRICK et al., 1998 Construction of a pine BAC library, pp. P18 in *Plant & Animal Genomes VI Conference*, San Diego, CA.

JENKINS, H., N. HARDY, M. BECKMANN, J. DRAPER, A. R. SMITH et al., 2004 A proposed framework for the description of plant metabolomics experiments and their results. Nature Biotechnology **22:** 1601-1606.

KODRZYCKI, B., S. CHANG, C. EAGLETON, K. R. FOUTZ, K. C. GAUSE et al., 2004 Functional genomics in loblolly pine and eastern cottonwood, pp. W93 in *Plant & Animal Genomes XII Conference*, San Diego, CA.

KOMULAINEN, P., G. R. BROWN, M. MIKKONEN, A. KARHU, M. R. GARCIA-GIL et al., 2003 Comparing EST-based genetic maps between Pinus sylvestris and Pinus taeda. Theoretical and Applied Genetics **107**: 667-678.

KRUTOVSKY, K. V., M. TROGGIO, G. R. BROWN, K. D. JERMSTAD and D. B. NEALE, 2004 Comparative mapping in the Pinaceae. Genetics **168**: 447-461.

LE DANTEC, L., D. CHAGNE, D. POT, O. CANTIN, P. GARNIER-GERE et al., 2004 Automated SNP detection in expressed sequence tags: statistical considerations and application to maritime pine sequences. Plant Molecular Biology **54**: 461-470.

LIPPERT, D., J. ZHUANG, S. RALPH, D. ELLIS, M. GILBERT et al., 2005 Proteome analysis of early somatic embryogenesis in *Picea glauca*. Proteomics **5**: 461-473.

LORENZ, W. W., and J. F. D. DEAN, 2002 SAGE Profiling and demonstration of differential gene expression along the axial developmental gradient of lignifying xylem in loblolly pine (Pinus taeda). Tree Physiology **22**: 301-310.

LORENZ, W. W., F. SUN, C. LIANG, Z. XIN, D. KOLYCHEV et al., 2005 Identification Of Drought Response Genes In Loblolly Pine (Pinus Taeda L.) By Analysis Of Expressed Sequence Tag Libraries, pp. W109 in *Plant & Animal Genomes XIII Conference*, San Diego, CA.

MCKEAND, S., T. MULLIN, T. BYRAM and T. WHITE, 2003 Deployment of genetically improved loblolly and slash pines in the south. Journal of Forestry **101**: 32-37.

PETERSON, D. G., S. R. SCHULZE, E. B. SCIARA, S. A. LEE, J. E. BOWERS et al., 2002 Integration of Cot analysis, DNA cloning, and high-throughput sequencing facilitates genome characterization and gene discovery. Genome Research **12**: 795-807.

PLOMION, C., C. PIONNEAU, J. BRACH, P. COSTA and H. BAILLERES, 2000 Compression wood-responsive proteins in developing xylem of maritime pine (Pinus pinaster Ait.). Plant Physiology **123**: 959-969.

POULIS, B., and P. V. ADERKAS, 2003 Identification of proteins present in Douglas-fir ovular secretions using proteomics, pp. S3.4 in *IUFRO Tree Biotechnology*, Umea, Sweden.

PRATT, L. H., M.-M. CORDONNIER-PRATT, C. LIANG, F. SUN, H. WANG et al., 2005 MAGIC DB As A Resource For Pine Functional Genomics, pp. W122 in *Plant & Animal Genomes XIII Conference*, San Diego, CA.

QUACKENBUSH, J., J. CHO, D. LEE, F. LIANG, I. HOLT et al., 2001 The TIGR Gene Indices: analysis of gene transcript sequences in highly sampled eukaryotic species. Nucleic Acids Research **29**: 159-164.

SEWELL, M. M., B. K. SHERMAN and D. B. NEALE, 1999 A consensus map for loblolly pine (Pinus taeda L.). I. Construction and integration of individual linkage maps from two outbred three-generation pedigrees. Genetics **151**: 321-330.

STASOLLA, C., P. V. BOZHKOV, T. M. CHU, L. VAN ZYL, U. EGERTSDOTTER et al., 2004 Variation an transcript abundance during somatic embryogenesis in gymnosperms. Tree Physiology **24**: 1073-1085.

SUMNER, L. W., P. MENDES and R. A. DIXON, 2003 Plant metabolomics: large-scale phytochemistry in the functional genomics era. Phytochemistry **62**: 817-836.

TEMESGEN, B., G. R. BROWN, D. E. HARRY, C. S. KINLAW, M. M. SEWELL et al., 2001 Genetic mapping of expressed sequence tag polymorphism (ESTP) markers in loblolly pine (Pinus taeda L.). Theoretical and Applied Genetics **102**: 664-675.

THIELLEMENT, H., M. ZIVY and C. PLOMION, 2002 Combining proteomic and genetic studies in plants. Journal of Chromatography B-Analytical Technologies in the Biomedical and Life Sciences **782**: 137-149.

TULSIERAM, L. K., J. C. GLAUBITZ, G. KISS and J. E. CARLSON, 1992 Single Tree Genetic-Linkage Mapping in Conifers Using Haploid DNA from Megagametophytes. Bio-Technology **10**: 686-690.

VAN ZYL, L., S. VON ARNOLD, P. BOZHKOV, Y. Z. CHEN, U. EGERTSDOTTER et al., 2002 Heterologous array analysis in Pinaceae: hybridization of Pinus taeda cDNA arrays with cDNA from needles and embryogenic cultures of P-taeda, P-sylvestris or Picea abies. Comparative and Functional Genomics **3**: 306-318.

VELCULESCU, V. E., L. ZHANG, B. VOGELSTEIN and K. W. KINZLER, 1995 Serial Analysis of Gene-Expression. Science **270**: 484-487.

WANG, X. R., and A. E. SZMIDT, 2001 Molecular markers in population genetics of forest trees. Scandinavian Journal of Forest Research **16:** 199-220.

WATKINSON, J. I., A. A. SIOSON, C. VASQUEZ-ROBINET, M. SHUKLA, D. KUMAR *et al.*, 2003 Photosynthetic acclimation is reflected in specific patterns of gene expression in drought-stressed loblolly pine. Plant Physiology **133:** 1702-1716.

WHEELER, D. L., C. CHAPPEY, A. E. LASH, D. D. LEIPE, T. L. MADDEN *et al.*, 2000 Database resources of the National Center for Biotechnology Information. Nucleic Acids Res **28:** 10-14.

WHETTEN, R., Y. H. SUN, Y. ZHANG and R. SEDEROFF, 2001 Functional genomics and cell wall biosynthesis in loblolly pine. Plant Molecular Biology **47:** 275-291.

ZHANG, H. B., and C. C. WU, 2001 BAC as tools for genome sequencing. Plant Physiology and Biochemistry **39:** 195-209.

CHAPTER 5

A NEW DIRECTION FOR CONIFER GENOMICS

KERMIT RITLAND[1], STEVEN RALPH[1,2], DUSTIN LIPPERT[1,2], DAINIS RUNGIS[1] AND JÖRG BOHLMANN[1,2]

[1]Department of Forest Sciences
[2]Michael Smith Laboratories
University of British Columbia
Vancouver B.C. V6T1Z4 Canada

Abstract. Extensive genome resources have now been developed for both spruce and pine. However as other technologies develop and costs drop, other resources become feasible. One such resource, not yet available in conifers, is collecting complete cDNA sequences (full length messenger RNA transcripts). We describe our approach for obtaining such sequences and the importance of this resource. Second, existing resources and new technologies are allowing new approaches in functional and comparative genomics. In functional genomics, new technologies and the integration of mRNA, protein and metabolite expression profiling promises to improve the identification of gene function, interaction and regulation. In comparative genomics, an apparent slowdown of evolutionary rate in the pine family should allow genome information to be easily transferred between spruce and pine (and other members of the pine family). Ultimately, genomics research should impact practical problems and traits in breeding and management programs. In our own program, we target spruce resistance mechanisms to the spruce weevil, a serious pest in British Columbia.

1. INTRODUCTION

Conifer genomics is now entering the arena of potential applications. The basic toolbox of genomic resources has been constructed for both spruce and pine. Large-scale expressed sequence tag (EST) databases of size 150-200K and associated cDNA clone stocks have been developed for spruce (white spruce or *Picea glauca* and Sitka spruce or *Picea sitchensis*) and loblolly pine (*Pinus taeda*). cDNA microarray chips have been designed and utilized for both species, and the development of these EST databases are allowing pioneering proteomic studies. A BAC library is being developed for loblolly pine and proposed for spruce. Many single nucleotide polymorphisms (SNPs), as well as microsatellites in untranslated regions of messenger RNA, can be identified in the EST databases, and additional SNPs can be easily identified using primers designed from ESTs.

It would seem that genomics can add a new dimension to tree breeding, providing accelerated breeding schedules and identification of novel traits upon which to focus breeding efforts. The most advanced tree breeding programs have progressed only three or four generations, as compared to thousands for many crop plants. New targets for breeding might be identified via better understanding of the genetics of disease and pest resistance, of cold and drought tolerance, even of CO_2 sequestration. As well, one can imagine that genomics might provide novel methods to monitor the health of existing forest stock and make revised predictions about the longer-term sustainability of conifer forests in our landscape. However, the recent dizzying advances of genomics may invite a too-fast rush into applications, as our actual level of genomics knowledge is still quite primitive, and like any revolutionary area, the ideas and hypotheses of interest in genomics will change rapidly. Even now it is apparent that the number of genes and their interdependencies guarantees that in the real world, changes of single genes result in a multitude of genetic effects, most unpredictable.

With these cautions, we will review some recent progress and prospects in the spruce portion of our genome project, *Treenomix*, funded by Genome Canada and Genome British Columbia. We focus on (1) the generation of full-length cDNA resources, (2) the technologies of mRNA, protein and metabolite expression profiling, and how these levels of gene expression might be integrated, and (3) the transfer of genomic information among members of the pine family. In our project, the realistic target of genomics activities is resistance mechanisms to the spruce weevil.

2. SPRUCE WEEVIL RESISTANCE AS A FOCUS FOR GENOMICS APPLICATIONS

The spruce shoot weevil (a.k.a. white pine weevil, *Pissodes strobi*), budworms, and certain bark beetle species are some of the most destructive insect pests of spruce forests worldwide (ALFARO et al., 2002). Larval feeding of the spruce weevil severely damages or kills the leading shoots of susceptible host trees, resulting in reduced growth and tree deformation (ALFARO et al., 2002). Weevil damage is a substantial risk factor in plantations of some of the most valuable timber trees in Canada, like Sitka spruce (*Picea sitchensis*) in coastal British Columbia (B.C.), white spruce (*P. glauca*) and interior spruce (*P. glauca x P. engelmanni*) in the interior of B.C., and Norway spruce (*P. abies*) in eastern Canada. Sitka spruce suffers the greatest damage. As a result, annual planting of Sitka spruce in B.C. has been reduced from 10 million seedlings to less than a million (KING et al.,1997).

Although genotypes of Sitka and interior spruce resistant to weevil damage have been described (ALFARO et al., 2004; KING and ALFARO 2004; KING et al., 2004), genomic mechanisms of conifer defence and resistance remain to be discovered and functionally characterized. To date, most mechanistic studies of conifer defence have been at the anatomical, chemical, and more recently at the biochemical/molecular levels

(TRAPP and CROTEAU 2001; HUBER et al., 2004). Species of spruce have undoubtedly evolved a plethora of anatomical and chemical, constitutive and inducible, direct and indirect defences against insect pests and insect-associated pathogens (HUBER et al., 2004). These include terpenoid and phenolic chemical defences, antimicrobial and potentially insecticidal peptides, a complex system of proteolytic enzymes and protease inhibitors, and other defences. Only the terpenoid oleoresin defence system has been thoroughly studied to date at the metabolic, biochemical and molecular genetic levels in a few conifer species. Terpenoid defences have been associated with resistance against weevils in Sitka and interior spruce, however, the genomic control of this resistance is not known. The genomic hardwiring and phenotypic plasticity of terpenoid, phenolic, and other chemical defences contributed, at least in part, to the successful evolution of long-lived conifer trees, which often survive for several hundred years in the same location, defeating faster evolving insect pests and pathogens. Both from an evolutionary and economic perspective, genomics will allow fruitful advances of our knowledge of how resistance evolves and is expressed, and how we might use these mechanisms to improve forest health.

3. NEW DIRECTIONS

3.1. Full-length cDNAs as the characterization of the expressed conifer genome

A major focus of our current *Treenomix* project has been the generation of high quality EST sequences. This requires sampling a diversity of tissue types under a variety of treatment conditions, and improved methods for constructing libraries from these tissues. To date (March 31, 2005), our sequence assembly of 113,612 3' ESTs has identified 39,280 unique transcripts, representing 16,688 contigs and 22,592 singletons (build #10 using CAP3 software at 95% identity and 40bp window). The average length of our *Treenomix* spruce quality- and vector-clipped EST sequences was 652bp. At the conclusion of our project we plan to have ca. 160,000 3' reads and ca. 41,000 5' reads. While we currently have nearly 40,000 unique transcripts, the actual number is about 2/3 this or 26,000, as estimated by performing CAP3 assembly a selected gene family, then completely sequencing all clone members in the family. Often different or unique transcripts were identified as members of the same gene (this can occur with sequencing errors, polymorphism, chimeras, and truncations of inserts).

However, in the absence of complete genome sequencing, sequences of full-length cDNA (FL-cDNA) clones, e.g., clones that contain the entire protein coding sequence, is the best way to fully characterize the expressed genome, as opposed to a large number of unique ESTs. The large size of conifer genomes currently precludes genome sequencing, and FL-cDNAs are the best alternative for deciphering the conifer genome, in terms of functional characterization and marker development. Also, as conifers are quite distantly related to angiosperms, FL-cDNAs are useful for gene annotation, as the likelihood of good BLAST hits to annotated angiosperm genes is more likely. Finally, even for sequenced organisms, construction of full-length cDNAs is a central focus in the post-sequence era of the various genome projects, such as *Arabidopsis*.

Over the last two years *Treenomix* has pioneered developments towards the first conifer ORFeome, FL-cDNA sequence database. There are two major steps in this activity. First, cDNA libraries that contain mainly full-length messenger RNA inserts are constructed. Among the 21 spruce cDNA libraries we have constructed, five have been full-length. In developing these libraries, a number of issues and methods pertaining to the synthesis and cloning of full-length cDNAs were identified and developed. In *Treenomix*, our methods involve modification of the RIKEN cap-trapper FL-cDNA protocol (CARNINCI et al., 2000). To maximize the recovery of clones with complete open reading frames, only cDNAs generated from these FL-cDNA libraries are appropriate in the following process of identifying FL-cDNAs.

Second, an efficient pipeline for sequencing full-length inserts is used. In earlier work with poplar, we laboriously estimated insert sizes by cDNA insert size by restriction digest and gel electrophoresis, and chose clones with the largest insert size for sequencing. With sequencing costs dropping from $4.79 CAN per read to $2.64 CAN per read over the three year course of our project, we recently adopted a different strategy which largely skips the insert sizing. In this strategy, the insert size distribution of a FL cDNA library is estimated from a small number of clones. If the library is validated as largely full-length as evidenced by the insert size distribution, we then proceed to sequence both 5' and 3' individuals from many randomly selected clones within the library. In our project, a total of 35,000 putative FL clones will be sequenced in this way from three FL libraries.

After sequencing both ends, valid full-length clones are identified by several selection criteria: those without potential contamination (e.g. bacteria, yeast and fungi) as determined by BLASTX searches, presence of the expected 5' cloning structure or cap, and the presence of a polyA tail at the 3' end. We also try to identify FL-cDNA sequences associated with defence and adaptation, utilizing transcript profiling along with a review of the literature to identify genes of interest. Now, often the 5' and 3' ends overlap (in roughly 60% of the clones), meaning immediate recovery of a FL-cDNA; otherwise primers are designed to attempt to bridge the two ends. In one rearray, we sequence inward from the inside ends of both the 5' and 3' initial reads. This captures most of the remaining complete cDNAs (up to size ca. 2500 bp). A small percentage (<5%) need further sequencing. The total success rate in our work with poplar was 4700 successes in 5100 clones. In spruce, our initial goal (by the end of the currently funded *Treenomix* project in September 2005) is to obtain the sequences of 5,000 FL-cDNAs, and over a longer term, to sequence an additional 10,000 FL-cDNAs.

Another issue in complete cDNA sequences is sequence quality. Most EST data are single-pass reads and hence not of the quality (>35 or 40 phred) acceptable by the current standards of the genome community. For this reason, we actually verify each sequence read above with two additional reads, one for verification and a second should one of the pair of reads fail.

3.2. mRNA and protein expression profiling, and their integration

The most immediate application of EST and cDNA resources is the study of gene expression via cDNA microarrays. Developing microarrays in the face of expanding EST and cDNA clone collections has been an iterative process. Our first spruce cDNA microarray consisted of 9,700 unique clones in 2002. This was followed a year later by a 16.7K array, and in 2005 we will complete a 22K array of unique clones. Baseline gene expression profiles have been established for Sitka spruce pest interactions, with greenhouse studies of spruce budworms and weevils, insect oral secretion elicitors, wounding, and methyl jasmonate (HUBER *et al.*, 2004; RALPH *et al.*, unpub.).

Previous studies of defense against insects or pathogens in spruce had a focus on a few, selected defence genes. This approach may have left many possible defence genes and mechanisms unnoticed. The new *Treenomix* resources for spruce genomics now allow for a much wider screening and discovery of constitutive and weevil-induced defences in resistant and susceptible genotypes. Our hypotheses are that (1) differences in resistant and susceptible trees are reflected in differences in constitutive or induced transcript, protein and metabolite profiles, (2) apparent differences in the resistance mechanisms of Sitka spruce (H898) and interior spruce (PG29) are reflected in differences in constitutive or induced transcript profiles, (3) defence and resistance mechanisms against weevils are associated with conifer-specific cell specialization of constitutive resin ducts, induced traumatic resin ducts, phloem parenchyma cells, and other specialized cell structures such as bark sclereids, (4) insect induced transcription factors control coordinated anatomical, cellular, and biochemical defences in spruce.

A second application of EST and full-length c-DNA (FL-cDNA) databases is for large-scale identification of expressed spruce proteins (LIPPERT *et al.*, 2005). Using 2-dimensional gel electrophoresis (2D-GE) coupled with liquid chromatography and tandem mass spectrometry (LC-MS/MS) we can reliably identify up to 80% of all proteins by using bioinformatics tools as described in LIPPERT *et al.* (2005). The success of this approach demonstrates the importance of deep EST and FL-cDNAs resources for proteome analysis in a species without a complete genome sequence.

Proteomics, a field in itself, has witnessed a number of technological developments in the past few years. By contrast, DNA sequencing methods have been largely static for a decade. Thus one can argue that most new advances will take place in the study of protein expression and interactions. However, currently there is no technology capable of measuring the entire proteome of an organism in a single analysis (unlike microarrays). To expand the coverage of proteome analysis it is important to include a methodical sample fractionation strategy (BRUNET *et al.*, 2003; STASYK and HUBER 2004). Separating each sample into a series of reduced-complexity fractions can greatly improve sensitivity and effectiveness of protein expression profiling. With further assembly of EST and FL cDNA sequences we may also be able to use global protein expression profiling via isotope ratioing (ICAT, iTRAQ) in spruce, which we have successfully applied in studies of the plant *Arabidopsis* in *Treenomix* (LIPPERT *et al.*, unpub.).

A final dimension of expression is the metabolome, which is the complete set of small molecules in a particular tissue or organism (WECKWERTH et al., 2003). Differences in genotype or changes of gene and protein expression can result in changes of the metabolome. Much control of metabolism occurs downstream of gene- and protein expression by regulation of enzyme activities, metabolic channelling, cell compartmentalization, or transport. Characterization of metabolites can provide an important indication of phenotype and genotype (FIEHN 2002).

However, plant metabolomes are more complex than most animals due to their extensive secondary metabolism: sessile organisms plants rely on secondary metabolites for interactions with their environment, including defence against herbivores or pathogens. With regard to conifer defence and resistance, secondary metabolites, e.g. terpenoids and phenolics, and signal molecules mediating insect-induced activation of secondary metabolisms are important (HUBER et al., 2004; MILLER et al., 2005). Metabolite profiles associated with insect-defence and resistance in conifers are largely unknown, with the exception of some terpenoid profiles (MARTIN et al., 2002, 2003; MILLER et al., 2005).

To pursue the metabolomics of insect resistance, our plan is to develop quantitative and qualitative profiles of (1) terpenoids, (2) phenolic secondary metabolites, and (3) signal molecules in constitutive and insect-induced tissues of resistant and susceptible Sitka spruce and interior spruce. A combination of different methods is needed to obtain profiles of complex mixtures of hundreds of secondary metabolites in spruce. Gas chromatography with mass spectrometry (GC/MS) or flame ionization detection (GC/FID), and liquid chromatography (LC/MS) methods allow high-resolution separation of very similar compounds, such as terpenoids, and their identification based on retention indices and fragmentation patterns in reference databases (MARTIN et al., 2002, 2003; MILLER et al., 2005). GC methods are ideal for analysis of many low molecular weight terpenoid and signal metabolites, but are limited to molecules that are volatile under the operating conditions of these instruments.

Now, an entry gate into genomewide systems biology would be the integration of mRNA, protein, and metabolite expression profiles. However, it is widely recognized that each level of expression profiling (gene-, protein-, and metabolite profiling) can only yield partly overlapping fingerprints (TIAN et al., 2004). At least, these approaches will lead to complementary sets of candidate genes, proteins, and metabolites for further investigations of their function in defence and resistance. Also, we have identified matching, insect-induced spruce transcript and protein IDs using microarrays and 2D-GE/LC-MS/MS. For example, dirigent proteins are among the most strongly up-regulated in bark tissues upon weevil attack, detected both by microarrays and qRT-PCR and at the proteome level (RALPH et al., unpub.; CHOWRIRA et al., unpub.).

Expression profiling at the genome, proteome and metabolite level, both for complex tissues and for micro-dissected tissues and cells, will lead to a comprehensive inventory of the weevil-induced defense system in resistant and susceptible Sitka and interior spruce. By comparison with susceptible trees we may be able to identify the most important induced defences in resistant trees.

3.3. Comparative genomics of the pine family

Despite their large size (6X human, 100X *Arabidopsis*), conifer genomes seem to be remarkably conserved. Across the Pinaceae (pine family — pine, spruce, fir, hemlock, Douglas-fir, larch), chromosome number is the almost same (N=12, except N=13 for Douglas-fir). In a pioneering conifer comparative genetic map, KRUTOVSKY et al. (2004) found extensive synteny and colinearity between pine and spruce, consistent with the hypothesis of conservative chromosomal evolution. This suggests that, as in the grass family, linkages and clusters of gene functions are preserved among species.

More interestingly, as inferred by sequence comparisons with *Pinus*, nucleotide substitution rates appear to be an order of magnitude lower in *Pinus* compared to angiosperms. Our own comparison of spruce vs. pine EST contigs has found an average synonymous substitution rate of about 4×10^{-10} per year (RITLAND et al., unpub.), assuming a divergence time of 140 MYA. This value is five times, or even ten times, less that observed for angiosperms, and is probably due to a lower mutation rate, as under the neutral theory, the substitution rate equals the mutation rate. Low levels of nucleotide heterozygosity have been found in recent studies of single nucleotide polymorphisms, a level consistent with a low mutation rate of 1.17×10^{-10} per year (BROWN et al., 2004). Longer generation times contribute to the lower mutation rate, but the absence of major chromosomal changes suggests that processes involving meiosis in conifers differ from angiosperms, and hence may also contribute to lower mutation rates.

This low rate of evolution implies that species such as spruce and pine would have the same degree of sequence similarity and microsynteny as angiosperms separated by 10-20 million years of evolution, which is typical for species within an angiosperm genus (such as *Populus*). As synteny is generally thought to be high between species of an angiosperm genus such as *Populus*, it should be possible to integrate genomic information (genetic maps, expression profiles, even SNPs) across the Pinaceae.

Among conifers, the largest effort in genetics and genomics has been devoted to loblolly pine, with large EST collections, rich genetic resources, and well developed genetic and QTL maps (NEALE and WHEELER 2004). Species of spruce rank second, largely because of Genome Canada funded projects, and spruce is also almost the most distant member of the Pinaceae from lobolly pine, hence loblolly pine and spruce are good bridges across the pine family. Joint genomic patterns in these two species might enable extension and integration of such genomic knowledge into other species, particularly lodgepole pine (being very proximal to loblolly pine) and Douglas-fir (being intermediate between loblolly pine and spruce).

Our first hypothesis is that the mechanisms of resistance against insects and pathogens are largely shared among members of the Pinaceae. While traits involved with resistance to biological agents are well known to evolve rapidly, the *suites* of genes involved are likely to be evolutionarily conserved, enabling transfer of knowledge among species. Specifically, gene discovery and annotation in one species can aid in the discovery and assignment of gene function in other related conifer species. This hypothesis is supported by our work on terpenoid synthase (TPS) defense gene families in Norway and Sitka spruce, loblolly pine, and grand fir (BOHLMANN et al., 2004;

MARTIN et al., 2004). In addition, conifers display overlapping defense systems, including terpenoid and phenolic defenses, against both insect pests and microbial pathogens (TRAPP and CROTEAU 2001; FRANCESCHI et al., 2000; HUBER et al., 2004). From the southern pine beetle and its associated fungal pathogens (the most serious pest of the coniferous forests of the southern USA) to the spruce weevil (the most serious pest of western Canadian spruce forests), we hypothesize that members of the pine family largely employ a similar suite of defenses.

Our second hypothesis is that the genetic maps of spruce and pine are largely co-linear. Use of EST databases also allows the identification of loci and the design of primers for loci that have the largest chance of mapping success within and across species: conserved orthologous `set (COS) markers (FULTON et al., 2002). These markers are identified by self-BLASTING an EST database, and identify ESTs that lack hits to related ESTs. Loblolly pine ESTs can identify a subset of these markers that are slowly evolving (conserved) as evidenced by low BLAST e-values. We designed and tested 73 COS markers using spruce, pine and Douglas-fir ESTs. Of these, 28 gave products that were mappable (segregating) in both a loblolly pine pedigree and in our spruce pedigrees (of the remaining, 6 were monomorphic and 37 had amplification problems). These are our start of a syntenic map of the pine family. For mapping candidate genes, we use celery endonuclease extract (CJE) to digest mismatches of heteroduplex DNA (COMAI et al., 2004). Sites which are heterozygous in individuals are easily and quickly identified and mapped (RUNGIS et al., submitted). COS marker development and the simplicity of the CJE genotyping system promises that hundreds of markers can be mapped in a syntenic comparison of spruce and loblolly pine.

4. FURTHER PROSPECTS

There are a number of other relevant conifer genome objectives not covered in this chapter; for many of these, see NEALE and WHEELER (2004). They also noted that support for genomics in conifers has lagged behind most major agricultural crops and model plant species, yet conifers are economically comparable to crop plant species. For the prospects of conifer genomics, the silver lining of this cloud is that it could be time for a comeback. In fact, there is even talk of sequencing the entire loblolly pine genome (NEALE and WHEELER 2004) which should be of benefit to all members of the pine family due to the above noted synteny. Also, the costs of genome research continually decline. As stated earlier, our sequencing costs have been reduced by nearly 50 % over the three years of our project. Recently developed array-based genotyping systems (Affymetrix, Illumina) cost substantially less than traditional methods (3-10 cents/genotype compared to $1/genotype for traditional) and have higher rates of genotyping success, information extraction and genotyping accuracy (THE INTERNATIONAL MULTIPLE SCLEROSIS GENETICS CONSORTIUM, 2004). This raises the real prospect that once a good suite of verified SNPs are obtained by the various conifer genome projects, natural SNP variation can be used effectively for marker-aided selection in operational tree breeding programs.

5. REFERENCES

ALFARO, R. I., J. H. BORDEN, J. N. KING, E. S. TOMLIN, R. L. MCINTOSH AND J. BOHLMANN, 2002 Mechanisms of Resistance in Conifers against shoot infesting insects. In: *Mechanisms and Deployment of Resistance in Trees to Insects*. M.R. Wagner, K.M. Clancy, F. Lieutier, and T.D. Paine (eds.). Kluwer Academic Press, Dordrecht, the Netherlands. Pages 101-126.

ALFARO, R. I., L. VAN AKKER, B. JAQUISH, AND J. N. KING, 2004 Weevil resistance of progeny derived from putatively resistant and susceptible interior spruce parents. Forest Ecology and Management **202**: 369-377.

BOHLMANN, J., D. MARTIN, B. MILLER B, AND D. P. W. HUBER, 2004 Terpenoid synthases in conifers and poplars. In: *Plantation FOREST Biotechnology for the 21st Century*. C. Walter and M. Carson (eds.).

BROWN, G. R., G. P. GILL, R. J. KUNTZ, C. H. LANGLEY, AND D. NEALE, 2004 Nucleotide diversity and linkage disequilibrium in loblolly pine. Proc. Nat. Acad. Sci. USA **101**: 15255-15260.

BRUNET, S., P. THIBAULT, E. GAGNON, P. KEARNEY, J. J. BERGERON, M. DESJARDINS, 2003 Organelle proteomics: looking at less to see more. Trends in Cell Biology **13**: 629-638.

CARNINCI, P., Y. SHIBATA, N. HAYATSU, Y. SUGAHARA, K. SHIBATA, M. ITOH, H. KONNO, Y. OKAZAKI, M. MURAMATSU, AND Y. HAYASHIZAKI, 2000 Normalization and subtraction of cap-trapper-selected cDNAs to prepare full-length cDNA libraries for rapid discovery of new genes. Genome Research **10**: 1617-1630.

COMAI, L., K. YOUNG, B. J. TILL, S. H. REYNOLDS, E. A. GREENE, C. A. CODOMO, L. C. ENNS, J. E. JOHNSON, C. BURTNER, A. R. ODDEN, AND S. HENIKOFF, 2004 Efficient Discovery of DNA Polymorphisms in Natural Populations by Ecotilling. The Plant Journal. **37**: 778-786

FIEHN, O., 2002 Metabolomics – the link between genotypes and phenotypes. Plant Molecular Biology **48**: 155-171.

FRANCESCHI, V. R., T. KREKLING, AND E. CHRISTIANSEN, 2002 Application of methyl jasmonate on *Picea abies* (Pinaceae) stems induces defence-related responses in phloem and xylem. Amer. J. Bot. **89**: 578-586.

FULTON, T. M., R. VAN DER HOEVEN, N. T. EANNETTA, AND S. D. TANKSLEY, 2002 Identification, analysis, and utilization of conserved ortholog set markers for comparative genomics in higher plants. The Plant Cell **14**: 1457-1467.

HUBER, D. P. W., S. RALPH AND J. BOHLMANN, 2004 Genomic hardwiring and phenotypic plasticity of terpenoid-based defences in conifers. Journal of Chemical Ecology **30**: 2401-2420.

KING, J. N., AND R. I. ALFARO, 2004 Breeding for resistance to a shoot weevil of Sitka spruce in British Columbia, Canada. In: *Plantation Forest Biotechnology for the 21st Century*. C. Walter and M. Carson (eds.).

KING, J. N., R. I. ALFARO, AND C. CARTWRIGHT, 2004 Genetic resistance of Sitka spruce (Picea sitchensis) populations to the white pine weevil (Pissodes strobi): distribution of resistance. Forestry **4**: 269-278.

KING, J. N., A. D. YANCHUK, G. K. KISS, AND R. I. ALFARO, 1997 Genetic and phenotypic relationships between weevil (*Pissodes strobi*) resistance and height growth in spruce populations of British Columbia. Can. J. Forest Res. **27**: 732-739.

KRUTOVSKY, K. V., M. TROGGIO, G. R. BROWN, K. D. JERMSTAD, AND D. B. NEALE, 2004 Comparative mapping in the Pinaceae. Genetics **168**: 447-461.

LIPPERT, D., J. ZHUANG, S. RALPH, D. ELLIS, M. GILBERT, R. OLAFSON, K. RITLAND, B. ELLIS, C. DOUGLAS, AND J. BOHLMANN, 2005 Proteome analysis of early somatic embryogenesis in *Picea glauca*. Proteomics **5**: 461-473.

MARTIN, D., J. FÄLDT, AND J. BOHLMANN, 2004 Functional characterization of nine Norway spruce *TPS* genes and evolution of gymnosperm terpene synthases of the *TPS-d* subfamily. Plant Physiology **135**: 1908-1927.

MARTIN, D., J. GERSHENZON, AND J. BOHLMANN, 2003 Induction of volatile terpene biosynthesis and diurnal emission by methyl jasmonate in foliage of Norway spruce (*Picea abies*). Plant Physiology **132**, 1586-1599.

MARTIN, D., D. THOLL, J. GERSHENZON, AND J. BOHLMANN, 2002 Methyl jasmonate induces traumatic resin ducts, terpenoid resin biosynthesis and terpenoid accumulation in developing xylem of Norway spruce (*Picea abies*) stems. Plant Physiology **129**: 1003-1018.

MILLER, B., L. L. MADILAO, S. RALPH, AND J. BOHLMANN, 2005 Insect-induced conifer defence: White pine weevil and methyl jasmonate induce traumatic resinosis, *de novo* formed volatile emissions, and accumulation of terpenoid synthase and octadecanoid pathway transcripts in Sitka spruce. Plant Physiology **137**: 369-382.

NEALE, D. B., AND N. C. WHEELER, 2004 The Loblolly Pine Genome Project. A Prospectus to Guide Planning and Funding of a USDA Forest Service Led Effort to Develop an Integrated Genomics Research Program in Loblolly Pine. (http://dendrome.ucdavis.edu/lpgp/pdf/prospectus.pdf)

RUNGIS, D., B. HAMBERGER, Y. BÉRUBÉ, J. WILKIN, J. BOHLMANN, AND K. RITLAND, 2005 Efficient genetic mapping of nucleotide polymorphisms based upon DNA mismatch digestion. Molecular Breeding (submitted).

STASYK, T., AND L. A. HUBER, 2004 Zooming in: Fractionation strategies in proteomics. Proteomics **4**: 3704-3716.

TIAN, Q., S. B. STEPANIANTS, M. MAO, L. WENG, M. C. FEETHAM, M. J. DOYLE, E. C. YI, H. DAI, V. THORSSON, J. ENG, D. GOODLETT, J. P. BERGER, B. GUNTER, P. S. LINSELEY, R. B. STOUGHTON, R. AEBERSOLD, S. J. COLLINS, W. A. HANLON, AND L. E. HOOD, 2004 Integrated genomic and proteomic analyses of gene expression in mammalian cells. Mol. Cell. Prot. **3**: 960-969.

THE INTERNATIONAL MULTIPLE SCLEROSIS GENETICS CONSORTIUM, 2004. Enhancing linkage analysis of complex disorders: an evaluation of high-density genotyping. Hum. Mol. Genet. **13**: 1943-1949.

TRAPP, S., AND R. CROTEAU, 2001 Defensive resin biosynthesis in conifers. *Ann. Rev. Plant Phys. and Plant Mol. Biol.* **52**: 689-724.

WECKWERTH, W., 2003 Metabolomics in systems biology. *Ann. Rev. Plant Biol.* **54**: 669-689.

CHAPTER 6

USING GENOMICS TO STUDY EVOLUTIONARY ORIGINS OF SEEDS

ERIC D. BRENNER AND DENNIS STEVENSON
The New York Botanical Garden
Bronx, NY, USA

Abstract. Today's gymnosperms are relicts from groups of plants whose history extends to nearly 300 million years ago when seeds first evolved. Consequently gymnosperms have been studied in relation to the fossil record to uncover important clues to understand the appearance and evolution of seeds and pollen. Extant gymnosperms are restricted to four surviving orders: the Gnetales, Coniferales, Ginkgoales and Cycadales, providing a marvelous biological source of data which has only recently been examined at the molecular level. Efforts to understand the molecular evolution of the gymnosperms has examined the expression of gymnosperms genes with similarity to genes in angiosperms involved in the development of leaves and reproductive structures via a degenerate based primer approach. Lately, genomic approaches have been engaged to understand what genes are involved in the evolution and development of gymnosperm structures. Below is a description of genomic and molecular studies currently to understand evolution in the gymnosperms.

1. THE RISE OF THE SEED PLANTS

The emergence of the seed plants is among one of the most significant developments to occur in the evolution of terrestrial plants (FOSTER and GIFFORD 1974)). Through dramatic changes in reproductive strategies, the seed plants (spermatophytes), which include the angiosperms and gymnosperms, were able to overcome one of the key barriers to reproductive efficiency. In non-seed plants, such as ferns, the gametophyte is photosynthetic, free living (exosporic) and not vascularized. Consequently fertilization can only occur in wet environments where flagellated gametes can swim from their source, the antheridia, to the archegonia to fertilize the egg. Because of their dependence on water mediated fertilization, hybridization in non-seed plants can occur only between close neighboring individuals as swimming sperm can only travel short distances. In contrast to non-seed plants, the seed plant megametophyte is retained within the spore wall (endosporic) and also within the maternal tissue. That is the megagametophyte is contained within the diploid integuments to form the ovule (Figure 1). Gamete transfer does not require water as a medium but rather it is the male gametophyte that is transferred. The male microgametophyte in seed plants is transported as a desiccation-resistant pollen grain directly to the ovule (or ovule enclosing carpel), often many miles away.

Upon reaching the female reproductive structure, a tube grows from the pollen, subsequently bringing the male gamete to the female gametophyte and ultimately to the egg. Similar to the pollen grain, the female gametophyte is also dessication resistant via protection by diploid tissue, the nucellus and integuments. The megagametophyte is no longer free-living, or photosynthetic and must receive nutrients from the vascularized, sporophytic parent during its development. The megagametophyte remains dependent on the maternal parent during pollination and later while the pollen tube is growing prior to fertilization. In most cases fertilization occurs while the megagametophyte is attached to the sporophyte, while in the oldest surviving seed plants fertilization occurs after the ovule has been shed, such as in cycads and gingko.

Upon fertilization the ovule becomes a seed. Within the seed, the megagametophyte function has been modified to serve primarily as nutrient storage tissue to sustain the developing embryo (In the angiosperms nutritive tissue typically comes from other tissue types including endosperm and cotyledons, while in gymnosperms nutrition primarily comes from the gametophyte). Upon maturity the seed becomes dehiscent and subsequently transported (by biotic or non-biotic sources) to a new locale with the possibility of finding favorable growing conditions. Consequently the mature seed consists of three generations: the maternal sporophyte, the gametophyte and the new embryonic sporophyte. Thus the ovule/seed is a major developmental innovation that enables long distance fertilization as well as dispersal of a diaspore containing both the embryo and storages tissue for its establishment in a process unique in land plants.

In addition to their important role enabling long distance fertilization, seeds are fundamentally important for humans. Seeds provide food, fiber and industrial goods. Considering their importance to society, a better understanding of the origins of seeds requires innovative studies to explain seed evolution. The oldest surviving seed plants are the gymnosperms, which exhibit a number of plesiomorphic (ancient) conditions reminiscent of the early fossil seed plants. Gymnosperms are hypothesized to have descended from the progymnosperms (ROTHWELL 1988), which date back in the fossil record to approximately 285-350 MYA (ROTHWELL and SCHECKLER 1988) to 359-385 MYA (TAYLOR and TAYLOR 1993).

One theory contends that the original gymnosperms were the seed ferns (FOSTER and GIFFORD 1974). The last common ancestor of extant seed plants and extant ferns is estimated by molecular clocks to have existed approximately 275-290 MYA (SAVARD et al., 1994). When molecular clock estimates include fossil calibration points, gymnosperms age ranges from 345 to 360 MYA depending on whether lycophyte or gymnosperm fossils, respectively, are used as calibration points (SOLTIS et al., 2002b). Thus, considerable variation in gymnosperm age estimations exists spanning nearly 100 million years. The discovery of new fossils from extinct gymnosperms and expanded molecular clock studies with deeper sequencing initiatives should help narrow this estimation gap.

1.1. Today's gymnosperms: remnants of early seed plants

Conflicting results between morphological and molecular data sets have made it difficult to define the precise phylogenetic hierarchy of the four extant gymnosperm clades: the Gnetales, Coniferales, Ginkgoales and Cycadales (DOYLE 1998; MAGOLLÓN and SANDERSON 2002; RYDIN et al., 2002). Among the gymnosperms, the Gnetales, according to the anthophyte theory, are argued to form a sister group to the angiosperms (CRANE 1985; DOYLE and DONOGHUE 1986; NIXON et al., 1994) as shown in Figure 2A. Based on morphological evidence, synapomophic characterstics shared between angiosperms and Gnetales include the presence of vessel elements, double fertilization, a double integument, and a reduction or loss of archegonia (FOSTER and GIFFORD 1974). Contrary to anthophyte morphological symnapomorphies, the majority of molecular phylogenies have placed the Gnetales as a sister group to the conifers (see Figure 2B) (BOWE et al., 2000; CHAW et al., 2000; DONOGHUE and DOYLE 2000; GOREMYKIN et al., 1996; SOLTIS et al., 2002a). However, some molecular evidence can be interpreted as supporting the anthophyte clade (RYDIN et al., 2002). Attempts to associate molecular expression data with morphological structures also place the Gnetales and conifers together with shared expression of orthologous genes indicating that the *Gnetum* strobilar collar and ovule are homologous to the conifer bract and ovule-ovuliferous scale complex (SHINDO et al., 2001).

The relative placement of the two other clades, Cycadales and Ginkgoales in the gymnosperms is also controversial. *Ginkgo* has been variably allied either with the Coniferales by some authorities (Figure 2A), partly due to similar characteristics such as axillary branching and simple leaves; or more commonly it is basal among the gymnosperms with the Cycadales (Figure 2B) (BHARATHAN et al., 1997; BOWE et al., 2000; CHAW et al., 2000; FROHLICH and PARKER 2000; NIXON et al., 1994). This basal position of cycads and *Ginkgo* is based upon common plesiomorphic (ancestral) characters similar to extinct seed ferns including pollen chambers that release motile male gametes, which swim to the egg (CHAMBERLAIN 1935; FRIEDMAN 1987; IKENO and HIRASE 1897). Also, the archegonium of cycads and *Ginkgo* consists of a large four-neck-celled chamber (CHAMBERLAIN 1935). Between cycads and *Ginkgo,* most molecular phylogenies place cycads as the sister group to all living seed plants (Figure 2B). Cycad fossils come from the Lower Permian at approximately 260-290 million years old (GAO and THOMAS 1989; MAMAY 1969).

Although such phylogenetic associations are intriguing, the absence of an overall, robust hierarchy that defines the relationship between the four extant gymnosperm clades will continue to create equivocal evolutionary interpretations on extant species relationships and the origins of seed plant structures. Genomics holds the potential to ultimately resolve this problem as already shown in smaller, more easily sequenced genomes. Such studies presage future advances. In a study of seven phylogenetically unresolved species of yeast, analysis of 106 common genes (a number far exceeding standard phylogenetic studies is only made possible with genomic high through-put sequencing), generated an unambiguous species relationship tree at 100% bootstrap support; irregardless of whether maximum likelihood or maximum parsimony was used (ROKAS et al., 2003). As such, genomic approaches are quickly becoming a way of obtaining high yielding sets of common genes to generate robust phylogenomic trees. For large genome species, which includes the gymnosperms, full sequencing is still too costly. In the absence of full genome sequence availability, EST (Expressed Sequence Tag) based approaches can yield considerable insight into the genome (RUDD 2003). EST based approaches are already being developed in at least one genus from each of the four gymnosperm orders, offering the possibility of resolving the phylogeny of extant gymnosperm. EST data sets are now available from the Coniferales; including *Pinus* (KIRST *et al.*, 2003; WHETTEN *et al.*, 2001), *Abies* (GENOME *et al.*, 2005) and *Cryptomeria* (UJINO-IHARA *et al.*, 2000), as well as from cycads (BRENNER *et al.*, 2003a). In addition, ESTs have been prepared from *Ginkgo* and *Gnetum* (Brenner *et al.*, unpublished results). These EST datasets promise to dramatically increase the number of molecular markers from each major gymnosperm node to generate a robust phylogenomic gymnosperm trees.

Robust phylogenetic trees can also be used to pinpoint evolutionary nodes associated with morphological adaptation and gene change. A common assumption is that changes in function and/or expression of genes are the driving force for structural evolutionary modification. Current studies in gymnosperms have attempted to identify genes with orthology to angiosperm regulators with defined roles in organogenesis and to explore their role in gymnosperm species representing key nodes in evolution. The most active area of research has explored conserved gene function during development of the male and female angiosperm reproductive organs, particularly pollen-bearing structures (microsporangia) and ovules (megasporangia) in gymnosperms, while other work has examined gene expression in structures associated with gymnosperm micro- and megasporangium, such as the ovuliferous bracts in conifers. Phylogenomic tree construction will allow quick identification and annotation of candidate genes to specifically focus on the development of the ovule in gymnosperms and to test classical theories regarding the origin of the ovule.

1.2. Theoretical origins of the ovule

By definition, an ovule is an integumented megasporangium and a seed is a fertilized ovule. Also, when describing an ovule, the term nucellus is used as a synonym for megasporangium. Thus, the megaspores originate in a megasporangium located in the nucellus, which in turn is enclosed within the integument. All of this becomes the seed. The origin of the integument is highly controversial with three main hypotheses concerning its origins (Figure 3). One of these considers the integument to be a new structure that developed as an outgrowth of the stalk (funiculus) of a terminal sporangium (Figure 3A-C). This is not unlike the development of the aril in *Taxus*, which is an outgrowth of the funiculus that develops after the integument in *Taxus* has been initiated. Thus, the sequence in *Taxus* is basipetal (BIERHORST 1971; MUNDRY 2000) and is similar to the sequence of initiation of the two integuments in the angiosperms (FOSTER and GIFFORD 1974). Another hypothesis concerning the origin of the integument of the seed plants considers it to be derived from sterilized sporangia (Figure 3 D-G). In this scenario, sporangia become clustered into sori Figure 3D) and then fused to form a synangium. The outer fused sporangia of a sorus become sterile with only one central sporangium remaining fertile. Thus, the sterile sporangia become the integument (Fig. 3G). This hypothesis is supported by the fact that the integument of many fossils as well as that of extant cycads have a series of vascular bundles arranged in a ring that are interpreted as the bundles of the stalks of the original fused sporangia (FOSTER and GIFFORD 1974). The third hypothesis would derive the integument from a truss of sterile telomes that become fused and enclose a sporangium (Fig. 3H-K).

Currently, data supporting any or all of these hypotheses is equivocal and often contradictory. The telome origin is primarily based upon transitional fossils that have little if any structural details. It should be noted that the ring of bundles in the integument described above as sterilized sporangial bundles are interpreted in the telome theory as the bundles of sterile telomes that are fused. The *de novo* hypothesis finds little support anatomically mainly because it can not be demonstrated to be new based on anatomical innovations. The sterile synangium hypothesis finds it support in vascular anatomy and the comparative morphology of synangial organization in the extant Marattiales, a group of eusporangiate ferns, and in the microsporophylls of the cycads. In both of these groups the sporangia are arranged in clusters (i.e. sori) that are either elliptical or circular. In the microsporophylls of cycads, the microsporangia (pollen producing sporangia) are arranged in circles with the sporangial stalks exhibiting varying degrees of fusion.

The Gnetales and the angiosperms have two envelopes (or integuments) enclosing the nucellus, in contrast to the single integument of the cycads, *Ginkgo*, and the conifers. In the Gnetales, the outer integument is initiated before the inner integument (TAKASO 1984; TAKASO 1986), but in the angiosperms the process is reversed so that the inner integument is initiated first. Like the first integument, the origin of the second integument is enigmatic. Moreover, it is not clear that when two integuments are present, if the inner or the outer one is homologous to the single integument of the cycads, *Ginkgo*, and conifers. Compounding this is the aril of Taxus as mentioned above, which could also be a precursor to the outer integument of either the Gnetales or the angiosperms or both.

Thus, an important goal to understand the evolution of the ovule will be to try to sort out the homologies of these ovular structures: the inner integument, the outer integument, and the aril and to determine what genes are involved in their development. Such analysis can utilize the rich source of genetic information available in the molecular model plant *Arabidopsis*.

1.3. Arabidopsis: the molecular reference plant for ovule evolutionary studies

Arabidopsis currently serves as the most definitive model system to explore ovule development in plants, as well as general reproductive development. Because of its simplified structure and facile genetics, *Arabidopsis* studies have helped define key genetic steps in ovule specification and development (for reviews on ovule molecular development in angiosperms, see SKINNER *et al.* (2004) and ANGENENT and COLOMBO (1996)). In angiosperms, the ovule emerges from an area of tissue called the placenta. Ovule primordia develop with three distinguishable regions including the funiculus, where the ovule is attached to the developing carpel; the chalazal region, where the two integuments emerge; and the nucellus, which is the site of meiocyte formation. The meiocyte (or megaspore mother cell) produces the haploid tissue from which the gametophyte is formed. Although similar in early development the ovule of gymnosperms becomes clearly distinct from angiosperms ovules later in development.

1.4. Comparative genomics, gymnosperms vs. angiosperms

A comparative genomics approach must consider the fact that the gymnosperm ovule (Figure 1A) is considerably different from the angiosperm structure (Figure 1B). Angiosperm ovules have undergone a number of major changes including an increase in integument number, the evolution of double fertilization to form an endosperm (as mentioned above, double fertilization exists in the Gnetales, but it is independently derived (CARMICHAEL and FRIEDMAN 1995)), development of specialized nuclei in the egg sac (polar, synergid, antipodal), a dramatically decreased time interval between pollination and fertilization, a reduction or loss of archegonia, and a dramatic decrease in the number of cells in the megagametophyte (for review see FOSTER and GIFFORD (1974)). By comparing transcriptomes between gymnosperm and angiosperm ovules and even ovule structures, genomics will revolutionize our understanding of seed evolution.

1.5. Molecular approaches to understand regulation of reproductive growth in gymnosperms

Of the molecular developmental studies in gymnosperms so far, most have focused mainly on the Coniferales, due to their importance in wood production and on the Gnetales because of their possible assignment as the sister group to angiosperms. Such studies have virtually overlooked the Ginkgoales or the Cycadales, despite their fascinating and fundamentally important plesiomorphic (ancient) characters—particularly those related to early seed plants. Fortunately recent work has begun to refocus attention towards this basal group of extant gymnosperms (BRENNER *et al.*, 2003b).

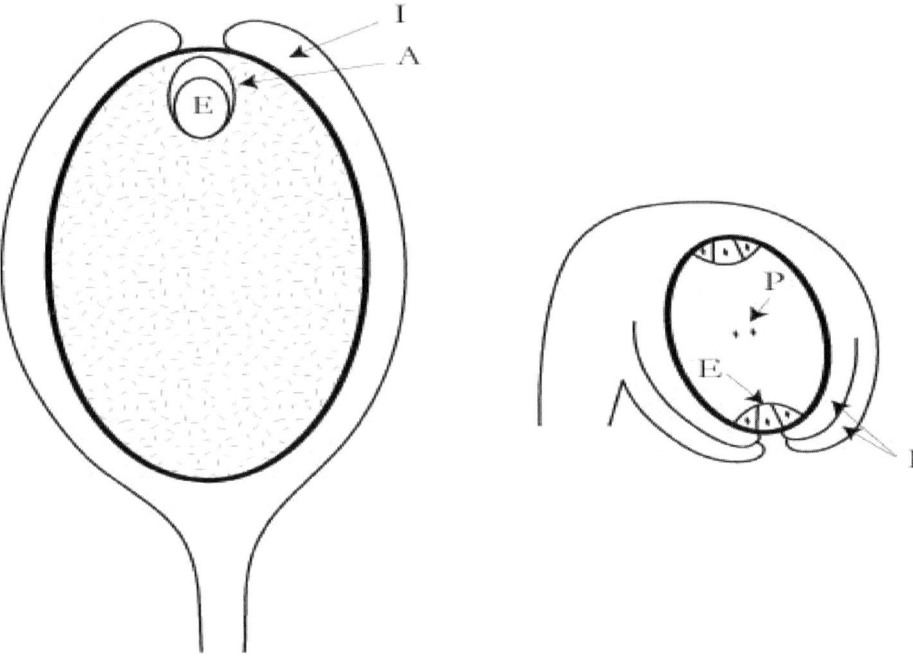

Figure 1. A simplified diagram illustrating basic differences between the gymnosperm (A) and angiosperm ovule (B) The developmental stage shown in both illustrations represents a point when each respective ovule is receptive to fertilization. The integuments of diploid sporophytic origins are shown surrounding the megagametophyte. The haploid megagametophyte region is delineated in bold. In comparison to angiosperms, the gymnosperm ovule is large comprising many more cells (shown with stippling). The typical angiosperm ovule megagametophyte cell number is reduced—shown here with 8 nuclei and seven cells. In gymnosperms the nutritive tissue is stored in the megagametophyte, much of which develops prior to fertilization. Endosperm development in angiosperms is tightly linked with fertilization. The endosperm is formed after a second fertilization event of one or more of the polar nuclei shown. Most gymnosperm ovules have several archegonia (one is shown here enclosing an egg). Archegonia are not present in angiosperms. Unitegmy is the ancestral condition in gymnosperms whereas most angiosperms have two integuments. E, egg; I, Integument; A, archegonia. P, polar nuclei.

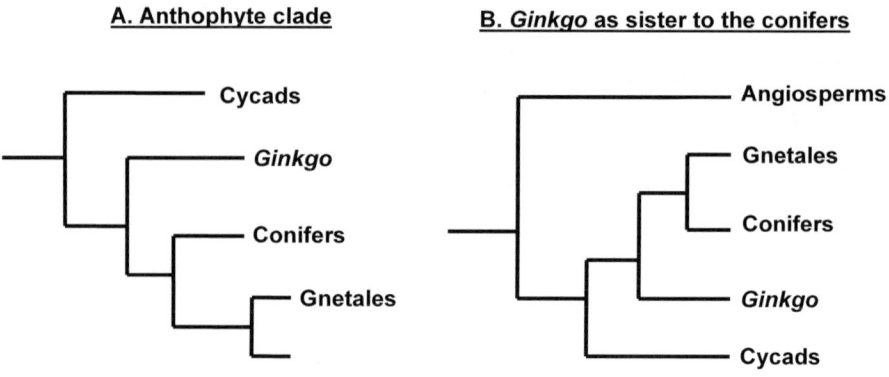

Figure 2. Two alternative scenarios regarding the phylogenetic hierarchy among the gymosperms: A) Placement of the Gnetales as the sister group to the angiosperms reflecting an "anthophyte" clade in agreement with morphological analysis (CRANE 1985; DOYLE and DONOGHUE 1986; NIXON et al., 1994). B) Molecular realignment showing an association between the Gnetales and conifers (BOWE et al., 2000; CHAW et al., 2000; DONOGHUE and DOYLE 2000; GOREMYKIN et al., 1996; SOLTIS et al., 2002a). Cycads are shown basal with Ginkgo potentially as sister to Conifers (BHARATHAN et al., 1997; BOWE et al., 2000; CHAW et al., 2000; FROHLICH and PARKER 2000; NIXON et al., 1994), however cycads and Ginkgo may also form a separate subclade within the gymnosperms (not shown).

In general, most molecular studies in gymnosperms have focused on the analysis of conserved gene families. Gene families of interest are cloned with degenerate primers designed to amplify conserved gene regions, and orthology determinations of the amplified and sequenced products are based on gene alignments and tree assembly. Genes with orthology to key regulators in angiosperms are then examined in organ/tissue localization studies and tested for complementation in mutants where their original developmental role was identified in the cognate ortholog. The majority of these molecular evolution studies in gymnosperms have examined developmental gene candidates within the context of their role in reproduction— including induction and development of reproductive organs.

Although the evolution of ovule development has not been a primary focus in gymnosperm molecular studies, information gained by examining expression of genes correlated with other aspects of gymnosperm reproduction can be used to further study the development of the ovule in gymnosperms. The following studies present reproductive development for the gymnosperms.

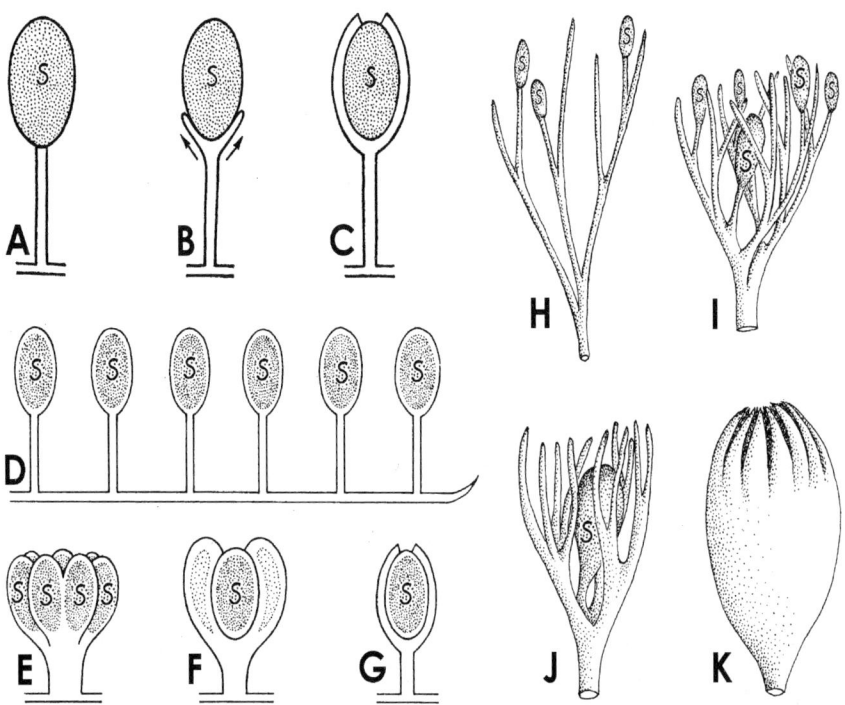

Figure 3. Three alternative hypotheses concerning the origin of the integument in seed plants. A-C. De novo origin of the integument from an outgrowth of the sporangial stalk (funiculus). D-G. Synangial hypotheses where sporangia become aggregated into a sorus with the outer sporangial eventually becoming sterilized and becoming the integument. H-K. Telome origin of the integument by the amalgamation of sterile telomes surrounding a fertile telome (sporangium). (A-G, after MEEUSE 1967 (MEEUSE 1966); H-K, after STEWART AND ROTHWELL 2000; STEWART and ROTHWELL 1983).

2. THE TRANSITION FROM VEGETATIVE TO REPRODUCTIVE GROWTH

In angiosperms many of the key regulators that control the dynamic step converting a vegetative reproductive shoot have been cloned. In *Arabidopsis*, *LEAFY* is one of the main initiators of reproductive growth (Also known as *FLOROCAULA* in

Antirrhinum (COEN *et al.*, 1990)). Loss of *LEAFY* expression in angiosperms prevents floral initiation resulting in a reiteration of vegetative growth (hence the "leafy" phenotype) (WEIGEL *et al.*, 1992). Gymnosperm *LEAFY* homologs have been cloned in all gymnosperm orders (FROHLICH and PARKER 2000; HIMI *et al.*, 2001; MELLEROWICZ *et al.*, 1998; MOURADOV *et al.*, 1998). In *Pinus radiata* the *LEAFY* homologue is expressed in primordia of the bract/ovuliferous scale complex and can complement the *Arabidopsis LEAFY* mutant, consequently reestablishing reproductive development (MELLEROWICZ *et al.*, 1998; MOURADOV *et al.*, 1998). Likewise, the *Gnetum parvifolium* homolog, *GpLFY* is expressed in vegetative and strobili primordia and can complement *Arabidopsis leafy* mutant (SHINDO *et al.*, 2001).

In most gymnosperms there are two *LEAFY* homologues: "Leaf" and "Needle" with potential evolutionary implications (FROHLICH and PARKER 2000; HIMI *et al.*, 2001). The Leaf form shows a closer phylogenetic relationship to angiosperm *LEAFY*. As mentioned above, gymnosperms produce separate male and female reproductive structures. The cycads, *Ginkgo* and Gnetales are dioecious (separate male and female structures-although some member of the Gnetales, produce sterile female ovules within the male cone). Conifers can be either dioecious or moneocious. Frolich and others (FROHLICH and PARKER 2000) hypothesized that the Leaf class controls male reproductive identity whereas the Needle controls female identity—each being expressed to form either a male or female reproductive shoot, respectively. According to Frolich's "Mostly Male Theory of Evolution" two events occurred during the evolutionary transition from the gymnosperms to the angiosperms to form a bisexual axis seen in early angiosperms: 1. formation of ectopic ovules on male structures to make a bisexual flower coupled with, 2. loss of the female identity gene Needle resulting in loss of the female flower. Contrary to this theory however, expression of *GpLFY,* the *LEAFY* homologue in *Gnetum* has been determined to be expressed not only in male, but also in the female shoot— consequently putting the genetic basis of this theory in doubt (SHINDO *et al.*, 2001). However, as a concept, the feminization of male reproductive organs to form a bisexual flower in gymnosperms could have happened through ectopic ovule formation, which has been observed in angiosperms (FAVARO *et al.*, 2003) and gymnosperm (HARA 1997) mutants, particularly in *Gingko*.

2.1. Revealing genes that determine reproductive organ identity in gymnosperms

Once reproductive growth has been induced in gymnosperms, presumably by genes such as *NEEDLY* and/or *LEAFY,* organ specification ensues, possibly by activation of a set of genes orthologous to angiosperm MADS box transcription factors. *MADS* genes in plants were first discovered because disruption in their function causes homeotic mutations during flower organ specification. The ABC theory of development in angiosperms, as originally demonstrated in *Arabidopsis* and *Antirrhinum,* hypothesizes that specific combinations of different MADS box genes leads to the development of each distinct flower whorl. Thus in the model plant *Arabidopsis*, expression of A (*APETALLA-1*) genes form sepals; A with B (*APETALLA-3* and *PISTILLATA-1)* genes from petals; B with C (*AGAMOUS)* genes form stamens and C genes alone form the carpel (BOWMAN et al., 1991; COEN and MEYEROWITZ 1991; SCHWARZ-SOMER et al., 1990). Furthermore, another class of MADS genes, known as the *SEPELLATA* interact with the ABC genes (BYRNE et al., 2001). In conifers degenerate PCR and *in situ* analysis has explored the role of some key *MADS* gene homologs within the context of plant evolution and have determined that of the ABC genes, the B and C gene lineage is conserved in nearly all seed plants examined (GOTO et al., 2001; NG and YANOFSKY 2001; THEISSEN et al., 2000; THEISSEN et al., 1996) as described below.

2.2. Gymnosperms and the origins of the MADS box ABC gene lineage

The *MADS*-box floral homeotic gene lineages have been estimated to be 340 to 385 MYA and may predate or possibly coincide with the origin of seed plants (PURUGGANAN 1997; PURUGGANAN et al., 1995). Unfortunately, the exact origin of the ABC genes is difficult to resolve due to the extinction of many key intermediates between early progymnosperms, the first seed plants and modern seed plants. Current molecular evidence describes horsetails and ferns as the extant sister group to the seed plants (PRYER et al., 2001). In order to study the origin of the ABC gene lineage, *MADS* box genes have been found in the fern *Ceratopteris richardii*. Phylogenetic comparisons have determined that the ABC *MADS* clade are not present in ferns (HASEBE et al., 1998; MUNSTER et al., 1997). Expression of *Ceratopteris MADS* genes is present in vegetative and reproductive tissue (HASEBE et al., 1998; MUNSTER et al., 1997) indicating an ancient role for MADS box genes in reproduction. The complete genomic sequencing of a non-seed vascular plant, the lycopod, *Selaginnella moellendorfii,* by the Joint Genome institute will reveal the full complement of *MADS* genes in a vascular seed-less plant, as well as other genes whose changes may be associated with evolutionary structural modifications.

2.3. The ABC theory applied to the specification of male vs. female organs in gymnosperms

Gymnosperms are the oldest living ovule/seed producing plants. In a search for genes that control the production of ovules, paralogs to the B and C lineage have been isolated from gymnosperms. The C gene ortholog (*AGAMOUS*) has been isolated from at least one species representing each gymnosperm order. Beginning with the conifers, the *AGAMOUS* ortholog from Black spruce *(Picea mariana)*, *SAG1* (RUTLEDGE et al., 1998) and closely related Norway spruce *(Picea abies)*, *DAL2* (TANDRE et al., 1998) have been cloned. Both *Picea AGAMOUS* orthologs are expressed exclusively in male and female reproductive tissues. In the compound female cone, expression is present in the developing ovule-bearing organ and to a smaller extent in the ovuliferous scale (RUTLEDGE et al., 1998; TANDRE et al., 1995; TANDRE et al., 1998). In Gnetales, the *Gnetum gnemon* C gene, *GGM3,* is expressed in both male and female reproductive structures, but not in the leaf. *GGM3* is initially present in the developing ovule and antherophore (pollen bearing unit) (BECKER et al., 2003; WINTER et al., 1999). Later in development *GGM3* appears in the outer envelope of the ovule and antherophore. In the basal gymnosperms

AGAMOUS orthologs have been cloned in *Ginkgo biloba*, *GBM5* (JAGER et al., 2003) and *Cycas edentata*, *CyAG* (ZHANG et al., 2004). As in the more derived gymnosperms, *Ginkgo* and *Cycas AGAMOUS* orthologs are expressed in both male and female reproductive organs. In cycads this includes not only the ovule, but also the megasporophyll—which is reminiscent of a derived leaf (for an exceptional review of cycad biology, see (NORSTOG and NICHOLLS 1997)). In male tissues, *CyAG* is present in the central axis of the male cone, microsporophyll and microsporangium. *CyAG* is not detected in vegetative structures including the root, stem and leaf (ZHANG et al., 2004). In *Ginkgo*, *in situ* analysis was not performed, however contrary to all gymnosperm studies, non-quantitative RT-PCR results indicate that the *Ginkgo AGAMOUS* gene, *GBM5*, also is expressed strongly in young leaves (JAGER et al., 2003), whereas in conifers, *Gnetum*, and *Cycas*, it is not expressed vegetatively. Overall, such studies strikingly reveal a strongly conserved expression pattern for *AGAMOUS* orthologs during ovule and ovule associated structure development from the earliest living seed plants through to modern plants.

Recently, it has been shown that *AGAMOUS* is not the sole gene responsible for the formation of the gynoecium in *Arabidopsis*. *AGAMOUS* is a member of a small clade of closely related genes that also includes *SHATTERPROOF1* and *2*, and *STICK* (*SHP1,2, STK1*), which play various, interacting roles during carpel and ovule development (PINYOPICH et al., 2003). The role of these closely related genes involved in ovule development reveals their redundant nature. In fact loss of all three mutants, *STK, SHP1* and *SHP2*, causes a conversion of integuments into carpel like structures (PINYOPICH et al., 2003). Other MADS box genes are necessary players in ovule development including *SEPETALA1, 2* and 3 (FAVARO et al., 2003), which is consistent with their role as components of the *MADS* transcriptional complex during flower development (HONMA AND GOTO 2001). Coordinate regulation of transcriptional MADS components within a protein complex argues that additional regulators besides the *AGAMOUS* orthologs induce ovule/seed formation. Future studies on ovule specification in gymnosperms will need to consider the importance of coordination of additional MADS players, as well as unknown genes as have been the case during studies of B genes.

In gymnosperms there is only one *AGAMOUS* ortholog in all clades examined as described above. Furthermore, gymnosperm *AGAMOUS* appears phylogenetically basal to the angiosperm genes, suggesting that a single gene is the ancestral state. Thus, duplications and specialization in angiosperms of *AGAMOUS* paralogs could have led to formation and elaboration of the gynoecium.

2.4. Determining the role of B genes in gymnosperms

Similar to angiosperms, in gymnosperms B gene coupled with C gene expression is believed to control the formation of male organs, which is thus a universal pattern in all spermatophytes (THEISSEN et al., 2000). However, expression of orthologous B genes in gymnosperms is complex. For example, in *Gnetum gnemon*, it is predicted that there is not one, but two B genes, *GGM2* and *GGM15*, as orthology designations between these two genes are ambiguous (BECKER et al., 2003). Both *GGM2* and *GGM15* are strongly expressed specifically in male organs and excluded from developing ovules or associated ovuliferous structures, which is consistent with B gene expression patterns in angiosperms. In conifers the situation is even more complex. In Northern spruce, there are three genes *DAL11, DAL12,* and *DAL13,* which are classified as B gene orthologs. As with *Gnetum,* all three genes show expression specifically in the developing pollen cone, with local variations between genes. *DAL11* and *DAL12* are present in the pollen cone bud, whereas *DAL13* is in the peripheral zone basal to the emerging microsporophyll primordia. Later in development *DAL11* expression is found in both the central pith and surrounding vascular tissue of the cone, whereas *DAL12* is only in the vascular tissues. *DAL13* is not expressed in the cone bracts at all but only in the microsporophylls (SUNDSTROM and ENGSTROM 2002). The presence of multiple B genes was also described in a separate conifer family, Taxodiaceae, in the species *Cryptomeria japonica.* In *Cryptomeria,* two B gene orthologs, *CrMADS1* and *CrMADS2,* were isolated, which into a subclade with Northern Spruce *DAL13, DAL11, Pinus PrDGL* and *Gnetum GGM2*. Expression of *CrMADS1* and *CrMADS2* also was expressed exclusively during male development (FUKUI et al., 2001). In *Pinus radiata* only one B homolog, *PrDGL* has been isolated. Expression of *PrDGL* is consistent with the theme of B expression in male organs as restricted to the male cone (MOURADOV et al., 1999).

The variable expression patterns of multiple B gene orthologs indicate that these related genes have specific and distinct roles during male reproductive development in *Gnetum* and conifers. To add an additional level of complexity, in *Gnetum,* additional MADS genes have been detected, which also show strong expression in the developing antherophore. However, these additional MADS genes do not belong to the B gene clade, but instead to other clades including *GGM9* and *GGM11,* which are members of the *AGL6* gene family (BECKER et al., 2000).

Some MADS genes from gymnosperms fall out separately from any known clade of the MADS gene superfamily from angiosperms, such as *GGM7,* which consequently forms a novel subclade (BECKER et al., 2000). Similarly, in Norway Spruce (*Picea abies*), the MADS box gene *DAL10* shows expression specificity in developing seed and pollen cones, and like *GGM7* has no orthology to known angiosperm MADS clades (CARLSBECKER et al., 2003). Intriguingly, prior to lateral organ initiation, expression of *DAL10* was detected in the peripheral zone of the meristem. Later in the developing cone, *DAL10* is expressed in the axis basal to the newly produced bract/ovuliferous-scale complex. Such early expression, prior to the appearance of the *AGAMOUS* homologue, *DAL2,* in the same tissue lead the authors to hypothesize that *DAL10* might act upstream of organ identity genes (CARLSBECKER et al., 2003).

2.5. Complementation studies assess gymnosperm MADS gene function

Functional complementation adds another piece of evidence revealing the conservation of function across plant groupings. Transgenic expression of the C gene from *Picea abies* (Tandre 1998) and *Picea mariana* (RUTLEDGE et al., 1998) in *Arabidopsis* leads to homeotic mutations in *Arabidopsis* similar to the ectopic expression of the native *AGAMOUS* gene in *Arabidopsis.* Such changes include the homeotic conversion of sepals to carpels and petals to stamens as well as more generalized effects including curling of leaves and dwarfism (RUTLEDGE et al., 1998). In *Cycas*, functional complementation of the C gene homologue, *CyAG*, leads to similar complementation.

The same approach using functional complementation of *Arabidopsis* B gene mutants supports a partially conserved role of *Gnetum* B gene *GGM2* (WINTER et al., 2002). Expression of *GGM2* in the *Arabidopsis* B gene mutants *ap3* and *pi* showed partial complementation of B function including formation of petaloid sepals and stamenoid carpels (WINTER et al., 2002) whereas complementation with wildtype *AP3* and *PI* causes complete transformation of sepals into petals and carpels into stamens (KRIZEK and MEYEROWITZ 1996).

To sum up, the overall expression of MADS genes in gymnosperm reproductive development reveals that signals regulating homologous male and female structures might have basic pattern development plan for B and C genes, but will not simply copy the developmental orthologs found in Angiosperms. Some roles may be ancestral and lost during the evolution of angiosperms, whereas others may have diverged after the separation of angiosperms and gymnosperms.

2.6. HOX genes in gymnosperms

Besides ABC genes, homeobox or "*HOX*" genes have been investigated to some degree in gymnosperms and play an important role in ovule development. *HOX* genes in higher plants are regulators of meristem maintenance, and organ determination and development and have been grouped into several families, which diverged prior to the separation of animals and plants. (BHARATHAN et al., 1997; CHAN et al., 1998). In plants, *KNOX-1* controls cell identity in the shoot meristem (for review see (BARTON 2001)). In most vascular plants, *KNOX1* expression is important in the meristem where it maintains meristematic activity. During leaf primordia formation, *KNOX1* is down-regulated and remains repressed whereas during complex leaf formation *KNOX1* is reactivated at a later stage in the primordia development (BHARATHAN and SINHA 2001). A cross survey of vascular plants confirmed this expression pattern (using a KNOX-1 universal plant antibody) in nearly all seed plants (with pea as the only exception (GOURLAY et al., 2000)) including the pinnately complex leaf in the cycad genus *Zamia* (BHARATHAN et al., 2002) Similar expression of *HBK1*, a *KNOX-1* homologue from Norway spruce revealed meristem localization and subsequent down regulation in leaf primordia (SUNDAS-LARSSON et al., 1998).

In gymnosperms, four *KNOX*-1 homologs have been reported from Norway spruce (SUNDAS-LARSSON et al., 1998) and black spruce (GUILLET-CLAUDE et al., 2004; HJORTSWANG et al., 2002). At least one of these genes has an important role in embryo development (HJORTSWANG et al., 2002). A possible role for *HOX* genes in conifers may also be ovule development, a subject waiting further study.

2.7. Gene specific approaches study ovule development in gymnosperms

The *KNOX-1* ortholog in *Arabidopsis, SHOOT MERISTEMLLESS (STM)*, along with *CUP-SHAPED COTYLEDONS1 (CUC1* and *CUC2)* (ISHIDA et al., 2000), are necessary for meristem maintenance in the shoot (TAKADA and TASAKA 2002) and later play a key role in ovule development by specifying the region where the placenta will form. Another homeodomain protein involved in determination of ovule identity is *BEL1* and plays an important role later during integument formation (ROBINSON-BEERS et al., 1992). BEL has been identified as forming a complex with STM, possibly involved in maintaining indeterminacy of the meristem (BELLAOUI et al., 2001). Virtually little is known about ovule primordial formation in gymnosperms. Recently, four *BEL*-like homologues, *MELBEL1-4* have been found in *Gnetum gnemon*. *MELBEL-1* based on northern analysis is expressed in leaves and both reproductive organs (BECKER et al., 2002). However, little is known about the specific expression of *BEL* during ovule formation in *Gnetum*.

Upon primordial formation, *AINTEGUMENTA (ANT)* is important for outgrowth of integument formation in *Arabidopsis* (ELLIOTT et al., 1996). One of the defining characteristics of *ANT* is that it has two *AP2* domains. *ANT* operates in a coordinated fashion with a number of genes to regulate ovule growth and development. *ANT* interacts with *LEUNIG* to direct ovule initiation and development as revealed in the *ANT /LUG* double mutant where ovule formation is completely abolished (LIU et al., 2000). *ANT* expression is reiterated during a later step in ovule development where it plays an important role in integument formation, along with *WUSCHEL* (GROSS-HARDT et al., 2002), and *NOZZEL (NOZ)* (SCHIEFTHALER et al., 1999). At the stage when the ovule primordial becomes distinct, *ANT* acts to promote integument formation, and *NZZ* restricts *ANT* activity to the integument forming chalazal region. *WUS* acts in concert with *ANT* to induce integument formation, but through a cell autonomous process as indicated by the fact that *WUS* is not expressed at the site of integument formation, but in the nucellus (GROSS-HARDT et al., 2002). As mentioned, in gymnosperms little work has been performed to examine the role of genes involved with ovule formation. Homologs to genes containing *AP2* domains have been cloned from conifers which include two *APETAL2*-like gene from Norway spruce *PaAP2L1* and *PaAP2L2* (VAHALA et al., 2001), and three from *P. thunbergii*, PtANTL1/PtAP2L1/PtAP2L2. PtANTL1 showed closest homology to *AINTEGUMENTA* (SHIGYO and ITO 2004). All three genes from *P. thunbergii* show expression in developing female cones and ovules. Similar to *Arabidopsis ANT, in situ* analysis of *PtANTL1* reveals its presence in developing pine nucellus and integuments. Future work to detect and examine potential key interacting genes with *ANT*, such as *LUG, WUS,* and *NOZ* orthologs, in gymnosperms will greatly help to understand the role of early molecular steps in ovule development.

In *Arabidopsis*, *ANT* activates the downstream gene *INNER-NO-OUTER (INO)*. *INO* is a *YABBY* gene family member, which is known to promote abaxial organ development. *INO* belongs to the five member *YABBY* family of zinc-finger transcription factors that regulate radial patterning (adaxial/abaxial) in *Arabidopsis* (ESHED et al., 1999; ESHED et al., 2001). *Arabidopsis INO* mutants have an inner but no outer integument. The *YABBY* gene *INNER NO OUTER (INO)* is necessary for outer integument development. Consequently *ino* mutants have an inner, but no outer integument enclosing the ovule (VILLANUEVA *et al.,* 1999). The *ino* phenotype is reminiscent of the condition in ancestral seeds which were unitegmic (one integument), as seen in all gymnosperms, except possibly *Gnetales* as described above. Besides *INO*, another *YABBY* gene family member is *CRABSCLAW*, which may also be important for ovule formation (BOWMAN and SMYTH 1999). In gymnosperms, a *YABBY* gene family paralog has been detected in a *Cycas rumphii* EST library from leaves (BRENNER *et al.,* 2003a). *INO* RNA levels in *Arabidopsis* integuments are monitored by *SUPERMAN (SUP)* through transcriptional attenuation (MEISTER *et al.,* 2002), an ortholog of *SUP* has not been reported in gymnosperms.

BEL1, described above, beside its role controlling the formation of the ovule primordia, is also a regulator of integument development, where it interacts with *INO* (MEISTER *et al.,* 2004). As indicated by the name, *BEL1* leads to the formation of a swollen bell-shaped structure in the place where integuments would normally be formed (REISER *et al.,* 1995). *MELBEL* genes from *Gnetum gnemon*, mentioned in a previous section, are *BEL* orthologs and need to be tested for their role in integument formation (BECKER *et al.,* 2002). Identification and characterization of these genes involved in integument formation is an important next step to understand gymnosperm reproductive biology.

3. CONCLUSIONS

The discovery of gymnosperm homologs and paralogs for *Arabidopsis* developmental regulators is an important step towards understanding the evolution and the genesis of ovules in gymnosperms. Cloning and characterization of these gene families remain incomplete, whereas other homologs have yet to be identified—if they exist. Thus, a molecular evolutionary development approach is able to target specific gene families of interest to only a certain degree. The approach is limited to cloning genes related to those with known function in model plants providing they have not diverged in sequence too greatly over time. The advent of genomics allows us to overcome such limitations and examine a variety of genes from multiple families simultaneously. A high through-put genomics approach from normalized EST libraries can identify a large proportion of the genes present in a selected tissue/organ for examination (RUDD 2003). Currently the only limit with genomics is the price involved in library construction and gene sequencing. However, as sequencing prices fall precipitously, genomic approaches become more attractive.

EST analysis in gymnosperms has already provided considerable insight into evolution. Comparative genomics methods have shown that the majority of genes found in *Pinus taeda* are also present in *Arabidopsis* when large contigs formed from deep EST sequencing are used in the comparison (KIRST *et al.,* 2003). However, a small subset of long contigs in this dataset has no matches to *Arabidopsis* genes and therefore could potentially be gymnosperms specific. In cycads comparative genomics followed by full cDNA sequencing identified a subset of *Cycas* genes, which are potentially specific to gymnosperms and not present in rice or *Arabidopsis*. In this same study, a variety of genes with homology to important development regulators in higher plants were identified including *CONSTANS*, two homeobox genes, *ARGONAUT* and *COP9* (BRENNER et al., 2003a).

Ultimately, with the decreasing price of sequencing and the improved methods of genome annotation, full genome sequencing will become available for larger genome size species. Approximately 25% of the gymnosperms have been assayed for genome size (LEITCH et al., 2005). Among the gymnosperms, the possibility of sequencing the genome of *Pinus* with a 1C value of 5.8 to 32.2 pg (LEITCH et al., 2005) which is large by comparison to *Arabidopsis* which has a 1C value of 0.16 pg is gaining popularity especially due to its economic importance. *Gnetum ula* has been reported to have the smallest gymnosperm genome with a 1C value of 2.3 pg (LEITCH et al., 2001), which although still substantial in size, might eventually put it within the range of acceptable costs for full genome sequencing.

In summary, the task that lies ahead is to efficiently collect, categorize and analyze these genes by developing new bioinformatics methods to understand gene change and evolution on a genome-wide level. Outside the MADS box family of gene regulators, only a few of the many genes expected to be involved in reproductive development have been cloned in gymnosperms—and of those isolated, even fewer have been explored at the functional level. EST-based and full genome sequencing approaches will greatly expand our access to genes that play critical roles in evolution and development in gymnosperms—including genes not only of known—but of unknown homology. Hence genomics approaches can deliver a comprehensive set of tools to understand the origin of the seed.

4. REFERENCES

The *Selaginella moellendorffii* genome sequencing project 2005 http://www.jgi.doe.gov/sequencing/why/CSP2005/selaginella.html.

ANGENENT, G. C., AND L. COLOMBO, 1996 Molecular control of ovule development. Trends in Plant Science **1**: 228-232.

BARTON, M., 2001 Leaving the meristem behind: regulation of *KNOX* genes. Genome Biol **2**: Reviews1002.

BECKER, A., M. BEY, T. R. BURGLIN, H. SAEDLER and G. THEISSEN, 2002 Ancestry and diversity of *BEL1*-like homeobox genes revealed by gymnosperm (*Gnetum gnemon*) homologs. Dev Genes Evol **212**: 452-457.

BECKER, A., H. SAEDLER and G. THEISSEN, 2003 Distinct MADS-box gene expression patterns in the reproductive cones of the gymnosperm Gnetum gnemon. Dev Genes Evol **213**: 567-572.

BECKER, A., K. U. WINTER, B. MEYER, H. SAEDLER and G. THEISSEN, 2000 MADS-Box gene diversity in seed plants 300 million years ago. Mol Biol Evol **17**: 1425-1434.

BELLAOUI, M., M. S. PIDKOWICH, A. SAMACH, K. KUSHALAPPA, S. E. KOHALMI et al., 2001 The Arabidopsis BELL1 and KNOX TALE homeodomain proteins interact through a domain conserved between plants and animals. Plant Cell **13**: 2455-2470.

BHARATHAN, G., T. E. GOLIBER, C. MOORE, S. KESSLER, T. PHAM et al., 2002 Homologies in leaf form inferred from *KNOXI* gene expression during development. Science **296**: 1858-1860.

BHARATHAN, G., B. J. JANSSEN, E. A. KELLOGG and N. SINHA, 1997 Did homeodomain proteins duplicate before the origin of angiosperms, fungi, and metazoa? Proc Natl Acad Sci USA **94**: 13749-13753.

BHARATHAN, G., and N. R. SINHA, 2001 The regulation of compound leaf development. Plant Physiol **127**: 1533-1538.

BIERHORST, D. W., 1971 *Morphology of Vascular Plants*. Macmillan.

BOWE, L. M., G. COAT and C. W. DEPAMPHILIS, 2000 Phylogeny of seed plants based on all three genomic compartments: extant gymnosperms are monophyletic and Gnetales' closest relatives are conifers. Proc Natl Acad Sci USA **97**: 4092-4097.

BOWMAN, J. L., and D. R. SMYTH, 1999 *CRABS CLAW*, a gene that regulates carpel and nectary development in Arabidopsis, encodes a novel protein with zinc finger and helix-loop-helix domains. Development **126**: 2387-2396.

BOWMAN, J. L., D. R. SMYTH and E. M. MEYEROWITZ, 1991 Genetic interactions among floral homeotic genes of Arabidopsis. Development **112**: 1-20.

BRENNER, E. D., D. W. STEVENSON, R. W. MCCOMBIE, M. S. KATARI, S. A. RUDD *et al.*, 2003a Expressed sequence tag analysis in *Cycas*, the most primitive living seed plant. Genome Biol **4**: R78.

BRENNER, E. D., D. W. STEVENSON and R. W. TWIGG, 2003b Cycads: evolutionary innovations and the role of plant-derived neurotoxins. Trends Plant Sci **8**: 446-452.

BYRNE, M., M. TIMMERMANS, C. KIDNER and R. MARTIENSSEN, 2001 Development of leaf shape. Curr Opin Plant Biol **4**: 38-43.

CARLSBECKER, A., J. SUNDSTROM, K. TANDRE, M. ENGLUND, A. KVARNHEDEN *et al.*, 2003 The *DAL10* gene from Norway spruce (*Picea abies*) belongs to a potentially gymnosperm-specific subclass of MADS-box genes and is specifically active in seed cones and pollen cones. Evol Dev **5**: 551-561.

CARMICHAEL, J. S., and W. E. FRIEDMAN, 1995 Double Fertilization in *Gnetum gnemon*: The Relationship between the Cell Cycle and Sexual Reproduction. Plant Cell **7**: 1975-1988.

CHAMBERLAIN, C. J., 1935 *Gymnosperms. Structure and Evolution*. University of Chicaco Press, Chicago.

CHAN, R. L., G. M. GAGO, C. M. PALENA and D. H. GONZALEZ, 1998 Homeoboxes in plant development. Biochim. Biophys. Acta. **1442**: 1-19.

CHAW, S. M., C. L. PARKINSON, Y. CHENG, T. M. VINCENT and J. D. PALMER, 2000 Seed plant phylogeny inferred from all three plant genomes: monophyly of extant gymnosperms and origin of Gnetales from conifers. Proc Natl Acad Sci USA **97**: 4086-4091.

COEN, E. S., and E. M. MEYEROWITZ, 1991 The war of the whorls: genetic interactions controlling flower development. Nature **353**: 31-37.

COEN, E. S., J. M. ROMERO, S. DOYLE, R. ELLIOTT, G. MURPHY *et al.*, 1990 *FLORICAULA*: a homeotic gene required for flower development in *Antirrhinum majus*. Cell **63**: 1311-1322.

CRANE, P. R., 1985 Phylogenetic analysis of seed plants and the origin of angiosperms. Annals of Missouri Botanical Gardens **72**: 716-793.

DONOGHUE, M. J., and J. A. DOYLE, 2000 Seed plant phylogeny: Demise of the anthophyte hypothesis? Curr Biol **10**: R106-109.

DOYLE, J. A., 1998 Molecules, morphology, fossils, and the relationship of angiosperms and Gnetales. Mol Phylogenet Evol **9**: 448-462.

DOYLE, J. A., and M. J. DONOGHUE, 1986 Seed plant phylogeny and the origin of angiosperms: an experimental cladistic approach. Botanical Reviews **52**: 321-431.

ELLIOTT, R. C., A. S. BETZNER, E. HUTTNER, M. P. OAKES, W. Q. TUCKER *et al.*, 1996 *AINTEGUMENTA*, an *APETALA2*-like gene of Arabidopsis with pleiotropic roles in ovule development and floral organ growth. Plant Cell **8**: 155-168.

ESHED, Y., S. F. BAUM and J. L. BOWMAN, 1999 Distinct mechanisms promote polarity establishment in carpels of Arabidopsis. Cell **99**: 199-209.

ESHED, Y., S. F. BAUM, J. V. PEREA and J. L. BOWMAN, 2001 Establishment of polarity in lateral organs of plants. Current Biology **11**: 1251-1260.

FAVARO, R., A. PINYOPICH, R. BATTAGLIA, M. KOOIKER, L. BORGHI *et al.*, 2003 MADS-box protein complexes control carpel and ovule development in Arabidopsis. Plant Cell **15**: 2603-2611.

FOSTER, A. S., and E. M. GIFFORD, 1974 *Comparative Morphology of Vascular Plants*. W. H. Freeman and Company, San Francisco.

FRIEDMAN, W., 1987 Morphogenesis and experimental aspects of growth and development of the male gametophyte of *Ginkgo biloba in vitro*. American Journal of Botany **74**: 1816-1830.

FROHLICH, M. W., and D. S. PARKER, 2000 The mostly male theory of flower evolutionary origins: from genes to fossils. Systematic Botany **25**: 155-170.

FUKUI, M., N. FUTAMURA, Y. MUKAI, Y. WANG, A. NAGAO et al., 2001 Ancestral MADS box genes in Sugi, *Cryptomeria japonica* D. Don (Taxodiaceae), homologous to the B function genes in angiosperms. Plant Cell Physiol **42**: 566-575.

GAO, Z., and B. A. THOMAS, 1989 A review of fossil cycad megasporophylls, with new evidence of *Crossozamia pomel* and its associated leaves from the lower Permian of Taiyuan, China. Review of Palaeobotany and Palynology **60**: 205-223.

GENOME, BRITISH, COLUMBIA, FORESTRY, GENOMICS et al., 2005 http://www.treenomix.com/cDNA-sequencing/default.aspx.

GOREMYKIN, V., V. BOBROVA, J. PAHNKE, A. TROITSKY, A. ANTONOV et al., 1996 Noncoding sequences from the slowly evolving chloroplast inverted repeat in addition to rbcL data do not support gnetalean affinities of angiosperms. Mol Biol Evol **13**: 383-396.

GOTO, K., J. KYOZUKA and J. L. BOWMAN, 2001 Turning floral organs into leaves, leaves into floral organs. Curr Opin Genet Dev **11**: 449-456.

GOURLAY, C. W., J. M. HOFER and T. H. ELLIS, 2000 Pea compound leaf architecture is regulated by interactions among the genes *UNIFOLIATA, COCHLEATA, AFILA*, and *TENDRIL-LESS*. Plant Cell **12**: 1279-1294.

GROSS-HARDT, R., M. LENHARD and T. LAUX, 2002 WUSCHEL signaling functions in interregional communication during Arabidopsis ovule development. Genes Dev **16**: 1129-1138.

GUILLET-CLAUDE, C., N. ISABEL, B. PELGAS and J. BOUSQUET, 2004 The Evolutionary Implications of knox-I Gene Duplications in Conifers: Correlated Evidence from Phylogeny, Gene Mapping, and Analysis of Functional Divergence. Mol Biol Evol **21**: 2232-2245.

HARA, N., 1997 Morphology and anatomy of vegetative organs in *Ginkgo biloba*, pp. 3-15 in *Ginkgo biloba-a Global Treasure*, edited by T. HORI, R. W. RIDGE, W. TULECKE, P. DEL TREDICI, J. TRÉMOUILLAUX-GUILLER et al., Springer-Verlag, Tokyo.

HASEBE, M., C. K. WEN, M. KATO and J. A. BANKS, 1998 Characterization of MADS homeotic genes in the fern Ceratopteris richardii. Proc Natl Acad Sci U S A **95**: 6222-6227.

HIMI, S., R. SANO, T. NISHIYAMA, T. TANAHASHI, M. KATO et al., 2001 Evolution of MADS-box gene induction by *FLO/LFY* genes. J Mol Evol **53**: 387-393.

HJORTSWANG, H. I., L. A. SUNDÅS, G. BHARATHAN, P. BOZHKOV, S. VON ARNOLD et al., 2002 *KNOTTED1*-like homeobox genes of a gymnosperm, Norway spruce, expressed during somatic embryogenesis. Plant Physiol Biochem **40**: 837-843.

IKENO, S., and S. HIRASE, 1897 Spermatozoids in gymnosperms. Ann. of Bot. **11**: 344-345.

ISHIDA, T., M. AIDA, S. TAKADA and M. TASAKA, 2000 Involvement of *CUP-SHAPED COTYLEDON* genes in gynoecium and ovule development in *Arabidopsis thaliana*. Plant Cell Physiol **41**: 60-67.

JAGER, M., A. HASSANIN, M. MANUEL, H. L. GUYADER and J. DEUTSCH, 2003 MADS-box genes in *Ginkgo biloba* and the evolution of the *AGAMOUS* Family. Mol Biol Evol **20**: 842-854.

KIRST, M., A. F. JOHNSON, C. BAUCOM, E. ULRICH, K. HUBBARD et al., 2003 Apparent homology of expressed genes from wood-forming tissues of loblolly pine (*Pinus taeda* L.) with Arabidopsis thaliana. Proc Natl Acad Sci U S A **100**: 7383-7388.

KRIZEK, B. A., and E. M. MEYEROWITZ, 1996 The Arabidopsis homeotic genes *APETALA3* and *PISTILLATA* are sufficient to provide the B class organ identity function. Development **122**: 11-22.

LEITCH, I. J., L. HANSON, M. WINFIELD, J. PARKER and M. D. BENNETT, 2001 Nuclear DNA C-values complete familial representation in gymnosperms. Ann Bot **88**: 843-849.

LEITCH, I. J., D. E. SOLTIS, P. S. SOLTIS and M. D. BENNETT, 2005 Evolution of DNA amounts across land plants (embryophyta). Ann Bot (Lond) **95**: 207-217.

LIU, Z., R. G. FRANKS and V. P. KLINK, 2000 Regulation of gynoecium marginal tissue formation by *LEUNIG* and *AINTEGUMENTA*. Plant Cell **12**: 1879-1892.

MAGOLLÓN, S., and M. J. SANDERSON, 2002 Relationships among seed plants inferred from highly conserved genes: sorting conflicting phylogenetic signals among ancient lineages. Am. J. Bot. **89**: 1991-2006.

MAMAY, S. H., 1969 Cycads: fossil evidence of late paleozoic origin. Science **164**: 295-296.
MEEUSE, A. D. J., 1966 *Fundamentals of Phytomorphology*. The Ronold Press Co., New York.
MEISTER, R. J., L. M. KOTOW and C. S. GASSER, 2002 *SUPERMAN* attenuates positive *INNER NO OUTER* autoregulation to maintain polar development of Arabidopsis ovule outer integuments. Development **129**: 4281-4289.
MEISTER, R. J., L. A. WILLIAMS, M. M. MONFARED, T. L. GALLAGHER, E. A. KRAFT *et al.*, 2004 Definition and interactions of a positive regulatory element of the Arabidopsis *INNER NO OUTER* promoter. Plant J **37**: 426-438.
MELLEROWICZ, E. J., K. HORGAN, A. WALDEN, A. COKER and C. WALTER, 1998 *PRFLL*–a *Pinus radiata* homologue of *FLORICAULA* and *LEAFY* is expressed in buds containing vegetative shoot and undifferentiated male cone primordia. Planta **206**: 619-629.
MOURADOV, A., T. GLASSICK, B. HAMDORF, L. MURPHY, B. FOWLER *et al.*, 1998 *NEEDLY*, a Pinus radiata ortholog of *FLORICAULA/LEAFY* genes, expressed in both reproductive and vegetative meristems. Proc Natl Acad Sci USA **95**: 6537-6542.
MOURADOV, A., B. HAMDORF, R. D. TEASDALE, J. T. KIM, K. U. WINTER *et al.*, 1999 A *DEF/GLO*-like MADS-box gene from a gymnosperm: *Pinus radiata* contains an ortholog of angiosperm B class floral homeotic genes. Dev Genet **25**: 245-252.
MUNDRY, I., 2000 Morphologische und morphogenetische Untersuchungen zur Evolution der Gymnospermen, E. Schweizbart'sche Verlagsbuchhandlung. N×§gele u. Obermiller.
MUNSTER, T., J. PAHNKE, A. DI ROSA, J. T. KIM, W. MARTIN *et al.*, 1997 Floral homeotic genes were recruited from homologous MADS-box genes preexisting in the common ancestor of ferns and seed plants. Proc Natl Acad Sci USA **94**: 2415-2420.
NG, M., and M. F. YANOFSKY, 2001 Function and evolution of the plant MADS-box gene family. Nat Rev Genet **2**: 186-195.
NIXON, K., W. CREPET, D. W. STEVENSON and E. FRIIS, 1994 A reevaluation of seed plant phylogeny. Annals of the Missouri Botanical Garden **81**: 484-583.
NORSTOG, K. J., and T. J. NICHOLLS, 1997 *The Biology of the Cycads*. Cornell University Press, Ithaca, N.Y.
PINYOPICH, A., G. S. DITTA, B. SAVIDGE, S. J. LILJEGREN, E. BAUMANN *et al.*, 2003 Assessing the redundancy of MADS-box genes during carpel and ovule development. Nature **424**: 85-88.
PRYER, K. M., H. SCHNEIDER, A. R. SMITH, R. CRANFILL, P. G. WOLF *et al.*, 2001 Horsetails and ferns are a monophyletic group and the closest living relatives to seed plants. Nature **409**: 618-622.
PURUGGANAN, M. D., 1997 The MADS-box floral homeotic gene lineages predate the origin of seed plants: phylogenetic and molecular clock estimates. J Mol Evol **45**: 392-396.
PURUGGANAN, M. D., S. D. ROUNSLEY, R. J. SCHMIDT and M. F. YANOFSKY, 1995 Molecular evolution of flower development: Diversification of the plant MADS-box regulatory gene family. Genetics **140**: 345-356.
REISER, L., Z. MODRUSAN, L. MARGOSSIAN, A. SAMACH, N. OHAD *et al.*, 1995 The *BELL1* gene encodes a homeodomain protein involved in pattern formation in the Arabidopsis ovule primordium. Cell **83**: 735-742.
ROBINSON-BEERS, K., R. E. PRUITT and C. S. GASSER, 1992 Ovule development in wild-type *Arabidopsis* and two female-sterile mutants. Plant Cell **4**: 1237-1249.
ROKAS, A., B. L. WILLIAMS, N. KING and S. B. CARROLL, 2003 Genome-scale approaches to resolving incongruence in molecular phylogenies. Nature **425**: 798-804.
ROTHWELL, G. W., 1988 Cordaitales in *Origin and Evolution of Gymnosperms*, edited by C. B. Beck Columbia University Press, New York.
ROTHWELL, G. W., and S. E. SCHECKLER, 1988 Biology of Ancestral Gymnosperms in *Origin and Evolution of Gymnosperms*, Edited by C. B. Beck. Columbia University Press, New York.
RUDD, S., 2003 Expressed sequence tags: alternative or complement to whole genome sequences? Trends Plant Sci **8**: 321-329.
RUTLEDGE, R., S. REGAN, O. NICOLAS, P. FOBERT, C. COTE *et al.*, 1998 Characterization of an *AGAMOUS* homologue from the conifer black spruce (*Picea mariana*) that produces floral homeotic conversions when expressed in Arabidopsis. Plant J **15**: 625-634.
RYDIN, C., M. KALLERSJO and E. M. FRIIS, 2002 Seed plant relationships and the systematic position of Gnetales based on nuclear and chloroplast DNA: conflicting data, rooting problems and the monophyly of conifers. Int. J. Plant Sci. **163**: 197-214.

SAVARD, L., P. LI, S. H. STRAUSS, M. W. CHASE, M. MICHAUD et al., 1994 Chloroplast and nuclear gene sequences indicate late Pennsylvanian time for the last common ancestor of extant seed plants. Proc Natl Acad Sci USA **91**: 5163-5167.

SCHIEFTHALER, U., S. BALASUBRAMANIAN, P. SIEBER, D. CHEVALIER, E. WISMAN et al., 1999 Molecular analysis of *NOZZLE*, a gene involved in pattern formation and early sporogenesis during sex organ development in *Arabidopsis thaliana*. Proc Natl Acad Sci U S A **96**: 11664-11669.

SCHWARZ-SOMER, Z., P. HUIJSER, W. NACKEN, H. SAEDLER and H. SOMMER, 1990 Genetic control of flower development: homeotic genes in *Antirrhinum majus*. Science **250**.

SHIGYO, M., and M. ITO, 2004 Analysis of gymnosperm two-AP2-domain-containing genes. Dev Genes Evol **214**: 105-114.

SHINDO, S., K. SAKAKIBARA, R. SANO, K. UEDA and M. HASEBE, 2001 Characterization of a *FLORICAUL/LEAFY* homologue of *Gnetum parvifolium* and its implications for the evolution of reproductive organs in seed plants. International Journal of Plant Science **162**: 1199-1209.

SKINNER, D. J., T. A. HILL and C. S. GASSER, 2004 Regulation of ovule development. Plant Cell **16 Suppl**: S32-45.

SOLTIS, D. E., P. S. SOLTIS and M. J. ZANIS, 2002a Phylogeny of seed plants based on evidence from eight genes. Am. J. Bot. **89**: 1670-1681.

SOLTIS, P. S., D. E. SOLTIS, V. SAVOLAINEN, P. R. CRANE and T. G. BARRACLOUGH, 2002b Rate heterogeneity among lineages of tracheophytes: integration of molecular and fossil data and evidence for molecular living fossils. Proc Natl Acad Sci U S A **99**: 4430-4435.

STEWART, W. N., and G. W. ROTHWELL, 1983 *Paleobotany and the Evolution of Plants*. Cambridge University Press, Cambridge.

SUNDAS-LARSSON, A., M. SVENSON, H. LIAO and P. ENGSTROM, 1998 A homeobox gene with potential developmental control function in the meristem of the conifer *Picea abies*. Proc Natl Acad Sci U S A **95**: 15118-15122.

SUNDSTROM, J., and P. ENGSTROM, 2002 Conifer reproductive development involves B-type MADS-box genes with distinct and different activities in male organ primordia. Plant J **31**: 161-169.

TAKADA, S., and M. TASAKA, 2002 Embryonic shoot apical meristem formation in higher plants. J Plant Res **115**: 411-417.

TAKASO, T., 1984 Structural-Changes in the Apex of the Female Strobilus and the initiation of the female reproductive organ (ovule) in *Ephedra-Distachya* L and *Ephedra-Equisetina* Bge. Acta Botanica Neerlandica **33**: 257-266.

TAKASO, T., F. Bouman, 1986 Ovule and seed ontogeny in *Gnetum gnemon* L. Botanical Magazine Tokyo **99**: 241-266.

TANDRE, K., V. A. ALBERT, A. SUNDAS and P. ENGSTROM, 1995 Conifer homologues to genes that control floral development in angiosperms. Plant Mol Biol **27**: 69-78.

TANDRE, K., M. SVENSON, M. E. SVENSSON and P. ENGSTROM, 1998 Conservation of gene structure and activity in the regulation of reproductive organ development of conifers and angiosperms. Plant J **15**: 615-623.

TAYLOR, T. N., and E. L. TAYLOR, 1993 *The Biology and Evolution of Fossil Plants*. Prentice Hall.

THEISSEN, G., A. BECKER, A. DI ROSA, A. KANNO, J. T. KIM et al., 2000 A short history of MADS-box genes in plants. Plant Mol Biol **42**: 115-149.

THEISSEN, G., J. T. KIM and H. SAEDLER, 1996 Classification and phylogeny of the MADS-box multigene family suggest defined roles of MADS-box gene subfamilies in the morphological evolution of eukaryotes. J Mol Evol **43**: 484-516.

UJINO-IHARA, T., K. YOSHIMURA, Y. UGAWA, H. YOSHIMARU, K. NAGASAKA et al., 2000 Expression analysis of ESTs derived from the inner bark of *Cryptomeria japonica*. Plant Mol Biol **43**: 451-457.

VAHALA, T., B. OXELMAN and S. VON ARNOLD, 2001 Two *APETAL2*-like genes from *Picea abies* are differentially expressed during development. Journal of Experimental Botany **52**: 1111-1115.

VILLANUEVA, J. M., J. BROADHVEST, B. A. HAUSER, R. J. MEISTER, K. SCHNEITZ et al., 1999 *INNER NO OUTER* regulates abaxial- adaxial patterning in *Arabidopsis* ovules. Genes Dev **13**: 3160-3169.

WEIGEL, D., J. ALVAREZ, D. R. SMYTH, M. F. YANOFSKY and E. M. MEYEROWITZ, 1992 *LEAFY* controls floral meristem identity in Arabidopsis. Cell **69**: 843-859.

WHETTEN, R., Y. H. SUN, Y. ZHANG and R. SEDEROFF, 2001 Functional genomics and cell wall biosynthesis in loblolly pine. Plant. Mol. Biol. **47**: 275-291.

WINTER, K. U., A. BECKER, T. MUNSTER, J. T. KIM, H. SAEDLER et al., 1999 MADS-box genes reveal that gnetophytes are more closely related to conifers than to flowering plants. Proc Natl Acad Sci USA **96:** 7342-7347.

WINTER, K. U., H. SAEDLER and G. THEISSEN, 2002 On the origin of class B floral homeotic genes: functional substitution and dominant inhibition in Arabidopsis by expression of an orthologue from the gymnosperm *Gnetum*. Plant J **31:** 457-475.

ZHANG, P., H. T. TAN, K. H. PWEE and P. P. KUMAR, 2004 Conservation of class C function of floral organ development during 300 million years of evolution from gymnosperms to angiosperms. Plant J. **37:** 566-577.

CHAPTER 7

METABOLIC PROFILING FOR TRANSGENIC FOREST TREES

HELY HÄGGMAN
Department of Biology, University of Oulu, Oulu, Finland

RIITTA JULKUNEN-TIITTO
Department of Biology, University of Joensuu, Joensuu, Finland

Abstract. In biological systems, the information flow goes from DNA to RNA to protein and then to metabolites which plants possess in high numbers. This information flow is not co-linear so metabolic profiling is an opportunity to better understand the connection between genotype and phenotype. Metabolic profiling or metabolomics is the fourth cornerstone of functional genomics which also includes genomics, transcriptomics and proteomics. Metabolic profiling analyses of transgenic plants is recent and mainly focused on primary metabolite studies of potato, tomato or *Arabidopsis*. The first report on the use of metabolic profiling to compare wild type and yeast invertase overexpressing in potato tubers was published by Roessner and co-workers as recently as 2000 and the integration of transcript and metabolic profiles i.e. co-response analyses in potato tubers, was published by Urbanczyk-Wochniak and co-workers in 2003. Metabolic profiling of transgenic trees is still in its infancy. This is partly due to the specific characteristics of tree species which make the metabolism complex but also emphasizes the necessity for technological improvements in metabolic profiling protocols. For trees, the expression of monolignol precursor genes and their effect on lignin content and composition have been studied in detail but this can not be regarded as a large-scale metabolic profiling effort. Metabolic profiling of transgenic trees hold potential for 1) linking changes in genotype to phenotype, 2) deepening understanding of tree metabolism and its regulation, 3) determining interactions between non-allelic genes (epistasis) and 4) pinpointing potential multiple effects of single transgenes (pleiotropic effects). Metabolic profiling data can also contribute to conventional and molecular tree breeding through characterization of commercially important lines or breeds or showing unexpected metabolic shifts. By integrating genomics, transcriptomics, proteomics and metabolomics data, we will gain a more comprehensive understanding of whole tree physiology and its regulation.

1. INTRODUCTION

The history of genetic modification of forest trees starts at late 1980s when the first report of stably transformed *Populus alba x grandidentata* was published by FILLATTI *et al.,* (1987). This was possible due to the previous challenges to create new combinations of genetic material from DNA molecules of different origin. Since these early days, genetic transformation techniques have been developed for several broad-leaved as well as coniferous species. However, the genetic improvement or molecular breeding of forest crops utilizing genetic transformation techniques, is today at an early stage. Most forest trees can still be regarded as undomesticated wild trees for the majority of our wood product needs. There is, however, a global shift towards tree plantations to meet the increasing need for fiber and to maximize both growth and yield and in this context also the potential of genetically modified tree crops will be evaluated.

The conventional tree breeding consists of several activities including selection, sexual and somatic hybridization, and testing but these techniques are limited by the sterility of the descents, the genetic barrier between species and the long life cycle and generation interval characteristics for several forest tree species (as reviewed by SEDEROFF 1995). To overcome this genetic barrier is now as it was then only possible via genetic transformation. The fundamental aspect of genetic modification to create new combinations of genetic material from DNA molecules of different origin was considered important for tree breeding purposes. In addition to these practical applications, transgenic approach has been widely used as a tool in tree and plant physiology, ecology, genetics and molecular biology (as reviewed by HERSCHBACH AND KOPRIVA 2002). These approaches have mostly been focused on regulating endogenous gene functioning either to study transgene expression or regulation or to consider effects of gene modification at a more holistic level which in forest trees still means microarray technology i.e. studies at a transcriptome level.

It is obvious that information flow from DNA to metabolites is not co-linear and there is a need to better understand the connection between genotype and phenotype. The recent studies indicate that this might be done by metabolic profiling (FIEHN 2002) and by comparing metabolomics, proteomics and microarray data for improved understanding of phenotypic characteristics controlled by genomics and modified by environmental constraints (as reviewed by PHELPS et al., 2002, EDWARDS AND BATLEY 2004). In trees, this might help to better understand metabolism and its regulation, to plan specific approaches for metabolic engineering, to characterize commercially valuable breeds, to explain the potentially unexpected changes i.e. pleiotrophic and epistatic effects of transgenes. To achieve this goal there is a need for more studies on genetically modified tree species with different traits from resistance characteristics to qualitative traits and a need for a more comprehensive research and technology outcome at metabolome level.

2. METABOLISM VERSUS DNA

In biological systems, information flows from DNA to RNA to protein and to metabolites. Today several plant genomes including *Arabidopisis,* rice and poplar have been sequenced and partially annotated and there is also considerable amount of partial sequence information (ESTs, expressed sequence tags) from other species available. Most of the current genomic research in trees has been carried out at the DNA and RNA levels. This is mostly due to the mature technologies that allow the sequencing and analysis of the expression levels of complete or partial genomes at the transcript level (URBANCZYK-WOCHNIAK et al., 2003). It is, however, obvious that metabolism is not co-linear with DNA sequence. Post-translational factors are functionally important in the cell and to find out interactions between genes and gene products and to look at their biological roles under different environmental conditions classical gene-by-gene approaches are not enough (as reviewed by WECKWERTH AND FIEHN 2002). The role of metabolomics as one of *the functional genomics tools* and as the essential link between genotypes and phenotypes has been discussed in several recent publications and reviews (FIEHN 2002; PHELPS et al., 2002; BINO et al., 2004).

Plant metabolism can be described as a complex network with crossroads and lonely streets. Conventionally and based on biochemistry, it has been divided into primary and secondary metabolism. Primary metabolism and the products of metabolism i.e. metabolites are considered the ones essential for plant growth and development (BRYANT et al., 1991; TEGELBERG et al., 2001; LAITINEN et al., 2004) or even regarded as overflow-components (e.g. LUCKNER et al., 1976; SEIGLER 1998). Typically, secondary compounds may be specific for certain plant species or tissues or they may be produced only at a certain developmental stage depending on the plant age and the prevailing environmental factors (e.g. BRYANT AND JULKUNEN-TIITTO 1995; TEGELBERG AND JULKUNEN-TIITTO 2001; LAITINEN et al., 2002). Many secondary metabolites are also inducible as a response to wounding or pathogens (e.g. WATERMAN AND MOLE 1994; JULKUNEN-TIITTO et al., 1995). Several secondary metabolites are important for the successful performance of plants in the ecosystem. Many of them are energy-rich and accumulate in high amounts indicating high production and maintenance costs for the plants. In some cases, the classical division to primary and secondary metabolism can, however, be arbitrary. In trees, lignin, which is the hydrophobic polymer network of phenylpropanoid units in cell walls, is a necessary component for secondary growth and development but biochemically lignin is considered as a product of secondary metabolism (e.g. SIEGLER 1998; WATERMAN AND MOLE 1994; BOERJAN et al., 2003)

The plant kingdom is rich in metabolites, the recent estimates exceeding 200,000 (as reviewed by WECKWERTH AND FIEHN 2002). This high number is due to the large number of genes, multiple substrate specificities for many enzymes, nonenzymic reactions and the fluidity of the metabolites i.e. convertion to other molecules, and subcellular compartmentation. As pointed out by STITT AND FERNIE (2003) in their review article, measuring metabolites is difficult. This is due to their fluidity and their chemistry leading to enormous number of single metabolites with wide range of structures when compared to nucleic acids whose chemical structure is relatively simple or proteins with a bit more complicated chemical structures. The controversial circumstances are also possible due to the chemical instability of the metabolites. For instance, willow salicylates, counting more than 15 individual components in living tissues may yield only three to four decomposition products if post-harvest or extraction procedures have not been carefully considered (ORIANS 1995; LINDROTH AND KOSS 1996; JULKUNEN-TIITTO AND SORSA 2001).

Terminologies used in classification of metabolomic approaches are still evolving and there are some overlaps as pointed out by GOODACRE *et al.*, (2004). Metabolomics is generally regarded as a comprehensive analysis of the whole metabolome under a given set of conditions. In metabolite profiling or metabolic profiling, which terms are often used interchangeably, the analysis is focused on a group of metabolites.

3. METHODS FOR METABOLIC PROFILING

Metabolic profiling means monitoring and quantification of known metabolites representing selective classes of the biochemical pathways in specific tissues. By profiling it is possible to differentiate genotypes and to identify loci involved in metabolic composition. In the metabolic-engineering projects, it is possible to reveal pleiotropic effects of transgenes and transgene function in the performance of complex organism and to achieve deeper understanding of plant metabolic networks.

The principal concern among analytical issues in metabolic profiling is the pre-handling of the plant material. The different metabolites, such as amino acids, lipids, sugars, phenolics, terpenoids and alkaloids having a variety of chemical characteristics result in difficulties if one aims to use the same plant material for extraction of many components. There are several different post-harvest methods (liquid nitrogen, at -20 C, drying with freeze-dryer, oven at elevated temperature, drying-room in low RH and temperature, and at room temperature in laboratory) and many of which are tested for most of the metabolites identified. Preservation in liquid nitrogen may be the best way for certain metabolites (proteins, amino acids, monoterpenes) but very deleterious for some others (such as salicylates), (STEINBRECHER *et al.*, 1999; JULKUNEN-TIITTO AND SORSA 2001). Freeze-drying is recommended for plant tannins but not for monoterpenes (e.g. HAGERMAN 1988). In the use of inappropriate post-harvest method the number of components will be either increased or decreased from that found in living-tissues.

The extraction of plant tissues for metabolite profiling is complicated and needs a serious concern because of a complex of the factors including metabolites of varying chemical nature and stabilities and a variety of physical conditions. The differences in solubility, compartmentalization and matrix effects are possible to overcome using a sequential extraction process with several solvent systems. This way the highest possible yield in quantity and quality of metabolites is achieved and easily conducted with semi- or automated extraction equipment available. Also the identification and quantification instrumentation offers reliable analysis of the metabolites. The liquid-chromatography (LC) and gas-chromatography (GC) with different mass selective (MS) detectors, such as quadrupol (Q), time-of-flight (TOF), ion-trap (ITR) combined with NMR (nucleo magnetic resonance) data are effective tools for the analyses of metabolites. The high-throughput screening with a high accuracy in mass and fragmentation patterns will be approached by two-dimensional instrumentation, such as GC/MS, LC/MS, GC/MS/MS, LS/MS/MS or LC/NMR/MS as proposed by BINO *et al.*, (2004).

Today, the most challenging task is to expand the technology to high-resolution, high-throughput analysis and the dissemination of metabolomics research data. To achieve these goals, BINO *et al.*, (2004) propose the following actions: improvement in the comprehensive coverage, facilitation of comparison of results between different laboratories and experiments and integration of metabolomics into systems biology. The International Committee on Plant Metabolomics (http://www.metabolomics.nl) represents a platform to accomplish these actions.

4. METABOLIC PROFILING AND TRANSGENIC PLANTS: MOVING TOWARDS SYSTEMS BIOLOGY

Investigations of metabolic network regulation upon genetic and environmental perturbations may be viewed as a necessity for pathway discovery and functional genomics (as reviewed by WECKWERTH AND FIEHN 2002). It can also be used for gene discovery and to ascribe their function in the context of the organism (TRETHEWAY 2001). Quite recently metabolic gene expression data in yeast (*Saccharomyces cerevisiae*) revealed a complex modular organization of co-expressed genes which was considered to be helpful in engineering of cellular metabolic functions (IHMELS *et al.*, 2002, 2004). IHMELS *et al.* (2002, 2004) pointed out that only a subset of genes in a given pathway displays a significant co-expression. These genes or transcriptional modules are composed of linear arrangements of enzymes in the metabolic network but also of genes from feeder pathways, transporters and transcription factors. The large-scale multiorganism research on metabolic pathways reveals an arrangement of functional associations that crosses the boundaries of classically defined pathways and thus emphasizes the modulocentric view in which modules rather than conventional pathways are the elementary functional units (KRIEGER *et al.*, 2004). SEGRÉ (2004) also indicates in his review article that understanding the complex set of regulatory strategies of metabolic pathways is like unravelling the "software" that is running on the better-known and characterized "hardware" components of the metabolic network.

During the history of transgenic plants there are some examples of unexpected results which might point to the modulocentric function of plant metabolism. For instance, BOEHMERT *et al.* (2000) succeeded in their transgenic approach in producing high levels of biodegradable plastic, polyhydroxybutyrate, in *Arabidopisis* leaves. Simultaneously, however, routine metabolic profiling experiments indicated rampant pleiotropic changes in organic acids, amino acids, and sugar alcohols. In another example, CHIKWAMBA et al. (2003) found out that a recombinant bacterial protein Lt-B), with a secretion orienting signal sequence, had an aberrant localization to amyloplasts, starch accumulating cell organelles of maize endosperm.

So far, metabolic profiling approaches of transgenic plants have mainly been focused on primary metabolite studies. The first report on the use of metabolic profiling in order to compare wild type and genetically modified higher plants was published by ROESSNER et al. (2000). They used and validated gas chromatography mass spectrometry (GC-MS) to analyse, both quantitatively and qualitatively, around 150 primary metabolite products such as sugars, sugar alcohols, amino acids and organic acids in wild type potato tubers and in transgenic ones overexpressing yeast invertase in tubers. Their data indicated and confirmed that the expression of invertase led to a shift in metabolism from starch biosynthesis to glycolysis and amino acid metabolism. They also found a strong and unexpected accumulation of maltose in transgenic tubers. In their subsequent study (ROESSNER et al., 2001) they were able to identify and quantify the major metabolic constituents of the potato tuber within a single chromatographic run. The profiles achieved from potato tubers incubated with various carbohydrates or from transgenic tubers with modified starch metabolism were then compared at the level of individual metabolites and subsequently by hierarchical cluster analysis and principal component analysis. By these means the authors were able to assign clusters to the individual plant systems and furthermore identification of the most important metabolites describing these clusters. Potato tuber system comparisons enabled further evaluation of potential phenocopies.

FIEHN et al. (2000) used the same methodology as ROESSNER et al. (2000) to analyze four distinct *Arabidopsis* genotypes including two background genotypes, Columbia and C24, and two mutant genotypes *dgd1* and *sdd1-1*. For the interpretation of GC-MS data the authors applied the principle component analysis and they found that there were 41 significant changes in the *sdd1*-1 mutant and 153 in the *dgd1* mutant when compared to the parental ecotypes C24 and Columbia, repectively.

Metabolite profiling approaches have also been used to study transgenic tomato plants overexpressing hexokinase by ROESSNER-TUNALI et al. (2003). In this study in addition to GC-MS analyses also conventional spectrophotometric and liquid chromatographic methodologies used allowed identification over 70 small–M metabolites in tomato fruit. Metabolic profiling data with point by point analyses and principal component analyses supported the enzyme activity data indicating that the metabolic phenotypes of the transgenics alongside with enzyme activities became less distinct from wild type during development.

The examples above indicate that the first steps in metabolite profiling have already been taken to analyze transgenic plants and to better understand the effect of changes at genotype level on phenotype. For the next step towards full-scale systems biology there is a need to technology outcome to allow high-throughput analyses already realized at transcriptome level. The integration of transcript and metabolic profiles, i.e. co-response analyses has been performed for the first time in potato tubers (URBANCZYK-WOCHNIAK et al., 2003). The authors used principal-component analysis to determine whether the developmental stages and different ectopically expressed transcript profiles of potato tubers could be discriminated. Interestingly, the results suggested that metabolic profiling has a higher resolution than expression profiling. This will need, as the authors point out, further studies to find out if this reflects the fact that changes at transcriptome level are less pronounced than at metabolite level or if this is simply due to low sensitivity of ESTs (expression signal tags) as probes. When applying statistical pairwise transcript-metabolite correlation analyses they found 363 positive and 208 negative correlations out of 26, 616 possible pairs. Several of the correlations found were novel ones. For instance, from the nutritionally important metabolites, tocopherol was negatively correlated with succinyl-co-enzyme-A synthetase, ascorbate was negatively correlated with homologue of the clock gene *CONSTANS,* and lysine was positively regulated by the transcription factor WRKY6. Although URBANCZYK-WOCHNIAK et al., (2003) agree that there may be several explanations for these correlations and there is certainly a need for a more direct experimentation to elucidate the mechanisms underlying these correlations they also emphasize the great potential of these unexpected correlations for biotechnological applications in which the goal is the modification of metabolite compositions by transgenic approach.

5. METABOLIC PROFILING AND TRANSGENIC TREES

Forest trees, such as pines, spruce, birches, poplars and willows contain a wide variety of different secondary metabolites such as phenolics and terpenoids (e.g. ORIANS 1995; PAN AND LUNDGREN 1995; LINDROTH AND KOSS 1996; SEIGLER 1998; HÄGGMAN et al., 2004; LAITINEN et al., 2004). In trees, especially secondary metabolites have been analysed in several cases to study different developmental stages, different stress-related situations, carbon allocation between growth and defence (e.g. BRYANT et al., 1991; WATERMAN AND MOLE 1994; TEGELBERG et al., 2001). In long-living trees the quantitative content of secondary metabolites may be high. For instance, the total concentration of phenolic compounds may reach in birch leaves and stem bark more than 10 and 20% of the dry weight, respectively, of which condensed tannins form the main part (30 to 80%) (e.g. LAITINEN et al., 2003). Accordingly, also the developmental stage specificity and/or qualitative specificity among tree species are found. For instance, young silver birch stems are covered by resin glands which contain several triterpenoids, such as papyriferic and pendulic acids (LAITINEN et al., 2003), while pine and spruce needles and young stems contain a number of different mono- and sesqiterpenes located mainly into resin ducts (e.g. PAN AND LUNDGREN 1995).

To date, there exist numerous stably transformed tree species as indicated in the recent reviews (PEÑA AND SÉGUIN 2001; HERSCHBACH AND KOPRIVA 2002; DIOUF

2003). The purpose of genetic transformation has in most cases been to improve knowledge in tree physiology and molecular biology (HERSCHBACH AND KOPRIVA 2002) or more practically oriented approaches such as reduction of generation time, production of sterile trees, pest or disease resistance or lignin manipulation (PEÑA AND SÉGUIN 2001; DIOUF 2003). Metabolic profiling approaches in transgenic trees have been scarce and none of them has really been focused on large scale metabolite profiling. The well studied lignin biosynthesis, mostly in model plants *Arabidopsis* and tobacco but also in trees, is not an exception. According to the recent review (BOERJAN et al., 2003) transgenic plants or mutants with modified expression of all monolignol biosynthesis genes, except HCT an acyltransferase controlling shikimate and quinate ester intermediates, have been studied in detail. Plants can tolerate large variations in lignin content and composition. In addition to p-coumaryl, coniferyl and sinapyl alcohols also other uncommon monomers can be incorporated into lignin polymer and this copolymerization may result in novel lignin structures (BOERJAN et al., 2003). It has been reported that especially reductions in lignin content by downregulation of C3H (*p*-coumarate 3-hydroxylase), CCoAOMT (caffeoyl-CoA *O*-methyltransferase) or CCR (cinnamoyl-CoA reductase) have been associated with altered growth and phenotypes that may significantly vary according to developmental and environmental conditions (as reviewed by BOERJAN et al., 2003). On the other hand, Hu and co-workers (1999), demonstrated that in aspen decrease in lignin content by down-regulation of 4CL (4-coumarate:CoA ligase) did not cause growth abnormalities and decrease in lignin content was substituted by increase in cellulose content. Experiments like these also provide an interesting approach for metabolic profiling to study e.g. carbon sequestration or more generally to find connections between genotypes and phenotypes.

Up to our knowledge, the first attempt to apply metabolic profiling to transgenic trees was our own work (HÄGGMAN et al., 2003). We are interested in the function of endogenous non-symbiotic haemoglobin genes in plants and we produced several hybrid aspen lines with heterologous *Vitreoscilla* haemoglobin (*vhb*) gene. Especially the cases where the specific function or molecular mechanisms of the transgene are not known metabolic profiling might provide new information. Based on previous experience on metabolite composition in aspen leaves (LINDROTH AND KOSS 1996), we analyzed salicylates, quercetin derivatives, kaempferol derivatives, myricetin derivatives, apigenins, phenolic acids, catechins, condensed tannins and total flavonoids in non-transgenic controls and transgenic hybrid aspen leaves expressing *vhb* (HÄGGMAN et al., 2003). The analyses were done both in optimal greenhouse conditions as well as after exposure to elevated UV-B radiation. We found that the number of significant differences between lines was higher after elevated UV-B illumination than under ambient UV-B illumination or at optimized greenhouse conditions. The differences were found in concentrations of total flavonoids, individual quercetin, kaempferol- and myricetin derivatives. These findings together with our other physiological and molecular data (HÄGGMAN et al., 2003) reflect the availability of extra energy resources for secondary metabolite production and a potential role of the transgene (*vhb*) in stress situation. The metabolic profiling data achieved subsequently need more detailed investigation including comparison with transcriptomics data.

Biosafety issues of transgenic plants have recently been emphasized for instance by establishment of European wide, web-based, public-access database (www.versailles.inra.fr/europe/gmorescom) to enhance communication regarding biosafety research. Environmental concerns of transgenic technology in plants have derived from the possibility of gene flow to close relatives of the transgenic plant, the possible undesirable effects of the transgenes or traits, and their possible effect on non-target organisms. Many economically important tree species are wind-pollinated, characterized by long life cycles, and key species within their respective ecosystems. Therefore the recognition of the unexpected (e.g. pleiotropic) effects of the transgenes is important and this effort might be achieved by metabolic profiling. However, to our knowledge, there have been no reports on unexpected changes discovered by metabolic profiling of transgenic forest trees at this time.

ZAMIR (2001) proposes improvement of crop plant breeding through the use of exotic libraries consisting of marker-defined genomic regions derived from wild species which are then introgressed onto the background of elite crop lines. He also specifically emphasizes the role of metabolic profiling to phenotype the newly generated genetic diversity. In most tree species, natural genetic variation (the tool of the breeder) is still quite large, as opposed to crop plants, but conventional breeding is slow due to long generation intervals. Most of the commercially important traits in trees such as wood density or growth are affected by more than one gene and by the environment. These traits vary quantitatively in tree species and the identified regions are called quantitative trait loci (QTL). With the development of informative genetic markers and marker maps it has become possible to map not only monogenic traits but also QTLs that affect phenotypic variation. In a genetic background with wide genetic diversity as in forest trees it would, however, be important to recognize marker genes with known function i.e. candidate genes for specific traits which co-locate with QTL. High-throughput metabolic profiling of phenotypes might be helpful in recognition of genes involved in quantitative traits, to get information about QTL interactions and phenotypic effects of QTL interactions as well as epistasis. Better understanding of quantitative traits at metabolome level could be utilized by tree breeding either by conventional means and / or in combination with transgene technology.

6. CONCLUDING REMARKS

Metabolic profiling of transgenic trees is still in its infancy. This is partly due to the specific characteristics of conifer species including long life span, long generation intervals and large genome sizes. All this means the organism's metabolism is sophisticated and also emphasizes the necessity for high-quality protocols for metabolic profiling. Protocol improvements must start with specific handling of the tree materials then continue to high-resolution and high-throughput analysis and finally be directed to better tools for data handling. The rewards for such improvements will be great; transgenic approaches combined with metabolic profiling of specific phenotypes will enable us to link changes in genotype to phenotype, to increase our understanding of plant metabolism and its regulatory networks. We will find new interactions between non-allelic loci or genes (epistasis) and potentially multiple metabolic effects caused by a single transgene (pleiotropy). Metabolic profiling data will also provide additional important information for both conventional and molecular tree breeding purposes and it will be useful for characterization of commercially important lines or breeds. It will certainly be important when considering biosafety aspects or unravelling potential unexpected changes especially if intended for production of commercial breeds or varieties. Finally, by integrating genomics, transcriptomics, proteomics and metabolomics data, we will gain a more comprehensive understanding of whole tree physiology and its regulation.

Acknowledgements: Research funding was provided by TEKES, the National Technology Agency (#52218 to HH) and Academy of Finland (#105214 to HH and #52784 to RJT). We are grateful to our colleagues Dr Anni Harju, Dr Tytti Sarjala and Dr Seppo Ruotsalainen from the Finnish Forest Research Institute, Finland to their valuable comments on this manuscript. We also acknowledge language revision by Maire Karjalainen.

7. REFERENCES

BINO, R.J., R.D. HALL, O. FIEHN, J. KOPKA, K. SAITO et al., 2004 Potential of metabolomics as a functional genomics tool. Trends in Plant Science **9**: 418-425.
BOEHMERT, K., I. BALBO, J. KOPKA, V. MITTENDORF, C. NAWRATH, Y. PORIER, G. TISCHENDORF, R.N.TRETHEWAY and L. WILLMIZER 2000 Transgenic *Arabidopsis* plants can accumulate polyhydroxybutyrate to up 4 % of their fresh weight. Planta **211**: 841-845.
BOERJAN, W., J. RALPH and M. BAUCHER 2003 Lignin biosynthesis. Annual Rev. Plant Biol. **54**: 519-546.
BRYANT, J.P., F.D. PROVENZA, J. PASTOR, P.B. REICHARDT, T.P. CLAUSEN and J.T. DU TOIT 1991 Interaction between woody plants and browsing mammals mediated by secondary metabolites. Annual Review of Ecology and Systematics **22**: 431-446.
BRYANT, J. AND R. JULKUNEN-TIITTO 1995 Ontogenetic development of chemical defense by seedling resin birch: Energy cost of defense production. J. Chem. Ecol. 21:883-896.
CHIKWAMBA, R.K., M.P. SCOTT, L.B. MEJIA, H.S. MASON and K. WANG 2003 Localization of a bacterial protein in starch granules of transgenic maize kernels. Proc. Natl. Acad. Sci. *U.S.A.* **100**: 11127-11132.
DIOUF, D. 2003. Genetic transformation of forest trees. African Journal of Biotechnology **2**: 328-333.
EDWARDS, D. and J. BATLEY 2004 Plant bioinformatics: from genome to phenome. Trends in Biotechnology **22**: 232-237.
FIEHN, O. 2002 Metabolomics – the link between genotypes and phenotypes. Plant Mol. Biol. **48**: 155-171.

FIEHN, O., J. KOPKA, P. DORMANN, T. ALTMANN, R.N. TRETHEWAY and L. WILLMITZER 2000 Metabolic profiling represents a novel and powerful approach for plant functional genomics. Nature Biotechnology **18**: 1157-1161.

FILLATTI, JJ., J. SELMER, B. MCCOWN, B. HAISSIG and L. COMAI 1987 *Agrobacterium* mediated transformation and regeneration of *Populus*. Mol. Gen. Genet. **206**: 192-199.

HAGERMAN, A. 1988 Extraction of tannin from fresh and preserved leaves. *J. Chem. Ecol.* **14**: 453-461.

HÄGGMAN, H., A.D. FREY, L. RYYNÄNEN, T. ARONEN, R. JULKUNEN-TIITTO, H. TIIMONEN, K. PIHAKASKI-MAUNSBACH, S. JOKIPII, X. CHEN and P.T. KALLIO 2003 Expression of *Vitreoscilla* haemoglobin in hybrid aspen (*Populus tremula* x *tremuloides*). Plant Biotechnology Journal **1**: 287-300.

HERSCHBACH, C. and S. KOPRIVA 2002 Transgenic trees as tools in tree and plant physiology. Trees **16**: 250-261.

IHMELS, J., G. FRIEDLANDER, S. BERGMANN, O. SARIG and N. BARKAI 2002 Revealing modular organization in the yeast transcriptional network. Nat.Genet. **31**: 370-377.

IHMELS, J., R. LEVY and N. BARKAI 2004 Principles of transcriptional control in the metabolic network of *Saccaharomyces cerevisiae*. Nature Biotechnology **22**: 86-92.

JULKUNEN-TIITTO, R., M. KIRSI and T. RIMPILÄINEN 1988 Methods of the analysis and the aroma composition of some species of herbal teas. Lebens. Wiss. Techn. **21**: 36-40.

JULKUNEN-TIITTO, R., J. BRYANT, H. ROININEN and P. KUROPAT 1995 Slight tissue wounding fails to induce consistent chemical defense in three willow (*Salix* spp.) clones. Oecologia **101**: 467-471.

JULKUNEN-TIITTO, R. and S. SORSA 2001 Testing the drying methods for willow flavonoids, tannins and salicylates. J. Chem Ecol. **27**: 779-789.

KRIEGER, C. J., P. ZHANG, L.A. MUELLER, A. WANG, S. PALEY, M. ARNAUD, J. PICK, S.Y. RHEE and P.D. KARP 2004 MetaCyc: a multiorganism database of metabolic pathways and enzymes. Nucleic Acids Res. **32**: 438-442.

LAITINEN, M.-L., R. JULKUNEN-TIITTO and M. ROUSI 2002 Foliar phenolic composition of European white birch during bud and leaf development. Physiologia Plantarum **114**: 450-460.

LAITINEN, M.-L., R. JULKUNEN-TIITTO, K. YAMAJI, J. HEINONEN AND M. ROUSI 2004 Variation in bark secondary compounds between and within clones: implication for herbivory. *Oikos* **104**: 316-326.

LINDROTH, R. and P. A. Koss 1996 Preservation of Salicaceae leaves for phytochemical analyses: Further assesment. J. Chem Ecol. **22**: 765-771.

LUCKNER, M.K., K. MOTHES and L. NOVER (eds) 1976 *Secondary Metabolism and Coevolution*, Nova Acta Leopoldina Suppl. No. 7, Deutsche Academie der Naturforscher Leopoldina

ORIANS, C.M. 1995 Preserving leaves for tannin and phenolic glycoside analyses. J. Chem Ecol. **21**: 1235-1243.

PAN, H. and L.N. LUNDGREN 1995 Phenolic extractives from root bark of *Picea abies*. Phytochemistry **39**: 1423-1428.

PEÑA, L. and A. SÉGUIN 2001 Recent advances in the genetic transformation of trees. Trends in Biotechnology **12**: 500-506.

PHELPS, T.J., A.V. PALUMBO and A. S. BELIAEV 2002 Metabolomics and microarrays for improved understanding of phenotypic characteristics controlled by both genomics and environmental constraints. Current Opinion in Biotechnology **13**: 20-24.

ROESSNER, U., C. WAGNER, J. KOPKA, R.N. TRETHEWEY AND L. WILLMITZER 2000 Simultaneous analysis of metabolites in potato tuber by gas chromatography- mass spectrometry. Plant J. **23**: 131-142.

ROESSNER, U., L. WILLMITZER and A.R. FERNIE 2000 High-resolution metabolic phenotyping of genetically and environmentally diverse potato tuber systems. Identification of phenocopies. Plant Physiology **127**: 749-764.

ROESSNER-TUNALI, U., B. HAGEMANN, A. LYTOVCHENKO, F. CARRARI, C. BRUEDIGAM, D. GRANOT and A. R. FERNIE 2003 Metabolic profiling of transgenic tomato plants overexpressing hexokinase reveals that the influence of hexose phosphorylation diminishes during fruit development. Plant Physiology **133**: 84-99.

SEDEROFF, RR. 1995 Forest trees. In *The transformation of plants and soil microorganisms* (Wang K, A. Herrera-Estrella and M. Van Montagu, eds.). Cambridge; Cambridge University Press, pp 150-163.

SEDEROFF, RR., J.J. MACKAY, J. RALPH and R.D. HATFIELD 1999 Unexpected variation in lignin. Current Opinion in Plant Biology **2**: 145-152.
SEGRÉ, D. 2004 The regulatory software of cellular metabolism. Trends in Biotechnology **22**: 261-265.
SEIGLER, D.S. 1998 *Plant Seconday Metabolism*. Kluwer Academic Publishers, Dordrecht, Netherlands.
STEINBRECHER, R., K. HAUFF, H. HAKOLA and J. RÖSSLER 1999 A revised parameteriation for emission modelling of isoprenoids for boreal trees. Biphorep Scientific final report. EC, Belgium, pp. 29-43.
STITT, M. and A.R. FERNIE 2003 From measurements of metabolites to metabolomics: an "on the fly" perspective illustrated by recent studies of carbon-nitrogen interactions. Current Opinion in Plant Biotechnology **14**: 136-144.
TEGELBERG, R., R. JULKUNEN-TIITTO and P. APHALO 2001 The effect of the long-term UV-radiation on birch (*Betula pendula* Roth) seedlings. Global Change Biology **7**: 839-848.
TEGELBERG, R. and R. JULKUNEN-TIITTO 2001 Quantitative changes in secondary metabolites of dark-leaved willow (*Salix myrsinifolia* Salisb.) exposed to enhanced ultraviolet-B radiation. Physiologia Plantarum **113**: 541-547.
URBANCZYK-WOCHNIAK, E., A. LUEDEMANN, J. KOPKA, J. SELBIG, U. ROESSNER-TUNALI, L. WILLMIZER and A.R. FERNIE 2003 Parallel analysis of transcript and metabolic profiles: a new approach in systems biology. EMBO Reports **4**: 989-993.
WATERMAN, P.G. and S.MOLE 1994 *Analysis of plant phenolic metabolites*. Blackwell Scientific Publications.
WECKWERTH, W. and O. FIEHN 2002 Can we discover novel pathways using metabolomic analysis? Current Opinion in Biotechnology **13**: 156-160.
ZAMIR, D. 2001 Improving plant breeding with exotic genetic libraries. Nature Rev. Genet. **2**: 983-989.

SECTION III

VIEWING TRANSGENIC CONIFER PLANTATIONS ON a LANDSCAPE SCALE

CHAPTER 8

DISPERSAL OF TRANSGENIC CONIFER POLLEN

GABRIEL KATUL[1,3], CLAIRE G. WILLIAMS[2], MARIO SIQUEIRA[1], DAVIDE POGGI [1,3,4], AMILCARE PORPORATO[3], HEATHER McCARTHY[1] AND RAM OREN[1]

[1]*Nicholas School of the Environment and Earth Sciences, Duke University, Durham, NC, USA*
[2]*Department of Biology, Duke University, Durham, NC, USA*
[3]*Department of Civil and Environmental Engineering, Duke University, Durham, NC, USA*
[4]*Dipartimento di Idraulica, Trasporti ed Infrastrutture Civili Politecnico di Torino, Torino, Italy*

Abstract. Long-distance dispersal (LDD) of pollen in conifers presents a risk for transgenic escape into unmanaged forests. Here, we report simulations of transgenic pollen dispersal and LDD from genetically modified forests using a mechanistic turbulent dispersal model. The dispersal model is based on coupled Eulerian-Lagrangrian closure (CELC) principles that model turbulent velocity excursions within the canopy. Contrary to recent studies and measurements from annual crop canopies, which reported maximum pollen dispersal distances ranging from 6 m to 800 m, conifer pollen LDD can readily exceed 8 km in less than 1 hour without escaping the atmospheric boundary layer. These LDD estimates were conducted using a conservative terminal velocity (V_t) of 0.07 m s^{-1}. When using a V_t of 0.03 m s^{-1} ± 0.02 m s^{-1}, which is characteristic of pine species pollen, LDD increased by almost a factor of 3, from about 8 to 21 km for a stand at its reproductive onset and from about 13 km to 33 km for a stand at near-harvesting age. The fact that pollen can travel such distances without escaping the ABL has important consequences about viability and ecological risk assessment and gene flow.

1. INTRODUCTION

Estimating probability of transgenic pollen escape is timely because recombinant DNA technology to genetically modify forest trees is being applied worldwide to commercially important species (FENNING AND GERSHENZON 2002), and pines are the major commodity species in both Northern and Southern Hemispheres. *Pinus taeda*, used here as a case study, is a long-lived woody perennial with delayed reproductive onset followed by abundant long-distance gene flow. *Pinus taeda* mating system is wind-pollinated and monoecious. Potential for transgenic escape is great because reproductive onset precedes harvest by a decade or more and there is natural hybridization and introgression among close relatives (RIGHTER AND DUFFIELD 1951).

Few mechanistic models track long-distance dispersal (LDD) and escape probability of pollen from the canopy. However, several mechanistic models have been developed that predict attributes of seed and pollen dispersal kernels, such as means and modes (OKUBO AND LEVIN 1989, DE HAAN AND ROTACH 1998, LOOS et al., 2003). LDD is often defined as the distance travelled by the 99% percentile of pollen (or seeds), though this definition is not be unique (NATHAN et al., 2002, CAIN et al., 2003, HIGGINS et al., 2003, NATHAN et al., 2003). Most released pollen, however, will fall within the perimeter of the source, often referred to as local neighbourhood diffusion (LND) (HENGEVELD 1989). Pollen dispersal can be viewed (and often modelled) as superposition of LND and LDD. Few stratified empirical diffusion models (HENGEVELD 1989 pp. 48-53) even proposed bi-normal distribution kernels to account for these fundamentally different dispersion mechanisms (NICHOLS AND HEWITT 1994, BULLOCK AND CLARKE 2000).

Pollen dispersal from conifer trees is different from that from agricultural crops in that tree height (h) accrues annually so released pollen is carried at increasing distances from a source, with a partial offset due to increased leaf area index (LAI) that attenuates mean wind speed and other turbulent statistics within the canopy. Windborne pollen LDD is associated with an uplifting process from within the canopy by turbulent eddies, comparable in size to h (RAUPACH et al., 1996, KATUL et al., 1998, POGGI et al., 2004a, POGGI et al., 2004b) and then transported to elevated regions within the atmospheric boundary layer (ABL), and even higher into the troposphere, by larger-scale eddy motion. Given the large mean wind speeds at such elevations in the atmosphere, dispersal distances as far as 600 to 1200 km from the source can readily occur (BESSEY 1883, ERDTMAN 1937, LANNER 1966, DI GIOVANNI et al., 1995, DI-GIOVANNI et al., 1996). Even a simple ballistic model calculation suggests that at a settling velocity range of 25.3 to 31.9 mm s^{-1}, conifer pollen at 300 m above the surface moving at a speed of 5 ms^{-1} (assumed constant with height and time) would drift some 47 to 60 km in about 3 hours (DI-GIOVANNI et al., 1996). However, P. taeda pollen that has traveled for several hours or has escaped the ABL into the troposphere may not be viable because of the high exposure to UV-B radiation, and cold temperature (TUSKAN et al., 1992). Most reports show that effective pollination distance of forest trees is more limited than recorded travel distances (SMOUSE et al., 2001). Similarly, studies on genetically modified (GM) corn pollen in Mexico found a slight viability loss after 1 hour of dispersion compared to a 100% viability loss after 2 hours of dispersion in wintertime (AYLOR et al., 2003).

An immediate consequence of the above findings is that LDD might be less consequential than previously thought given the reduced pollen viability potential associated with its occurrence. Hence, quantifying the distances of pollen LDD for events in which pollen did not escape from the top of the atmospheric boundary layer (~ 1 km high during daytime) or experienced long travel times (~ 1 hour) is central to ecological risk assessment. Yet, this combination of viability and distances are difficult to ascertain via field experiments.

To address this question, we adapt a three-dimensional coupled Eulerian-Lagrangian model (HSIEH et al., 2000, NATHAN et al., 2002) to compute pollen dispersal kernels from a *P. taeda* plantation. AYLOR AND FLESCH (2001) already used a simplified two-dimensional version of this model to compute spore concentrations of *Lycopodium* and *V. inaequalis* from a wheat canopy. The model parameters were estimated independently from meteorological, canopy and reproductive data collected at the Free Air CO_2 Enrichment (FACE) experiment at Duke University near Durham, North Carolina and used as a sample case study in the simulations.

2. METHODS

The CELC model calculations require the flow statistics above the canopy during the pollen release period along with the amounts and location of pollen released. *Pinus taeda* has male strobili within the lower 33% of the canopy so the release height was set at one-third the canopy height in all model calculations. Pollen release date was approximated using a heat sum model (BOYER 1978) for the Duke Forest plantation from 1999 to 2002 (Figure 1).

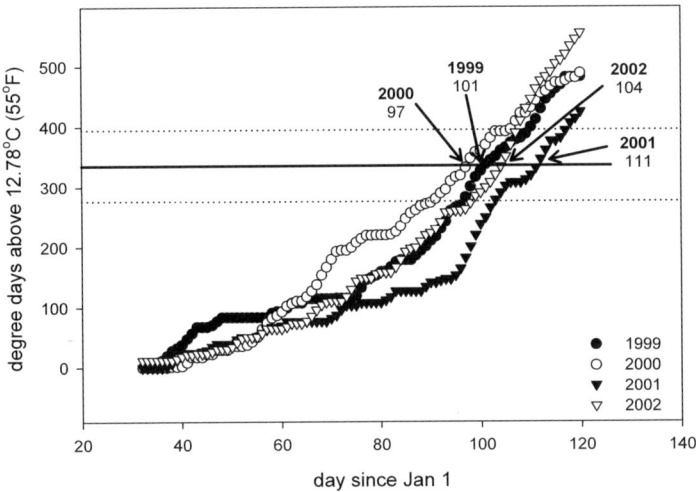

Bold line represents the critical heat sum that results in peak pollen shed. Dotted lines are 95% confidence intervals

Figure 1. Heat sum for pollen indicating Julian dates of production. Mean air temperature measurements collected at the Duke Forest FACE site are used.

Based on the heat-sum model, pollen release occurs between April 1 and April 25. The concomitant velocity measurements needed by CELC are described later.

Pollen dispersal quantity was based on pollen yield per tree per day. On average, one *P. taeda* tree at 16 m is estimated to produce 1133.9 grams of pollen over a 14-day pollen shed, roughly 81 g per tree per day on average (PARKER AND BLUSH 1996). Pollen grains (g^{-1}) usually range from 1.3 to 1.5 x 10^6. Hence, pollen production is about 113.4 x 10^6 pollen d^{-1} $tree^{-1}$. Using a tree density of a typical southern pine plantation (1750 trees ha^{-1}), the amount of pollen produced per 1 ha plantation is about 2 x 10^{11} pollen d^{-1}.

We used the long-term wind velocity data collected at the FACE facility, in Duke Forest (35° 97'N 79° 9'W) near Durham, North Carolina for driving the CELC model. The three velocity components and virtual potential temperature above the canopy were collected at 10 Hz using a CSAT3 (Campbell Scientific, Logan UT) triaxial sonic anemometer. These on-going velocity measurements commenced in 1998 and are part of a long-term CO_2 flux monitoring initiative (*Ameriflux*) within the United States (BALDOCCHI et al., 2001). The setup of the sonic anemometer, data collection, and post-processing are described elsewhere (SIQUEIRA et al., 2000, SIQUEIRA AND KATUL 2002, SIQUEIRA et al., 2002). In addition, detailed leaf area density measurements collected during the pollen production period permit us to estimate how the flow statistics inside the canopy vary in relation to the measurements collected above the canopy (KATUL AND ALBERTSON 1998, KATUL AND CHANG 1999).

3. THEORY

After Thomson's seminal work (THOMSON 1987), Lagrangian stochastic models for the trajectories of tracer-particles in turbulent flows are now routinely used in computational fluid mechanics and turbulence research (Pope 2000). These models are derived using the so-called well-mixed condition (*wmc*), which states that if a concentration of a scalar material is initially uniform at some time t_o it will remain so at any future time t in the absence of sources and sinks. The well-mixed condition is the most rigorous theoretical framework for computing Lagrangian trajectories and ensures consistency with prescribed Eulerian velocity statistics. Using the *wmc*, Thomson (1987) showed that the Lagrangian velocity of a tracer-particle is described by a generalized Langevin equation, of the form:

$$du_i = a_i(\vec{x},\vec{u},t)dt + b_{ij}(\vec{x},\vec{u},t)d\Omega_j$$

where $\vec{x}$ and $\vec{u}$ are the position and velocity vectors of a tracer-particle at time t, respectively. The terms a_i and b_{ij} are known as the drift and diffusion coefficients, respectively. The quantities $d\Omega_j$ are increments of a Wiener process with independent components, zero mean, and variance dt. Throughout, subscripts denote components of Cartesian tensors though both meteorological and index notations are also used interchangeably (i.e. the components of $\vec{x}$ are $x_1 = x$, $x_2 = y$, and $x_3 = z$) with x, y, and z representing the longitudinal, lateral, and vertical axes, respectively. As discussed in Thomson (1987), b_{ij} can be uniquely determined by requiring that the Lagrangian velocity structure function in the so-called inertial subrange match the well-known Kolmogorov's similarity theory (RODEAN 1996). The determination of a$_i$ remains a theoretical challenge requiring the use of *wmc*. For high Reynolds numbers, the *wmc* ensures that the Eulerian probability density function of the tracer particles $P(\vec{x},\vec{u},t)$ satisfies a Fokker-Planck equation,

$$\frac{\partial P}{\partial t} + \frac{\partial}{\partial x_i}(u_i P) = -\frac{\partial}{\partial u_i}(a_i P) + \frac{\partial^2}{\partial u_i \partial u_j}\left(\frac{1}{2}b_{ik}b_{jk}P\right).$$

The corresponding Langevin model of the above Fokker-Planck equation for a Gaussian, vertically inhomogeneous turbulence, consists of a set of three stochastic differential equations for the velocity components, given by (THOMSON 1987):

$$du_1' = \left[-\frac{C_o <\varepsilon>}{2}(\lambda_{11}u_1' + \lambda_{13}u_3') + \frac{\partial <\overline{u_1}>}{\partial x_3}u_3' + \frac{1}{2}\frac{\partial <\overline{u_1'u_1'}>}{\partial x_3}\right]dt + \left[\frac{\partial <\overline{u_1'u_1'}>}{\partial x_3}(\lambda_{11}u_1' + \lambda_{13}u_3') + \frac{\partial <\overline{u_1'u_3'}>}{\partial x_3}u_1' + \lambda_{33}u_3'\right]\frac{u_3'}{2}dt + \sqrt{C_o <\varepsilon>}\,d\Omega$$

$$du_2' = \left[-\left(\frac{C_o <\varepsilon>}{2} + \frac{1}{2}\frac{\partial <\overline{u_2'u_2'}>}{\partial x_3}u_3'\right)(\lambda_{22}u_2')\right]dt + \sqrt{C_o <\varepsilon>}\,d\Omega$$

$$du_3' = \left[-\frac{C_o <\varepsilon>}{2}(\lambda_{13}u_1' + \lambda_{33}u_3') + \frac{1}{2}\frac{\partial <\overline{u_3'u_3'}>}{\partial x_3}\right]dt + \left[\frac{\partial <\overline{u_1'u_3'}>}{\partial x_3}(\lambda_{11}u_1' + \lambda_{13}u_3') + \frac{\partial <\overline{u_3'u_3'}>}{\partial x_3}(\lambda_{13}u_1' + \lambda_{33}u_3')\right]\frac{u_3'}{2}dt + \sqrt{C_o <\varepsilon>}\,d\Omega$$

where u_i' are the (instantaneous) turbulent velocities at position x_i and time t, C_0 (≈ 5.5) is a similarity constant (related to the Kolmogorov constant) and λ_{11}, λ_{13}, λ_{22}, and λ_{33} can be derived by inverting the Reynolds stress tensor, and are given by:

$$\lambda_{11} = \frac{1}{<\overline{u'_1 u'_1}> - \frac{<\overline{u'_1 u'_3}>^2}{<\overline{u'_3 u'_3}>}}$$

$$\lambda_{22} = <\overline{u'_2 u'_2}>^{-1}$$

$$\lambda_{33} = \frac{1}{<\overline{u'_3 u'_3}> - \frac{<\overline{u'_1 u'_3}>^2}{<\overline{u'_1 u'_1}>}}$$

$$\lambda_{13} = \frac{1}{<\overline{u'_1 u'_3}> - \frac{<\overline{u'_1 u'_1}><\overline{u'_3 u'_3}>}{<\overline{u'_1 u'_3}>}}$$

Here, $<\overline{u_1}>$ is the mean longitudinal velocity (defined so that $<\overline{u_2}>=0$), $<\overline{u'_1 u'_1}>(=\sigma_u^2)$, $<\overline{u'_2 u'_2}>(=\sigma_v^2)$, $<\overline{u'_3 u'_3}>(=\sigma_w^2)$, are the standard deviations of the three velocity components, $<\overline{u'_1 u'_3}>$ ($=<\overline{w'u'}>$) is the Reynolds stress, and $<\overline{\varepsilon}>$ is the mean turbulent kinetic energy dissipation rate, $<.>$ is spatial averaging (RAUPACH AND SHAW 1982, FINNIGAN 2000), and over-bar is time averaging. The vertical distribution of the flow statistics $<\overline{u_1}>$, $<\overline{u'_1 u'_1}>$, $<\overline{u'_2 u'_2}>$, $<\overline{u'_3 u'_3}>$, $<\overline{u'_1 u'_3}>$, and $<\overline{\varepsilon}>$ needed to drive the Thomson (1987) model can be readily computed from Eulerian second order closure models (KATUL AND ALBERTSON 1998, AYOTTE et al. 1999, KATUL AND CHANG 1999, MASSMAN AND WEIL 1999, KATUL et al., 2001) as described in the Appendix. The above formulation, along with modeled $<\overline{u_1}>$, $<\overline{u'_1 u'_1}>$, $<\overline{u'_2 u'_2}>$, $<\overline{u'_3 u'_3}>$, $<\overline{u'_1 u'_3}>$, and $<\overline{\varepsilon}>$ using second order closure principles constitutes the formulation of the mean and turbulent velocity excursions within the CELC model (NATHAN et al., 2002). With these velocity statistics, and for the purposes of estimating dt, we define the relaxation time scale by

$$T_L = \frac{0.5 \times \left(\sigma_u^2 + \sigma_v^2 + \sigma_w^2\right)}{<\overline{\varepsilon}>}$$

and set $dt = 0.01 T_L$ in all model calculations. This estimate of dt satisfies all the theoretical constraints discussed elsewhere (REYNOLDS 1998a, b, d, c). As stated earlier, AYLOR AND FLESCH (2001) used a 2-D version of the above model to estimate spore dispersion from a wheat canopy. Their model includes empirically specified profiles of $<\overline{u_1}>$, $<\overline{u'_1 u'_1}>$, $<\overline{u'_3 u'_3}>$, and $<\overline{u'_1 u'_3}>$ and a simplified parameterisation for $<\bar{\varepsilon}>(=\dfrac{\sigma_w^2}{C_o T_L}$, with T_L assumed constant inside the canopy, but for the particle trajectory equations, it was reduced to account for the inertia effects of heavy-particle when compared to the no-mass particle transport).

Recent experiments by POGGI et al., (2005) and theoretical arguments by MASSMAN AND WEIL (1999) suggest that a constant length scale is more appropriate to modelling $<\bar{\varepsilon}>$ than a constant T_L. Also, a recent $K-\varepsilon$ study showed that either a constant length scale or the relaxation time scale (which varies inside the canopy) better describes the individual components of the mean turbulent kinetic energy dissipation rate budget (KATUL et al., 2004). Notwithstanding these simplifications, the Aylor-Flesch comparisons between model calculations and measurements suggested that such class of Lagrangian dispersion models could realistically reproduce diaspore spread. We note here that the meteorological, terminal velocity, and leaf area density inputs to the CELC approach are identical to the AYLOR-FLESCH model (with the exception of the σ_y, which is required here). LOOS et al. (2002) used a 1-D simplified analytical version of this model that is based on the localized near field (LNF) theory (RAUPACH 1983, 1987, RAUPACH 1989B, RAUPACH 1989A, PHILLIPS et al., 1997, SIQUEIRA et al., 2000) to successfully estimate the spread of GM pollen from transgenic maize. They also report good agreement between model calculations and measurements, at least within the 10 m from the source. However, the emphasis of the LNF method is based on a local homogenization of the near-field kernel so as to reproduce concentration variation neighbouring a source; the far-field kernel retains the usual gradient-diffusion approach. Hence, it is unlikely that LNF will be able to realistically simulate LDD for a canopy. Using an analytical approach, a solution for the dispersal kernel starting from a 3-dimensional Brownian motion for constant drift and dispersion coefficients from a point source can be derived (STOCKMARR 2002). STOCKMARR (2002) showed that upon radial averaging this solution, the power-law decay of the dispersal kernel tails is sensitive to whether a 2-D or a 3-D model is used. This perhaps motivates the retention of lateral dispersion in pollen dispersal calculations. Other approaches used to model LDD, such as the Puff model (AYLOR et al., 03), lack the detailed description of canopy turbulence needed to ensure realistic escape probabilities from the canopy. The simulations proposed here are among the most detailed three-dimensional simulations for pollen release inside a southern pine forest and consider the effect of age, through its effect on tree height, on *P. taeda* pollen dispersal kernels. We emphasize that the same approach can be used for calculating LDD of seeds (e.g., NATHAN et al., 2002).

4. MODEL SIMULATIONS

The pollen simulations were performed for the *P. taeda* plantation at reproductive onset (16 years; 14.6 m tall), and at a typical harvest age (25 years; 21.3 m tall).

Table 1. CELC model input and output variables for P. taeda pollen dispersal using measured meteorological and stand characteristics. The pollen release height, canopy height, and the normalized mean wind speed at the release height are also shown. The statistics pertaining to the dispersal kernels in Figures 4 and 5 are also presented. These statistics include the mode, the mean distance of the furthest 1% pollen ($D_{99\%}$), the fraction of pollen that travelled more than 1 km, the probability that pollen escapes the canopy volume, and the mean travel time corresponding to the pollen particle (D_{99}) are presented for the two age classes. The dispersal calculations are shown for mean terminal velocities of 0.03 ± 0.02 and 0.07 ± 0.02.

Variables	Dispersal Statistics			
	16 years		25 years	
Release Height (m)	4.87		7.1	
Canopy Height (m)	14.6		21.3	
LAI ($m^2\ m^{-2}$)	2.3		3.1	
U/u_* (at $h/3$)	1.14		0.78	
u_* for dislodging (m s^{-1})	0.44		0.64	
Terminal Velocity (m s^{-1})	0.07	0.03	0.07	0.03
Mode (km)	0.31	1.58	0.41	2.47
$D_{99\%}$ (km)	8.62	21.0	13.5	33.5
Fraction of pollen > 1 km (%)	4.3	24.9	4.8	28.2
Probability of escape from the canopy volume (%)	6.4	27.5	4.4	24.3
Travel Time of D_{99} (h)	0.35	0.73	0.51	0.81
D_{max} (km)	26.8	49.6	43.6	60.0

The release height and leaf area index (LAI) for these two age classes are presented in Table 1. The estimates of canopy height and LAI for the March-April period are for the moderately low fertility site of the Duke Forest (OREN et al., 2001).

The model computes the trajectory of a pollen particle with time from the release height until it is absorbed at the ground for a given u_* ($=\sqrt{-\overline{u_1'u_3'}}$), the friction velocity above the canopy. Pollen particles that reach 1 km elevation or travel the flow domain for more than 1 hour are considered unviable (or dead) and excluded. The model releases pollen when the local mean wind speed ($<\overline{u_1}>$) at the release height (see Table 1) exceeds 0.5 m s^{-1} (AYLOR AND PARLANGE 1975). This velocity was defined as a minimum velocity necessary to dislodge pollen by minor leaf fluttering or ventilation. Similarly, AYLOR et al. (2003) found that corn pollen dislodges at a $<\overline{u_1}>$ between 0.25 and 0.5 m s^{-1}. This threshold necessitates that u_* above the canopy must exceed a certain rate before pollen release occurs, which can vary among the two age classes considered here (see Table 1). To extrapolate the mean wind conditions for extreme scenarios not recorded within the 5-year velocity record at the site, the 30-minute 5-year sonic anemometer measured u_* distribution collected was fitted to a Weibull function. The Weibull probability density function (pdf) for u_* is given

$$pdf(u_*) = \frac{c u_*^{c-1}}{b^c} \exp\left(-\left[\frac{u_*}{b}\right]^c\right)$$

Here, b and c are shape and scale parameters, respectively, and can be determined from the sonic-anemometer measured u_* using the Method of Maximum Likelihood (MML). Using the MML, b and c can be determined by solving the two nonlinear equations:

$$b = \left[\frac{1}{n}\sum_{i=1}^{n}[u_*(i)]^c\right]$$

$$c = \frac{n}{\left(\frac{1}{b}\right)^c \sum_{i=1}^{n}[u_*(i)]^c \log(u_*(i)) - \sum_{i=1}^{n} \log(u_*(i))}$$

where $u_*(i)$ ($i = 1, 2, ...n$) are the sonic anemometer measured u_* time series, and n (= 39,273) is the number of u_* measurements used in the estimation of b (=0.456) and c (=2.11) as shown in Figure 2. To insure fully-developed turbulence conditions inside the canopy, all $u_*(i) < 0.1$ m s^{-1} were excluded from the b and c calculations (Lai et al., 2002). Furthermore, these small u_* do not contribute to pollen dislodging. As evidenced from Figure 2, the agreement between the Weibull function and the measured $pdf(u_*)$ is reasonably well for $u_*(i) > 0.1$ m s^{-1}, especially at the measured tails.

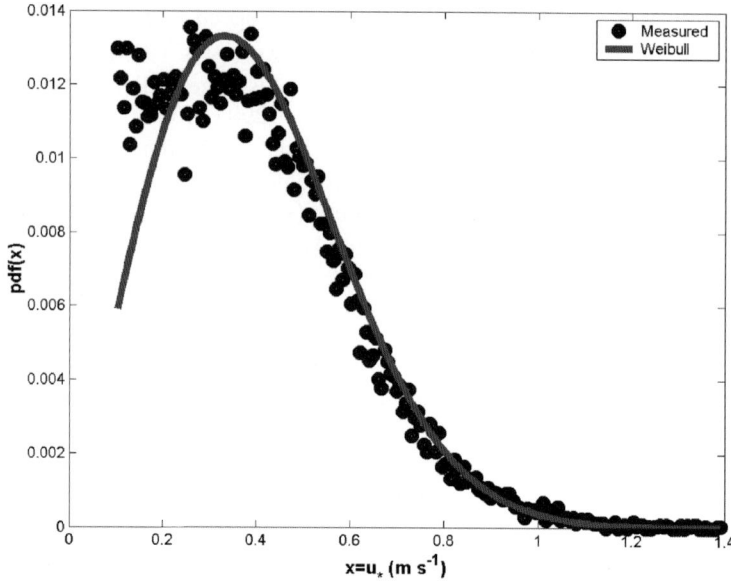

Figure 2. Measured and modelled probability distribution functions (pdf) of sonic anemometer measured friction velocity (u_*) above the pine canopy (1998-2003). The Weibull parameters are b =0.456 and c=2.11 as determined from MML.

It is desirable to use the widest possible range in u_* variability for estimating the parameters of the Weibull function; however, it is not clear whether this distribution is representative of the wind conditions during the pollination period shown in Figure 1. Hence, we conducted a Kolmogorov-Smirnov test (PRESS et al., 1992) to assess whether the histogram of the entire u_* record (divided into 30 bins) is consistent with the u_* distribution for the March-April months from 1998-2003. We found that the two distributions were not statistically different at the 95% confidence level and the Weibull parameters obtained from the entire record of $u_*(i)$ statistically represent the wind conditions during the pollination months. The pollen particle trajectories were computed as follows:

(i) Using the leaf area density and index (see Table 1), the Eulerian second-order closure model in the Appendix is used to compute the normalized $<\overline{u_1}>$, $<\overline{u'_1 u'_1}>$,

$\overline{<u'_2u'_2>}$, $\overline{<u'_3u'_3>}$, $\overline{<u'_1u'_3>}$, and $<\overline{\varepsilon}>$. The normalizing variables in the second-order closure model are u_* above the canopy and h.

(ii) A random u_* from the Weibull distribution was selected using

$$u_* = b\left(\log\left[\frac{1}{1-\mu}\right]\right)^{1/c}$$

where $\mu \in [0,1]$ is a uniform deviate random number, generated using the algorithm in Press et al. (1992). If u_* does not exceed the threshold u_* necessary to generate a $<\overline{u_1}>> 0.5$ m s^{-1} at the release height (see Table 1), no pollen particle is released and a new u_* is generated. Note that the maximum u_* generated from this approach can exceed the maximum recorded u_* within the 5-year record.

(iii) Using the generated u_*, h, and the normalized flow statistics from step (i), $<\overline{u_1}>$, $\overline{<u'_1u'_1>}$, $\overline{<u'_2u'_2>}$, $\overline{<u'_3u'_3>}$, $\overline{<u'_1u'_3>}$, and $<\overline{\varepsilon}>$ along with their vertical gradients are computed.

(iv) The three stochastic differential equations are simultaneously solved using the inputs from step (iii) for the turbulent velocity excursions along with the trajectory equation

$$x_i(t+dt) = x_i(t) + \int_t^{t+dt}\left(<\overline{u_i}>\delta_{i1} - V_t\delta_{i3} + u'_i\right)dt$$

where δ_{ij} is the Kronecker delta tensor, and V_t is the pollen terminal velocity. The estimation of V_t and dt are considered next. DI-GIOVANNI et al. (1995) reported a mean V_t for black spruce (*Picea mariana*) of 0.0319 m s^{-1} and for jack pine (*Pinus banksiana*) of 0.0253 m s^{-1}. They did not find any statistical difference in V_t between clustered and un-clustered pollen for these two species (DI-GIOVANNI et al., 1995). Notwithstanding statistical significance, their V_t measurement for black spruce and for triple grains (clumped) is 0.0626 ± 0.013 m s^{-1}. This estimate, which is an upper limit for conifer V_t, motivated us to choose a conservative V_t of

0.07 ± 0.02 ms^{-1} in the model calculations. Formal sensitivity analysis on V_t will be described in our discussions later.

Regarding the estimates of dt from T_L, it has been argued that a heavy particle has a smaller time scale T_L when compared to a weightless (or passive) particle (SAWFORD AND GUEST 1991, WILSON AND SAWFORD 1996, REYNOLDS 2000, REYNOLDS AND COHEN 2002). SAWFORD AND GUEST (1991) estimate that the difference between the passive and heavy particle time scale is given by

$$T_L^{heavy\ particle} = \frac{T_L^{passive\ partlice}}{\sqrt{1 + \beta' \left(\frac{V_t}{\sigma_w}\right)^2}}$$

where β' relates the Lagrangian and Eulerian time scales. The β' ranges from 1.0 to 1.5 (RAUPACH 1989a, c). With a conservative estimate of $\beta' \approx 1.5$, and noting that 1) $\sigma_w \sim u_*$ near the canopy top, 2) a minimum $u_* = 0.44$ m s^{-1} is needed to dislodge pollen from the bottom third of the canopy, the time scale correction (with $V_t = 0.07$ m s^{-1}) is about 1.019 or 1.9%. Hence, to a first order, coniferous pollen can be treated as a passive particle for the purposes of time scale calculations. The pollen release is initiated from rest ($u_i = 0$) at the release height ($= h/3$) from the ground. The radial distance from the source, the maximum elevation, and the travel time duration are recorded when the pollen particle arrives at the forest floor, assumed to be an absorbing boundary condition. An absorbing boundary condition prohibits re-suspension of the pollen into the air after falling on the ground. These trajectory calculations do not consider many other factors such as adsorption (and re-suspension) of pollen on canopy elements, or the effect of heavy rain that produce pollen washout from the air (AYLOR AND FLESCH 2001). This latter point will also be discussed within the context of the measured precipitation at the site.

5. RESULTS

As described earlier, the simulations for the reproductive and near-harvesting stages primarily differ in LAI and pollen release height. Difference in LAI between the early reproductive age (2.3 m^2 m^{-2}) and near-harvesting age (3.2 m^2 m^{-2}) of *P. taeda* is sufficiently large to impact the flow statistics inside the canopy (Figure 3). It is clear from Figure 3 that a 30% increase in LAI impacts primarily the second-order turbulence statistics (e.g. $<\overline{u_1'u_1'}>$, $<\overline{u_3'u_3'}>$, and $<\overline{u_1'u_3'}>$) and not as much $<\overline{u_1}>$.

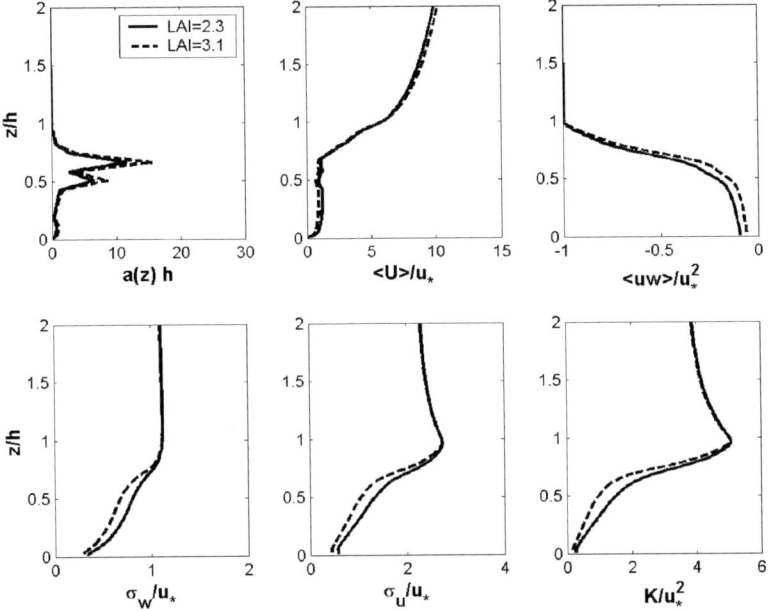

Figure 3. Modelled velocity statistics profiles needed to drive CELC model for the 16-year old (solid) and 25 year old (dashed) plantations. The normalized leaf area density is also shown.

Whether the increased attenuation of the second-order flow statistics with increasing LAI can be compensated for by increase in pollen release height will be considered. The dispersal kernels for the reproductive onset age and the typical harvest age were compared in Figure 4a. The key statistical attributes of these kernels are summarized in Table 1. Both modelled kernels suggest that pollen LDD are in excess of 8 km, and can be achieved without pollen escaping from the ABL. The mean travel time, corresponding to D_{99}, does not exceed one hour for all particles reported in Figures 4a (and 4b). While the increase in LAI at harvest age dampens the turbulence statistics within the canopy (see Figure 3), the increase in pollen release height more than compensates - resulting in a greater mode and larger LDD (Table 1). Regarding the escape probabilities ($\Pr(z > h)$) from the canopy volume, the results in Table 1 appear counter-intuitive. We expected that for the 25-year old stand, $\Pr(z > h)$ would be larger than the 16-year old stand because of a longer settling (or residence) time within the canopy, and hence, higher probability of encountering an ejection event. However, although the turbulent kinetic energy is about the same within the region $h/3$ and h in both stands, the smaller $<\bar{\varepsilon}>$ in the younger stand makes the flow appear more organized (i.e large T_L).

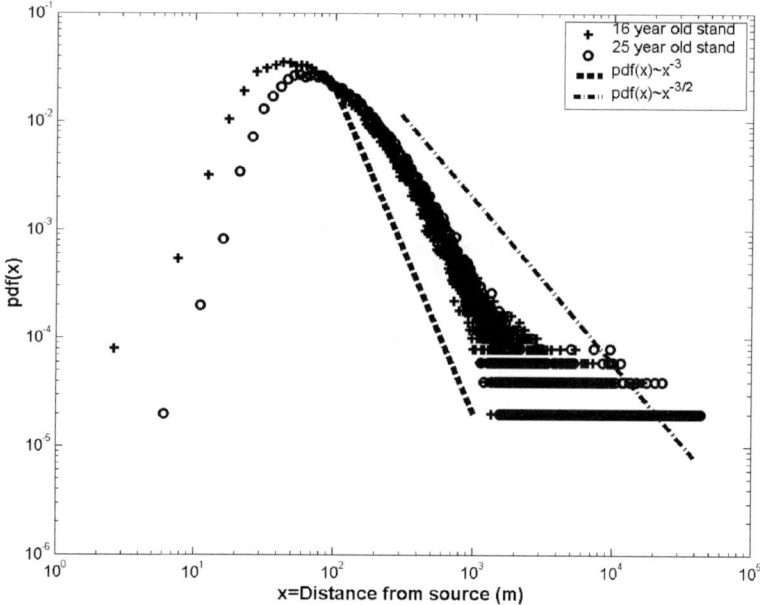

Figure 4a. Modelled dispersal kernels for the 16-year old and 25-year old P. taeda stands using 10^6 particles per age class. The -3 and $-3/2$ power-law decay of the dispersal kernels are shown for reference.

More organized flow leads to more coherent upward vertical motion, and hence, greater escape probability. Thus, while the pollen in the 25-year old-stand has a higher probability of encountering an updraft because of the longer residence time inside the canopy volume, its upward trajectory is less coherent when compared to the pollen trajectories in the 16-year old stand. The kernels in Figure 4a also exhibit approximate power-law decay between 200 m and 800 m. The power-law decay rate within this range has an approximate exponent = −3, which is steeper than the -1.78 reported by AYLOR et al. (2003) for corn pollen (that was bounded between 10 m and 60 m). As discussed in AYLOR et al. (2003), power-law tails suggest fat-tailed distributions. Interestingly, Figure 4a also suggests that beyond 1 km, the kernels clearly exhibit fat-tails that decay slower than −1.5. Kernel tails might exhibit different behaviour beyond 1 km due to pollen escape from the canopy volume encountering a turbulent flow field with different statistical properties than the flow inside the canopy. The flow above the canopy resembles a rough-wall boundary layer and has a larger integral time scale (i.e. also more coherent), a larger σ_w and $\bar{u}$. To illustrate this point further, we recomputed the kernels in Figure 4a after removing all distances associated with pollen that escaped from the canopy

(Figure 4b). It is clear from this figure that much of the dispersal events that yielded distances > 3 km can be attributed to pollen escaping the canopy volume.

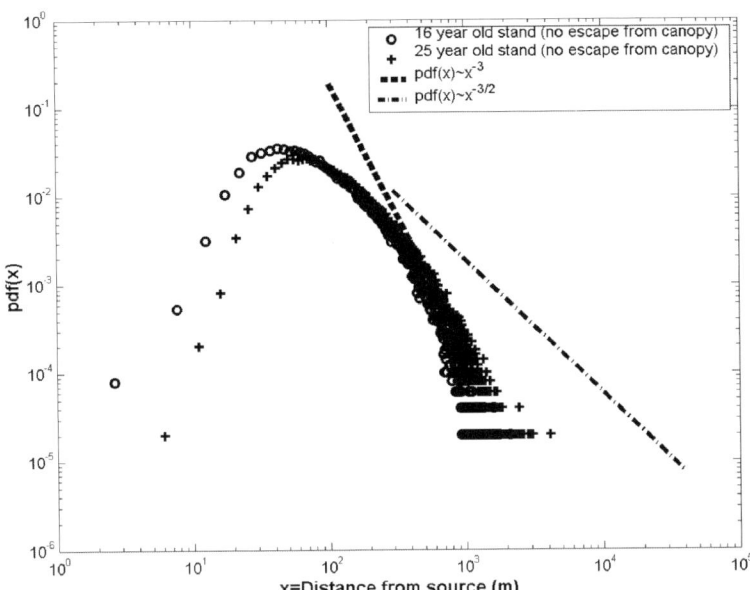

Figure 4b. Same as Figure 4a except that model excludes pollen that escaped the canopy volume.

To assess how sensitive the results in Table 1 are to the assumed V_t, we repeated all the calculations in Figure 4 for a V_t consistent with the mean (single-grained) V_t for black spruce and jack pine. This V_t is about 80% smaller than the V_t value used in Figure 4 (0.03 ± 0.02 m s^{-1} versus 0.07 ± 0.02 m s^{-1}). We found that the overall dispersal statistics in Table 1 are sensitive to V_t. Not surprisingly, the mode, LDD, and fraction of pollen escaping the canopy all increased with decreasing V_t. In fact, LDD increased nearly 3-fold for such an increase in V_t. Even the exponent of the power-law decay (see Figure 5a) has shifted closer to −1.5 for the reduced V_t, consistent with STOCKMARR (2002) solution for zero gravity. Simulations with $V_t \rightarrow 0$ are analogous to simulations in zero gravity. Again, for this reduced V_t, we find that the heavy-tails can be attributed to the pollen escape from the canopy volume as evidenced by Figure 5b.

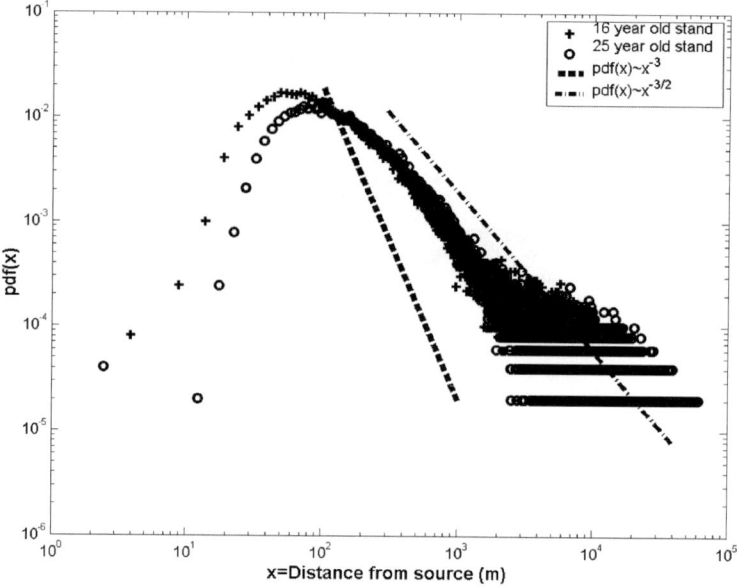

Figure 5a. Same as Figure 4a except that $V_t = 0.03 \pm 0.02$ m s^{-1}.

As mentioned, rainfall intensity (R_I) can impact pollen trajectories by removing or washing pollen from the air. AYLOR AND FLESCH (2001) found, for example, that a rainfall rate of 10 mm h^{-1} is too small for any washout to occur for *Venturia inaequalis* pollen. However, for larger intensities, they found that the loss of pollen from the air varies as $R_I^{0.787}$. Using the 30 minute-precipitation measurements collected at the site from 1998-2003 along with their threshold for washout, less than 10% of the 30-minute runs have rainfall events exceeding 0.1 mm h^{-1} (includes day and night), and we also found that 95% of the rainfall events have precipitation intensity not exceeding 10 mm h^{-1} (Figure 6). Hence, it is likely that for this site washout will not play a first-order effect on LDD in this southeastern pine forest. We note that the calculations leading to the summary results in Table 1 and Figures 4 and 5 neglect key processes known to affect pollen dispersal such as 1) pollen deposition and re-suspension on vegetation organs, 2) re-suspension from the forest floor, and 3) pollen washout from vegetated surface by rain. These processes tend to reduce pollen viability because of the increase in transport time. Hence, in a first order analysis, the distances reported in Table 1 are likely to serve as an upper limit on LDD for the wind statistics shown in Figure 2. Nonetheless, such upper limits are needed when gene flow and concomitant ecological risk assessments are being evaluated.

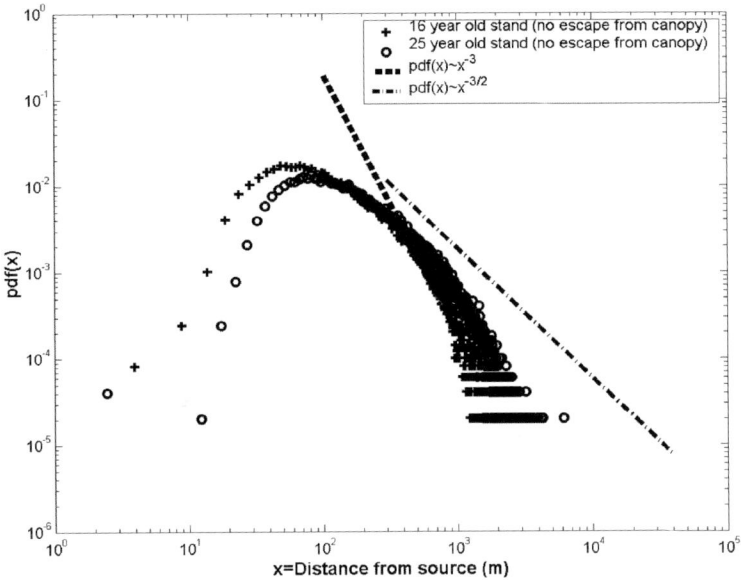

Figure 5b. Same as Figure 4b except for $V_t = 0.03 \pm 0.02$ m s^{-1}.

6. DISCUSSION AND CONCLUSIONS

Globally, public concern about risks associated with GM releases in geographic areas richly populated with diverse plants is rising exponentially (e.g. HAILS, 2000; STRAUSS et al., 2001; MANN AND PLUMMER, 2002). It was proposed that gene flow be managed using physical isolation distances, often determined from empirical farm scale experiments or subjective risk assessments relying on expert opinions (e.g. OGTR, 2001; GURA, 2001). For GM agricultural crops, we note that expert opinions for forest trees need not be entirely subjective as studies of gene flow for GM pollen in the past decade are numerous, in sharp contrast to studies of transgenic pollen of forest tree species. For example, pollen dispersal studies for maize *(Zea mays* L.) suggest a 7.5–30 m maximum distance (JAROSZ et al., 2003, LOOS et al., 2003).

Table 2. Recommended isolation distances for several genetically modified (GM) crops (as discussed in review by Messeguer, 2003) and/or maximum distance from the source in which pollen grains were detected.

Crop	Comments	Reference
Oilseed rape	Frequency of hybrid formation (GM and non-GM) was 0.156% at 200 m and 0.0038% at 400 m.	Messeguer (2003)
Maize (corn) (Zea mays L.)	Most pollen fell within 5 m, 98% of the pollen remained within a 25–50 m.	Same
Cultivated rice (Oryza sativa L.)	Pollen horizontal movement was limited to 10 m.	Same
Potato (Solanum tuberosum L.)	Isolation distance of 20 m appears adequate for transgenic potatoes. However, maximum distances for which pollen particles were detected is 80 m.	Messeguer (2003) and Moyes and Dale (1999).
Cotton (Gossypium hirustum L.)	Pollen trapped decreased to 0.03% at 50 m from the source. Other studies concluded that isolation distances of 50 m from the sources were adequate.	Messeguer (2003)
Apples	Maximum distance pollen particles were detected is 56 m	(Moyes et al., 2002)
Wheat	Maximum distance pollen were detected is 20 m	Same
Eucalypts	Minimum modelled distance to minimize gene flow between GM and wild forest stands is >> 100 m	Lincare and Ades (2004)
Loblolly pine (Pinus taeda)	LDD ranges from 26 – 60 km. Distances are for pollen particles that travelled < 1 hour and remained within the atmospheric boundary layer.	This study

Pollen experiments for canola (*Brassica rapa*) found viable pollen at distances greater than 1 km (RIEGER et al., 2002, MESSEGUER 2003). Isolation distances for several annual GM crops are summarized in Table 2 along with estimates of the maximum distances for transgenic pollen (CHRISTEY AND WOODFIELD 2001). Our model results show that transgenic conifer pollen is likely to disperse at least 2 orders of magnitude more than GM crops. Hence, an isolation distance based on the minimum limit established for seed production (i.e. 0.1% of pollen arriving at a location) would lead to bio-containment or isolation zones for *P. taeda* well in excess of 1 km. These model results are consistent with modelled gene flow results for *Eucalyptus* plantation, obtained from recent cellular automaton calculations that included both pollen and seed dispersal, mortality effects, disturbances, and fecundity (LINACRE AND ADES 2004).

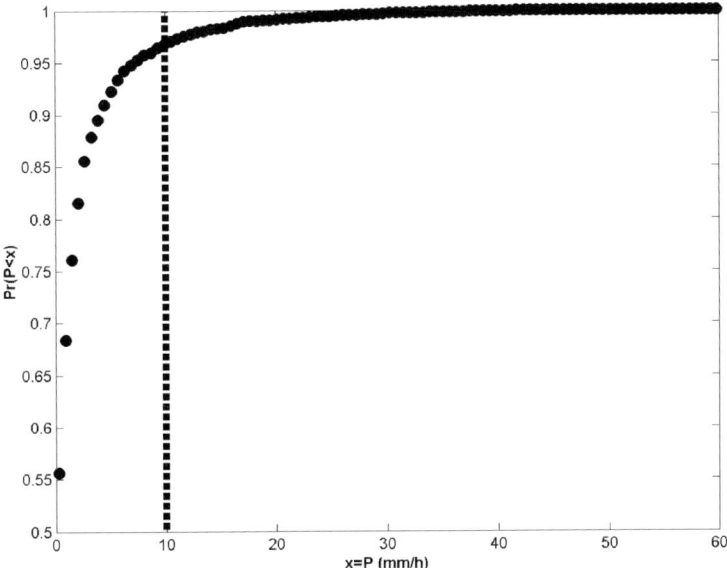

Figure 6. Cumulative precipitation (P) probability distribution function (Pr) for P>0. More than 95% of the precipitation events have intensities not exceeding 10 mm h-1 (vertical line).

Furthermore, our simulations suggest that conifer pollen is not likely to be of negligible viability at those distances due to excess UV-B, cold air temperatures, or dehydration, because of the short time interval for such long-distance movement. Given the long dispersal distances reported here, a regulatory framework that distinguishes between perennial crops and forests is only logical and has been advocated for almost a decade (e.g. STRAUSS *et al.,* 1995). LINACRE AND ADES (2004) stated that *"a less subjective and more transparent method is required for the proper evaluation of risk reduction measures"* if the general public is to accept the use of GM trees. Here, we show that stochastic models of particle dispersion in turbulent flows can augment risk assessments because model assumptions are explicit (*or transparent*) and the capacity to simulate ensemble of worst-case scenarios for a wide range of parameter space (*or proper evaluation of risk*) are now computationally possible.

Acknowledgements: Katul and Oren acknowledge support from the Center on Global Change at Duke University. Williams acknowledges the support of the USDA-Forest Service and the National Science Foundation (DEB #0454650). This project was supported by the US National Science Foundation (NSF-EAR), the Biological and Environmental Research, (BER) Program, U.S. Department of Energy, through the Southeast Regional Center (SERC) of the National Institute for Global Environmental Change (NIGEC), and through the Terrestrial Carbon Processes Program (TCP) and the FACE project.

APPENDIX 1: SECOND-ORDER CLOSURE MODEL FOR THE EULERIAN FLOW STATISTICS

In this appendix, the estimation of the Eulerian flow statistics needed to drive the THOMSON (1987) Lagrangian dispersion model are described, followed by the numerical scheme, closure constants, and boundary conditions. Upon temporal and spatial averaging the conservation of momentum equations, and following the closure approximations of canopy flows (WILSON AND SHAW 1977), the standard second order closure model of WILSON AND SHAW simplifies to the following set of ordinary differential equations (ODEs):

Mean Momentum:

$$0 = -\frac{d<\overline{u'w'}>}{dz} - C_d\, a(z) <\overline{u}>^2$$

Tangential Stress Budget:

$$0 = -<\overline{w'^2}>\frac{d<\overline{u}>}{dz} + 2\frac{d}{dz}\left(q\lambda_1 \frac{d<\overline{u'w'}>}{dz}\right) - \frac{q<\overline{u'w'}>}{3\lambda_2} + Cq^2\frac{d<\overline{u}>}{dz}$$

Longitudinal Velocity Variance:

$$0 = -2<\overline{u'w'}>\frac{d<\bar{u}>}{dz} + \frac{d}{dz}\left(q\lambda_1 \frac{d<\overline{u'^2}>}{dz}\right) + 2C_d a(z)<\bar{u}>^3$$

$$-\frac{q}{3\lambda_2}\left(<\overline{u'^2}> - \frac{q^2}{3}\right) - \frac{2}{3}\frac{q^3}{\lambda_3}$$

Lateral Velocity Variance:

$$0 = \frac{d}{dz}\left(q\lambda_1 \frac{d<\overline{v'^2}>}{dz}\right) - \frac{q}{3\lambda_2}\left(<\overline{v'^2}> - \frac{q^2}{3}\right) - \frac{2}{3}\frac{q^3}{\lambda_3}$$

Vertical Velocity Variance:

$$0 = \frac{d}{dz}\left(3q\lambda_1 \frac{d<\overline{w'^2}>}{dz}\right) - \frac{q}{3\lambda_2}\left(<\overline{w'^2}> - \frac{q^2}{3}\right) - \frac{2}{3}\frac{q^3}{\lambda_3}$$

Where $q = \sqrt{<\overline{u_i'u_i'}>}$ is a characteristic velocity scale, C_d is a drag coefficient, $a(z)$ is a leaf area density, $\lambda_j = a_j L_{ws}$, with L_{ws} a characteristic length scale specified using the formulation in KATUL AND ALBERTSON (1998) and is not permitted to increase at a rate larger than k, the Von Karman constant, and a_1, a_2, a_3 and C are determined so that the flow conditions well above the canopy reproduce well-established surface layer similarity relations. With estimates of the five constants (a_1, a_2, a_3, C, and α), the above five ODEs can be solved iteratively for the five flow variables $<\bar{u}>$, $<\overline{u'w'}>$, $<\overline{u'^2}>$, $<\overline{v'^2}>$, $<\overline{w'^2}>$, which are used to drive the Lagrangian model. Note that this approach neglects dispersive fluxes, which appears to be reasonable for dense canopies (POGGI et al., 2005).

A.1 The Computational Grid

The computational flow domain was set from zero to 50×h. The grid node spacing is $\Delta z = 0.05$ m. This grid density was necessary due to rapid variability in leaf area density close to the canopy top. Parameter values at the exact location of the pollen are calculated by interpolation between the grid nodes.

A.2 The Numerical Scheme

The five ODEs for the WILSON AND SHAW (1977) model were first discretized by central differencing all derivatives. An implicit numerical scheme was constructed for each ODE with boundary conditions to be discussed in section A.3. The tridiagonal system resulting from the implicit forms of each discretized equation was solved using the *Tridag* routine in PRESS et al., (1992, pp. 42-43) to produce the turbulent statistic profile. Profiles for all variables were initially assumed, and a relaxation scheme was used for all computed variables (WILSON 1988). Relaxation factors as small as 5% were necessary in the iterative scheme because of the irregularity in the leaf area density profile. The measured leaf area density was interpolated at the computational grid nodes by a cubic-spline to insure finite second derivatives of $a(z)$. Convergence is achieved when the maximum difference between two successive iterations in $<\bar{u}>$ did not exceed 0.0001%. We checked that all solutions were independent of Δz.

A.3 Boundary Conditions and Closure Constants

Typically, the well-established flow statistics in the atmospheric surface layer provide convenient upper-boundary conditions for closure models. The boundary conditions used are:

$$z = 0 \begin{cases} \sigma_u = 0 \\ \sigma_v = 0 \\ \sigma_w = 0 \\ u_* = 0 \\ <\bar{u}> = 0 \end{cases}$$

$$z = 50 \times h \begin{cases} \sigma_u = A_u u_* \\ \sigma_v = A_v u_* \\ \sigma_w = A_w u_* \\ u_* = 1 \\ \dfrac{d<\bar{u}>}{dz} = \dfrac{u_*}{k_v(z-d)} \end{cases}$$

Where σ_θ is the standard deviation of any flow variable $\theta (= <\theta'^2>^{1/2})$, A_u=2.2, A_v=2.0, and A_w=1.4 (KATUL AND CHANG 1999). The closure constants are dependent on the choice of the boundary conditions and are determined by assuming that in the atmospheric surface layer $(z > 2h)$, the flux-transport term is negligible and that $<\overline{u'w'}>$, $<\overline{u'^2}>^{1/2}$, and $<\overline{w'^2}>^{1/2}$ become independent of z for near-neutral conditions. These simplifications result, after some algebraic manipulations, in the following relationships between A_u, A_v, and A_w and a_2, a_3, and C:

$$a_2 = \frac{A_q(A_u^2 - A_w^2)}{6}$$

$$a_3 = \frac{-A_q^3(A_u^2 - A_w^2)}{A_w^2 - \frac{A_q^2}{3}}$$

$$C = \left(\frac{A_w}{A_q}\right)^2 - \frac{2}{A_q^2(A_u^2 - A_w^2)}$$

Where $A_q = (A_u^2 + A_v^2 + A_w^2)^{1/2}$. The closure constant a_1 is determined by noting that the eddy-diffusivity is $k(z-d)u_*$ in the surface layer. Hence, $q\lambda_1$ becomes identical to $k(z-d)u_*$ leading to $a_1 = 1/A_q$. The above equations are the first analytic expressions relating closure constants to ASL boundary conditions for the WILSON AND SHAW (1977) model. Table A1 summarizes the closure constants used resulting from our choice of A_u, A_v, and A_w.

Table A1. Closure constants for the second-order closure model for $A_u=2.2$, $A_v=2.0$, and $A_w=1.4$.

WILSON AND SHAW (1977)	Value
a_1	0.30
a_2	1.58
a_3	20.8
α	0.07
C	0.12
C_d	0.20

6. REFERENCES

AYLOR, D., AND J. PARLANGE. 1975. Ventilation required to entrain small particles from leaves. Plant Physiology **56**:97-99.

AYLOR, D. E., AND T. K. FLESCH. 2001. Estimating spore release rates using a Lagrangian stochastic simulation model. Journal of Applied Meteorology **40**:1196-1208.

AYLOR, D. E., N. P. SCHULTES, AND E. J. SHIELDS. 2003. An aerobiological framework for assessing cross-pollination in maize. Agricultural and Forest Meteorology **119**:111-129.
AYOTTE, K. W., J. J. FINNIGAN, AND M. R. RAUPACH. 1999. A second-order closure for neutrally stratified vegetative canopy flows. Boundary-Layer Meteorology **90**:189-216.
BALDOCCHI, D., E. FALGE, L. H. GU, R. OLSON, D. HOLLINGER, S. RUNNING, P. ANTHONI, C. BERNHOFER, K. DAVIS, R. EVANS, J. FUENTES, A. GOLDSTEIN, G. KATUL, B. LAW, X. H. LEE, Y. MALHI, T. MEYERS, W. MUNGER, W. OECHEL, K. T. P. U, K. PILEGAARD, H. P. SCHMID, R. VALENTINI, S. VERMA, T. VESALA, K. WILSON, AND S. WOFSY. 2001. FLUXNET: A new tool to study the temporal and spatial variability of ecosystem-scale carbon dioxide, water vapor, and energy flux densities. Bulletin of the American Meteorological Society **82**:2415-2434.
BESSEY, C. 1883. Remarkable fall of pine pollen. American Naturalist **17**:658.
BOYER, W.D. 1978. Heat accumulation: An easy way to anticipate the flowering of southern pines. Journal of Forestry **76**:20-23
BULLOCK, J. M., AND R. T. CLARKE. 2000. Long distance seed dispersal by wind: measuring and modelling the tail of the curve. Oecologia **124**:506-521.
CAIN, M. L., R. NATHAN, AND S. A. LEVIN. 2003. Long-distance dispersal. Ecology **84**:1943-1944.
CHRISTEY, M., AND D. WOODFIELD. 2001. Coexistence of genetically modified and non-genetically modified crops. 427, Crop & Food Research Confidential Report, Riverdale, MD.
DE HAAN, P., AND M. W. ROTACH. 1998. A novel approach to atmospheric dispersion modelling: The Puff-Particle Model. Quarterly Journal of the Royal Meteorological Society **124**:2771-2792.
DI-GIOVANNI, F., P. KEVAN, AND J. ARNOLD. 1996. Lower planetary boundary layer profiles of atmospheric conifer pollen above a seed orchard in northern Ontario, Canada. Forest Ecology and Management **83**:87-97.
DI-GIOVANNI, F., P. K. PG, AND M. NASR. 1995. Settling velocities of some pollen and spores and their variability. Grana **34**:39-44.
ERDTMAN, G. 1937. Pollen grains recovered from the atmosphere over the Atlantic. Medd. Göteborgs Bot. Trädg **12**.
FENNING, T. M., AND J. GERSHENZON. 2002. Where will the wood come from? Plantation forests and the role of biotechnology. Trends in Biotechnology **20**:291-296.
FINNIGAN, J. 2000. Turbulence in plant canopies. Annual Review of Fluid Mechanics **32**:519-571.
GURA, T. 2001. The battlefields of Britain. Nature **412**:760-763
HAILS, R.S. 2000. Genetically modified plants - the debate continues. Trends Ecol. Evol. **15**:14-18.
HENGEVELD, R. 1989.*Dynamics of biological invasions*. Chapman and Hall, London. 160 p.
HIGGINS, S. I., J. S. CLARK, R. NATHAN, T. HOVESTADT, F. SCHURR, J. M. V. FRAGOSO, M. R. AGUIAR, E. RIBBENS, AND S. LAVOREL. 2003. Forecasting plant migration rates: managing uncertainty for risk assessment. Journal of Ecology **91**:341-347.
HSIEH, C. I., G. KATUL, AND T. CHI. 2000. An approximate analytical model for footprint estimation of scaler fluxes in thermally stratified atmospheric flows. Advances in Water Resources **23**:765-772.
JAROSZ, N., B. LOUBET, B. DURAND, A. MCCARTNEY, X. FOUEILLASSAR, AND L. HUBER. 2003. Field measurements of airborne concentration and deposition rate of maize pollen. Agricultural and Forest Meteorology **119**:37-51.
KATUL, G., B. VIDAKOVIC, AND J. ALBERTSON. 2001. Estimating global and local scaling exponents in turbulent flows using discrete wavelet transformations. Physics of Fluids **13**:241-250.
KATUL, G. G., AND J. D. ALBERTSON. 1998. An investigation of higher-order closure models for a forested canopy. Boundary-Layer Meteorology **89**:47-74.
KATUL, G. G., AND W. H. CHANG. 1999. Principal length scales in second-order closure models for canopy turbulence. Journal of Applied Meteorology **38**:1631-1643.
KATUL, G. G., C. D. GERON, C. I. HSIEH, B. VIDAKOVIC, AND A. B. GUENTHER. 1998. Active turbulence and scalar transport near the forest- atmosphere interface. Journal of Applied Meteorology **37**: 1533-1546.
KATUL, G. G., L. MAHRT, D. POGGI, AND C. SANZ. 2004. One and two equation models for canopy turbulence. Boundary-Layer Meteorology **113**:81-109.
LANNER, R. 1966. Needed: a new approach to the study of pollen dispersion. Silv. Genetica **15:** 50-52.
LINACRE, N., AND P. ADES. 2004. Estimating isolation distances for genetically modified trees in plantation forestry. Ecological Modelling **179**:247-257

Loos, C., R. Seppelt, S. Meier-Bethke, J. Schiemann, and O. Richter. 2003. Spatially explicit modelling of transgenic maize pollen dispersal and cross-pollination. Journal of Theoretical Biology **225**:241-255.

Mann, C., and M.L. Plumer. 2002. Forest biotech edges out of the lab. Science **295**:1626-1629.

Massman, W. J., and J. C. Weil. 1999. An analytical one-dimensional second-order closure model of turbulence statistics and the Lagrangian time scale within and above plant canopies of arbitrary structure. Boundary-Layer Meteorology **91**:81-107.

Messeguer, J. 2003. Gene flow assessment in transgenic plants. Plant Cell, Tissue and Organ Culture. **73**:201-212.

Moyes, C. L., J. M. Lilley, C. A. Casais, S. G. Cole, P. D. Haeger, and P. J. Dale. 2002. Barriers to gene flow from oilseed rape (Brassica napus) into populations of *Sinapis arvensis*. Molecular Ecology **11**:103-112.

Nathan, R., G. G. Katul, H. S. Horn, S. M. Thomas, R. Oren, R. Avissar, S. W. Pacala, and S. A. Levin. 2002. Mechanisms of long-distance dispersal of seeds by wind. Nature **418**:409-413.

Nathan, R., G. Perry, J. T. Cronin, A. E. Strand, and M.L. Cain. 2003. Methods for estimating long-distance dispersal. Oikos **103**:261-273.

Nichols, R., and G. Hewitt. 1994. The genetic consequences of long-distance dispersal during colonization. Heredity **72**:312-317.

OGTR, 2001. Risk analysis framework for license applications to the office of the gene technology regulator. Office of the Gene Technology Regulator, Canberra, Australia.

Okubo, A., and S. A. Levin. 1989. A Theoretical Framework for Data-Analysis of Wind Dispersal of Seeds and Pollen. Ecology **70**:329-338.

Oren, R., D. S. Ellsworth, K. H. Johnsen, N. Phillips, B. E. Ewers, C. Maier, K. V. R. Schafer, H.McCarthy, G. Hendrey, S. G. McNulty, and G. G. Katul. 2001. Soil fertility limits carbon sequestration by forest ecosystems in a CO2-enriched atmosphere. Nature **411**:469-472.

Parker, S., and T. Blush. 1996. Quantifying pollen production of loblolly pine (*Pinus taeda* L.) seed orchard clones. 163, Westvaco Forest Research Report.

Phillips, N., A. Nagchaudhuri, R. Oren, and G. Katul. 1997. Time constant for water transport in loblolly pine trees estimated from time series of evaporative demand and stem sapflow. Trees, Structure and Function **11**:412-419.

Poggi, D., J. D. Albertson, and G. G. Katul. 2005. Scalar dispersion within a model canopy: Measurements and Lagrangian models. Advances in Water Resources. In press.

Poggi, D., A. Porporato, L. Ridolfi, J. Albertson, and G. G. Katul. 2004a. The effect of vegetation density on canopy sublayer turbulence. Boundary-Layer Meteorology **111**:565-587.

Poggi, D., G. G. Katul, and J. D. Albertson. 2004b. Momentum transfer and turbulent kinetic energy budgets within a dense model canopy. Boundary Layer Meteorology **111**:589-614.

Poggi, D., J. D. Albertson, and G. G. Katul. 2004c. A note on the contribution of dispersive fluxes to momentum transfer within canopies. Boundary Layer Meteorology **111**: 615-621.

Pope, S. B. 2000. Turbulent flows. Cambridge University Press, Cambridge.

Press, W., S. Teukolsky, W. Vetterling, and B. Flannery. 1992. Numerical Recipes in Fortran: The Art of Scientific Computing. Cambridge University Press, Cambridge.

Raupach, M. R. 1983. Near-Field Dispersion from Instantaneous Sources in the Surface-Layer. Boundary-Layer Meteorology **27**:105-113.

Raupach, M. R. 1987. A Lagrangian analysis of scalar transfer in vegetation canopies. Quarterly Journal of the Royal Meteorological Society **113**:107-120.

Raupach, M. R. 1989a. Applying Lagrangian Fluid-Mechanics to Infer Scalar Source Distributions from Concentration Profiles in Plant Canopies. Agricultural and Forest Meteorology **47**:85-108.

Raupach, M. R. 1989b. A practical lagrangian method for relating scalar concentrations to source distributions in vegetation canopies. *Quarterly Journal of the Royal Meteorological Society* **115**: 609-632.

Raupach, M. R. 1989c. A Practical Lagrangian Method for Relating Scalar Concentrations to Source Distributions in Vegetation Canopies. Quarterly Journal of the Royal Meteorological Society **115**:609-632.

Raupach, M. R., J. J. Finnigan, and Y. Brunet. 1996. Coherent eddies and turbulence in vegetation canopies: The mixing-layer analogy. Boundary-Layer Meteorology **78**:351-382.

Raupach, M. R., and R. H. Shaw. 1982. Averaging Procedures for Flow within Vegetation Canopies. Boundary-Layer Meteorology **22**:79-90.

REYNOLDS, A. M. 1998a. Comments on the 'Universality of the Lagrangian velocity structure function constant (C-0) across different kinds of turbulence'. Boundary-Layer Meteorology **89**:161-170.

REYNOLDS, A. M. 1998b. A Lagrangian stochastic model for the trajectories of particle pairs and its application to the prediction of concentration variance within plant canopies. Boundary-Layer Meteorology **88**:467-478.

REYNOLDS, A. M. 1998c. Modelling particle dispersion within a ventilated airspace. Fluid Dynamics Research **22**:139-152.

REYNOLDS, A. M. 1998d. On the formulation of Lagrangian stochastic models of scalar dispersion within plant canopies. Boundary-Layer Meteorology **86**:333-344.

REYNOLDS, A. M. 2000. On the formulation of Lagrangian stochastic models for heavy-particle trajectories. Journal of Colloid and Interface Science **232**:260-268.

REYNOLDS, A. M., AND J. E. COHEN. 2002. Stochastic simulation of heavy-particle trajectories in turbulent flows. Physics of Fluids **14**:342-351.

RIEGER, M. A., M. LAMOND, C. PRESTON, S. B. POWLES, AND R. T. ROUSH. 2002. Pollen-mediated movement of herbicide resistance between commercial canola fields. Science **296**:2386-2388.

RIGHTER, F., AND J. DUFFIELD. 1951. A summary of interspecific crosses in the genus *Pinus* made at the Institute of Forest Genetics. J. Hered. **42**:75-80.

RODEAN, H. 1996. Stochastic Lagrangian Models of Turbulent Diffusion. American Meteorological Society, Boston, 83 pages.

SAWFORD, B. L., AND F. M. GUEST. 1991. Lagrangian Statistical Simulation of the Turbulent Motion of Heavy-Particles. Boundary-Layer Meteorology **54**:147-166.

SIQUEIRA, M., AND G. KATUL. 2002. Estimating heat sources and fluxes in thermally stratified canopy flows using higher-order closure models. Boundary-Layer Meteorology **103**:125-142.

SIQUEIRA, M., G. KATUL, AND C. T. LAI. 2002. Quantifying net ecosystem exchange by multilevel ecophysiological and turbulent transport models. Advances in Water Resources **25**:1357-1366.

SIQUEIRA, M., C. T. LAI, AND G. KATUL. 2000. Estimating scalar sources, sinks, and fluxes in a forest canopy using Lagrangian, Eulerian, and hybrid inverse models. Journal of Geophysical Research-Atmospheres **105**:29475-29488.

SMOUSE, P. E., R. J. DYER, R. D. WESTFALL, AND V. L. SORK. 2001. Two-generation analysis of pollen flow across a landscape. I. Male gamete heterogeneity among females. Evolution **55**:260-271.

STOCKMARR, A. 2002. The distribution of particles in the plane dispersed by a simple 3-dimensional diffusion process. Journal of Mathematical Biology **45**:461-469.

STRAUSS, S., W. H. ROTTMAN, A. BRUNNER, L.A. SHEPPARD. 1995. Genetic engineering of reproductive sterility in forest trees. Molecular Breeding **1**:5-26.

STRAUSS, S., M.M. CAMPBELL, S. N. PRYOR, P. COVENTRY, AND J. BURLEY. 2001. Plantation Certification and Genetic Engineering: Ban on Research is Counter Productive. Journal of Forestry December: 4-7.

THOMSON, D. J. 1987. Criteria for the selection of stochastic-models of particle trajectories in turbulent flows. Journal of Fluid Mechanics **180**:529-556.

TUSKAN, G., T. TSCHAPLINSKI, B. FRANC, AND N. EDWARDS. 1992. UV-B radiation reduces germination capacity of loblolly pine and red spruce pollen. *in* In Proc. 12th North American Forest Biology Workshop. Ontario Ministry of Natural Resources, Sault Ste. Marie, Ontario.

WILSON, J. 1988. A Second Order Closure Model for Flow through Vegetation. Boundary-Layer Meteorology **42**:371-392.

WILSON, J. D., AND B. L. SAWFORD. 1996. Review of Lagrangian stochastic models for trajectories in the turbulent atmosphere. Boundary-Layer Meteorology **78**:191-210.

WILSON, N., AND R. SHAW. 1977. A Higher Order Closure Model for Canopy Flow. Journal of Applied Meteorology **16**:1198-1205.

CHAPTER 9

GENE FLOW IN CONIFERS

JEFFRY B. MITTON
Department of Ecology and Evolutionary Biology
University of Colorado

CLAIRE G. WILLIAMS
Department of Biology
Duke University

Abstract. The question of gene flow from transgenic conifers will draw heavily upon population genetics theory for answers about gene flow to surrounding forests. For this, we provide a brief summary of the conifer mating system and present some population genetics basics. As reviewed, several DNA-based polymorphism systems are available for conifers, particularly for the Pinaceae, but have not yet been applied to the question of transgenic escape. Future gene flow studies will benefit from the biparental organelle inheritance in the Pinaceae which allows measuring gene flow distances independently for male and female lineages. If nuclear, codominant markers are needed, then biologists will have the choice of the stable and conservative allozymes or rapidly-evolving nuclear microsatellites. Because allozymes produce enzymes that regulate metabolism, they might occasionally be the targets of selection, while the microsatellites are mostly neutral markers. Mitochondrial (mt) DNA and chloroplast (cp) DNA polymorphisms provide the opportunity to identify and track maternal and paternal lineages, respectively. These organellar markers provide the best tools for dissecting gene flow with high precision. Because most conifer seeds tend to have shorter dispersal distances compared to pollen, mtDNA is useful for making inferences about numbers and geographic location of outliers in addition to tracing advection.

1. GENE FLOW

Gene flow has a major impact on the distribution of genotypes within and among populations. Gene flow is not a constant, but is dependent on the mating system and the dispersal distances of pollen and seeds, all of which vary with environmental factors and geographic localities.

Isolated populations will diverge genetically. Allelic frequencies drift over time in finite populations, due to stochastic variation in survival and reproduction. The incidental loss of alleles may differ between populations, increasing the genetic distance between them. In addition, mutations may introduce new alleles into populations, further distinguishing them. With sufficient time, perfectly isolated populations will become completely differentiated, so that they do not share any alleles.

Gene flow among populations retards or prevents divergence. Low levels of gene flow between populations will retard their divergence and make them more similar, while high gene flow will homogenize them. The impact of gene flow on population structure can be illustrated quantitatively by modeling genetic variation at a single gene. Consider gene flow into a population from populations that have different allelic frequencies. If the proportion of migrants into a population is m, and the frequency of A in the migrants is $\bar{p}$, then p', the new frequency of A in the population, will be as follows:

$$p' = \bar{p}m + p(1-m).$$

If gene flow were unopposed by other forces, the populations connected by gene flow would ultimately share the same alleles, at the same frequencies.

1.1. The mating system in conifers

The importance of mating systems has long been apparent to theoreticians, field biologists, and plant breeders (WRIGHT 1931, 1969; ALLARD et al., 1968; LANDE AND SCHEMSKE 1985; BROWN 1990). Mating systems determine the distribution of genotypes within populations, and influence the degree of differentiation among populations. While outcrossing promotes gene flow, and brings genotypic proportions to Hardy-Weinberg equilibrium, selfing reduces gene flow, and brings genotypic distributions to an equilibrium described by Wright's equilibrium law (WRIGHT 1931, 1969). Selfing reduces heterozygosity by half each generation, reducing the effective rate of recombination, and thereby promoting genetic organization within populations (ALLARD et al., 1968).

Conifers have mixed mating systems (BROWN 1990). That is, some proportion of seed is produced by selfing (s), and the complementary proportion (t = 1-s) is produced by outcrossing. There is dramatic variation among species, with most species almost completely outcrossed (MULLER 1977; MITTON et al., 1977, 1981; ADAMS AND JOLY 1980; MORAN et al., 1980; SHAW AND ALLARD 1981, 1982; SHEN et al., 1981; EL- KASSABY et al., 1985, 1987; EPPERSON AND ALLARD 1984; FARRIS AND MITTON 1984; KING et al., 1984; CHELIAK et al., 1985; SNYDER et al., 1985; FRIEDMAN AND ADAMS 1985a; NEALE AND ADAMS 1985a,b: FURNIER AND ADAMS 1986; SHEA 1987; DENTI AND SCHOEN 1988; ERICKSON AND ADAMS 1989, 1990; GIBSON AND HAMRICK 1991). A meta-analysis of 52 conifers from eight genera and three conifer families (O'CONNELL 2003) confirmed that conifers in the family Pinaceae were predominantly outcrossed. However, some species, such as tamarack, *Larix laricina,* eastern white cedar, *Thuja occidentalis,* and Chihuahua spruce, *Picea chihuahuana,* have population outcrossing rates as low as t = 0.53 (KNOWLES et al., 1987), t = 0.51 (PERRY AND KNOWLES 1990), and t = 0.07 (LEDIG et al., 1997), respectively. Some outcrossing rates in conifers are summarized in Table 1.

Table 1. Outcrossing rates for conifers. All taxa except Thuja plicata are members of the Pinaceae family. Picea chihuahuana, a species which occurs in small, isolated populations is included as the only conifer with predominantly selfing.

Species	t	Reference
Tamarack (*Larix laricina*)	.729	(KNOWLES et al., 1987)
Douglas-fir (*Pseudotsuga menziesii*)	.752	(STAUFFER AND ADAMS 1993)
White spruce (*Picea glauca*)	.730	(INNES AND RINGIUS 1990)
Loblolly pine (*Pinus taeda*)	.994	(FRIEDMAN AND ADAMS 1985)
Norway spruce (*Picea abies*)	.956	(MORGANTE et al., 1991)
Scots pine (*Pinus sylvestris*)	.940	(MUONA AND HARJU 1989)
Black spruce (*Picea mariana*)	.924	(BOYLE AND MORGENSTERN 1986)
Ponderosa pine (*Pinus ponderosa*)	.960	(MITTON et al., 1977, 1981)
Limber pine (*Pinus flexilis*)	.980	(SCHUSTER AND MITTON 2000)
Noble fir (*Abies procera*)	.940	(SIEGISMUND AND KJAER 1997)
Radiata pine (*Pinus radiata*)	.900	(MORAN et al., 1980)
Silver fir (*Abies alba*)	.890	(SCHOEDER 1989)
Serbian spruce (*Picea omorika*)	1.00	(KUITTINEN AND SAVOLAINEN 1992)
Western red cedar (*Thuja alicata*)	.770	(O'CONNELL 2003)
Chihuahua spruce (*Picea chihuahuana*)	.076	(LEDIG et al., 1997)

In most conifers, each individual has separate male and female reproductive structures on one tree (monecy or 'one house'), although in some species, individuals are either male or female (dioecy or 'two houses'). If male and female strobili occur on the same plant then geitonomy or selfing via separate strobili on the same plant is possible (RICHARDS 1997). Autogamy, or selfing within the same flower, is not possible for conifers or other gymnosperms because male and female strobili are separate rather than combined into a single structure. For outcrossing monecious plants, selfing must be minimized at one or more stages despite the presence of male and female strobili on the same plant. In conifers, selfing is avoided at pollination or excluded after fertilization.

Mating systems, like other traits, are not constants, but are influenced by both environmental and genetic variation. Although there is still much variation that we do not yet understand, there are some patterns that we can predict. For example, the rate of outcrossing is expected to increase with stand density. Truly isolated trees, separated by many kilometers from conspecifics, can receive only their own pollen and set only selfed seed (FOWLER 1965). Similarly, in stands with low density, there will be low levels of outcrossing, for each tree stands in a mist of its own pollen, only partly diluted by the pollen of nearest neighbors. For this reason, lower levels of selfing are expected in years of low pollen production. In stands with normal or high density, the abundance of pollen sources necessarily dilutes selfing pollen, and outcrossing climbs to nearly 100%. An example of the association between stand density and outcrossing was found in ponderosa pine at the forest-grassland ecotone in eastern Colorado (FARRIS AND MITTON 1984). In a stand of normal density within the continuous forest, the average of estimates of outcrossing from allozymes and from albino seedlings was 0.96, and none of the estimates different from 1.00 (MITTON et al., 1981). Just a few kilometers to the east, a stand with nearest neighbor distances of 25-200 m had a level of outcrossing of only 0.80 (FARRIS AND MITTON 1984). However, the expected relationship between stand density and rate of outcrossing is not always observed (NEALE AND ADAMS 1985b).

The timing of pollen release and female receptivity will also affect variation in rates of outcrossing, and can impose limits to gene flow. In a study of Douglas- fir growing in a seed orchard, Erickson and Adams (1989, 1990) found considerable clone to clone variation in the timing of pollen release and female receptivity. Outcrossing was highest in clones whose pollen release followed female receptivity, and was lowest in clones in which male and female activities were coincident. A study of limber pine, *Pinus flexilis*, revealed that the timing of pollen release and female receptivity imposes limits to gene flow among populations at different elevations (SCHUSTER et al., 1989). Limber pine occurs at elevations ranging from 1,500 m to 3475 m on the Great Plains and the Front Range of Colorado, where spring arrives first on the plains, and then proceeds to higher elevation. Strong winds and steep terrain guaranteed that pollen released from the highest elevation could reach trees at the lowest elevation. But pollination phenology responds to the seasons, so that pollen released at high elevations may reach lower elevations where female surfaces are no longer receptive. The timing of release and receptivity suggests that successful pollen dispersal might be limited to a mere 1,000 feet in elevation.

1.1.1. Maternal haploid gametophyte

Within a fertilized pine seed, the embryo is surrounded by female gametophyte which serves as storage tissue that provides the nutrition for early growth and development. This tissue is haploid, and of maternal origin, and it provides several opportunities for geneticists and population biologists. A sample of seeds from a single tree can be used to test for segregation at an allozyme or microsatellite locus. Similarly, a sample of tissue can be used to measure the rate of recombination among a set of nuclear markers. Thus, although conifers might live centuries or even millennia, and are somewhat difficult to cross, the haploid maternal tissue makes them handy for genetic analyses.

1.1.2. Paternity analyses

In each pine seed, a diploid embryo is enveloped in maternal haploid tissue that carries the maternal contribution to that embryo. Simple subtraction of the maternal contribution from the offspring genotype yields the paternal contribution, or the set of genes carried by the pollen. The capability to clearly identify the paternal contribution provides more resolution to paternity analyses than can be obtained in angiosperms, which have triploid endosperm. With a sufficient number of loci, each with ample genetic variation, the paternal parent can be identified with moderate to high confidence.

A paternity analysis of limber pine, *Pinus flexilis*, examined gene flow within an isolated population on the eastern plains of Colorado (SCHUSTER AND MITTON 2000). This population is composed of 371 trees partially isolated among four small canyons and an adjacent hillside. All 277 trees producing pollen, the potential paternal parents, were scored for ten polymorphic allozyme loci. Ten polymorphic allozyme loci were scored in 474 embryos and their associated gametophytes. For 19 seeds, or 4% of the total, the paternal parent was identified with certainty. The average distance of pollen flow for these 19 seeds was 174 meters; similar estimates of pollen flow were obtained from most likely and fractional methods, which use probabilistic methods of identifying paternal parents and make fuller use of the data. These results indicated near panmixia within this population, which covers 15 hectares. Approximately 6.5% of the pollen came from outside the population; potential sources of pollen were 2-40 km away on the plains, and over 100 km away at lower elevations in the mountains.

1.2. Wind pollination

Because pollen is carried by the wind, the wind conditions at the time of pollen release will influence direction and distance of gene flow and levels of outcrossing. Total lack of wind would cause each tree to be enveloped in selfing pollen, and to receive little outcrossing pollen. At the other extreme, very turbulent air disperses pollen to exceptional distances and this promotes outcrossing.

1.3. Polyembryony

Gymnosperms have long been thought to control the genetic composition of their offspring by producing multiple embryos then selecting one to mature but there is little experimental support for this idea. The concept of polyembryony as a selective sieve was first proposed as developmental selection (BUCHHOLZ 1922). Developmental selection rests on a mistaken interpretation of *Pinus* embryology (BUCHHOLZ 1918) later extended to a comparative study of conifer genera (BUCHHOLZ 1920). The number of egg cells per archegonium was over-estimated by a factor of eight (BUCHHOLZ 1918). The rosette tier in the proembryo stage was mistakenly thought divided to form four rosette embryos which meant eight egg cells were present in one archegonium, not one (SKINNER 1992). If so, competition within an archegonium would have been far more intense with eight egg cells, naturally elevating the importance of developmental selection.

Later, based on correct morphology, simple polyembryony was re-considered as a potential selective force reducing fitness costs (SORENSEN 1982; KLEKOWSKI and KAZARINOVA-FUKSHANSKY 1984). Competition between selfed and outcrossed embryos was hypothesized to arise from simple polyembryony (SORENSEN 1982). Only in the case of simple polyembryony would weaker selfed embryos will be less competitive against outcrossed embryos. Cleavage polyembryony was excluded from consideration as a selective sieve because embryos share a common genotype and are thus selectively equivalent to each other.

Simple polyembryony does not act as a selective sieve for many reasons. First, few archegonia are actually fertilized. Even fewer or even none of these fertilized archegonia arise from simple polyembryony or monozygotic embryos. Second, monozygotic embryos share a high degree of genetic similarity so the genetic differences among embryos within an ovule are slight. Perhaps the third and most important reason comes from the observation that a dead embryo, selfed or outcrossed, is rarely replaced by another embryo unless its death occurs in early stages of embryo development. Rather than a selective sieve, polyembryony is considered to be a vestigial character of the highly conserved female gametophyte.

1.4. Inbreeding depression

Inbreeding decreases heterozygosity, and relative to the benchmark of randomly mating populations, the fitness of inbred individuals is typically depressed. Inbreeding depression is manifest as a depression of growth rate, viability, developmental stability, fecundity, and fertility (LERNER 1954; WRIGHT 1977; CHARLESWORTH AND CHARLESWORTH 1987; WILLIAMS AND SAVOLAINEN 1996). Inbreeding depression can be severe in species which are typically outcrossing, (CRUMPACKER 1967; SIMMONS AND CROW 1977; FRANKLIN 1972; SORENSEN AND MILES 1982; CROW AND SIMMONS 1983), and it may be less intense (LANDE AND SCHEMSKE 1985) but not necessarily absent (CHARLESWORTH AND CHARLESWORTH 1987) in selfing species. Inbreeding depression and heterosis have been under investigation for decades (WRIGHT 1977; FRANKEL 1983), but there is still no consensus concerning the genetic mechanisms underlying these phenomena. Two contending hypotheses, the dominance hypothesis and the overdominance hypothesis, are compatible as the genetic explanation for inbreeding depression.

1.4.1. Partial dominance hypothesis

The partial dominance hypothesis focuses upon the phenotypic expression of recessive alleles which are either lethal or detrimental. Some groups of species, such as conifers and *Drosophila*, have relatively high frequencies of deleterious alleles, and the rate of mutation to deleterious alleles per gamete is surprisingly high (LYNCH 1988). Inbreeding causes rare recessive alleles to occur more frequently in the homozygous condition, increasing the frequency of aberrant phenotypes. From the perspective of the dominance hypothesis, inbreeding depression is the expression of deleterious alleles that had previously been masked in the heterozygous condition. Because completely inbred strains tend to be fixed for deleterious alleles at different loci, crosses between strains produce genotypes bearing deleterious alleles in the heterozygous condition, masking the great majority of deleterious alleles, producing heterosis.

1.4.2. Overdominance hypothesis

The overdominance hypothesis supposes the performance of heterozygotes to be superior to that of homozygotes. From this perspective, inbreeding depression results when inbreeding decreases the frequency of heterozygous genotypes and increases the frequency of homozygous genotypes. The proponents of this hypothesis typically base their reasoning on the fact that the majority of enzyme kinetic studies have revealed that the genotypes at a locus have different biochemical properties (MITTON 1997; EANES 1999). Biochemical studies frequently reveal that the alternate homozygous genotypes at a locus reach their optimal performance under different conditions. Heterozygotes usually have intermediate optima, and perform efficiently over a broader range of conditions. This general (but not universal) pattern of kinetic variation provides a mechanism that might explain why higher heterozygosity is favored in fluctuating environments (MITTON 1997).

1.4.3. Interpretation of population surveys

Population-based surveys have been used to argue in favor of partial dominance. Molecular markers are assayed at the population level then correlated with phenotypic traits including fitness components. This approach constitutes a weak argument for partial dominance in conifers, aside from high embryo genetic loads. Positive correlations between a small number of heterozygous loci and fitness in conifers were interpreted as evidence for overdominance (BUSH AND SMOUSE 1991). A lack of positive correlation reported for other conifer species caused others to later conclude that partial dominance, not overdominance, is prevalent for conifer species (STRAUSS 1986; STRAUSS AND LIBBY 1987; SAVOLAINEN AND HEDRICK 1995). A third group reported negative correlations for conifers (Mitton and Grant 1984), perhaps because there is no genetic explanation (DENG AND FU 1998).

The weakness of the population heterozygosity argument lies in the untested assumptions about any given population under scrutiny. First, the population must be in gametic disequilibrium so that a positive correlation or functional overdominance can be detected. Second, random mating must be assumed by default yet deviations from random mating or panmixia is the major cause of incongruent results. Populations have undetected deviations from these idealized assumptions and these small deviations can generate any type of correlation (CHARLESWORTH et al., 1991; DENG AND FU 1998). As a result, marker-based population surveys do not provide consistent findings even within a species because every local population has a different phylogeographic history (DENG AND FU 1998).

The majority of evidence is consistent with the partial dominance hypothesis, and the majority of forest biologists believe that the partial dominance hypothesis best describes the mechanism underlying inbreeding depression in conifers.

However, comparison of selected trees with a random sample yields data consistent with the overdominance hypothesis. When foresters established orchards of Engelmann spruce, *Picea engelmannii* (MITTON AND JEFFERS 1989), Norway spruce, *Picea abies*, (BERGMANN AND RUETZ 1991), Sitka spruce, *Picea sitchensis* (CHAISURISRI AND EL-KASSABY 1994), and Douglas-fir, *Pseudotsuga menziesii* (EL-KASSABY AND RITLAND 1996), they selected trees according to the normal criteria used by foresters, but they unknowingly chose trees with higher individual heterozygosities (Mitton and Pierce 1980) than would be found in a random sample of trees. It appears that foresters use the advantageous traits associated with highly heterozygous genotypes when they select trees for breeding.

1.5. Avoidance of inbreeding

For outcrossing monecious plants, selfing must be minimized at one or more stages despite the presence of male and female strobili on the same plant. In conifers, selfing is avoided at pollination or excluded after fertilization. Selfing avoidance is influenced by strobili position and morphology yet it is still highly subject to environmental vagaries. External factors such as wind speed, wind direction or even stand density determine how much pollen from non-self trees reach receptive female strobili (MITTON 1992; DYER AND SORK 2001). Proportions of selfed pollination vary widely from tree to tree, from year to year.

Separation between pollen release timing and female strobilus receptivity on the same tree (dichogamy) lowers self-pollination rates (ERICKSON AND ADAMS 1989). Selfing is avoided with spatial separation of female and male strobili in a single crown. Younger trees tend to have a prevalence of female or male strobili throughout the crown and older trees have female strobili clustered in the top of the crown and male strobili in the lower branches. Age of the tree also affects degree of selfing avoidance. Older *Pinus ponderosa* trees produce more cones and less pollen. Young *Pinus ponderosa* trees act as pollen donors, producing few cones yet providing pollen to older trees loaded with female strobili (MITTON 1992).

1.6. Selection against inbred genotypes

When self-pollination does occur, pre-zygotic barriers such as self-incompatibility systems do not eliminate self-pollinated ovules prior to fertilization. Self-exclusion in the *Pinaceae* and possibly other monoecious conifers occurs *after* fertilization. To date, there are no direct estimates of the proportion of archegonia pollinated or fertilized by a tree's own pollen. The proportion of self-fertilizations in terms of zygotes ranges from 10 to 25% in *Pinus sylvestris* (SARVAS 1962; KOSKI 1971). Self-pollination rates can be as high as 34% in the middle of the crown yet only 5% of viable seed is a product of self-fertilization in *Pinus taeda* (FRANKLIN 1969).

Similarly, the proportion of full seeds after self-fertilization compared to cross-fertilization in most conifers on average is only 39% (Griffin and Lindgren 1985). The viability gap between self-pollination and selfed seed maturity is the result of post-fertilization exclusion. Self-exclusion is collectively defined as the embryo lethal system, a selective sieve against self-pollinated embryos soon after fertilization (BRAMLETT AND POPHAM 1971; KOSKI 1971).

Comparisons of seeds collected from field sites and the mature trees at the site typically provide evidence of selection against inbred genotypes. For seeds of Washoe pine, *Pinus washoensis*, the multilocus estimate of outcrossing was t = 0.86. The outcrossing rate, *t*, can be related to the equilibrium level of the inbreeding coefficient, F_{is}, as

$$F_{is} = (1-t)/(1+t),$$

so the outcrossing rate of t = 0.86 would produce an equilibrium value of F_{is} = 0.11. However, the value of F in the maternal seed trees was F = -.10 (MITTON, LATTA AND REHFELDT, 1997). These results indicate that the inbred genotypes detected in the seedlings are eliminated in natural populations. Similarly, the outcrossing rate in a stand of ponderosa pine was t = 0.96 (MITTON *et al.*, 1981; LINHART *et al.*, 1981), which would produce an equilibrium value of F_{is} = 0.02. However, the observed value in the mature trees at the site was F_{is} = -011 ± 0.02. The shift from F_{is} = 0.02 to F_{is} = -0.11 indicates not only that inbred genotypes were eliminated, but also that selection had favored highly heterozygous trees.

2. MOLECULAR MARKERS

2.1. Nuclear markers

Numerous types of molecular markers might be used to study genetic variation in conifers. Large numbers of RAPDs and AFLPs can be developed quickly. However, these markers are most commonly presence/absence of bands, and without further analysis, it is not possible to determine whether presence of a band in an individual reflects heterozygosity or homozygosity of the dominant allele. Geneticists have usually favored codominant markers, historically allozyme loci, and more recently, the more variable microsatellite loci.

2.1.1. Allozymes

Surveys of electrophoretically detectable genetic variation of proteins, or allozyme variation, have been used to measure genetic variation and describe geographic variation of many species of conifers (HAMRICK *et al.*, 1979; HAMRICK AND GODT 1989; LOVELESS AND HAMRICK 1984; MITTON 1983; MITTON AND DURAN 2004). These surveys have identified conifers as the most genetically variable group of species (HAMRICK *et al.*, 1979; HAMRICK AND GODT 1989; HAMRICK *et al.*, 1992). For example, the proportion of loci polymorphic in gymnosperms, dicots and monocots was .71, .59, and .45, respectively (P < 0.001; Hamrick and Godt 1989).

Similarly, the proportion of loci heterozygous within populations in gymnosperms, dicots, and monocots was 0.16, 0.14, and 0.10, respectively ($P < 0.001$). A survey of genetic variation within populations of annuals, short-lived herbaceous species, short lived woody species, long-lived herbaceous species and long-lived woody plants reported expected heterozygosities of 0.101, 0.098, 0.096, 0.082, and 0.148, respectively ($P < 0.001$; HAMRICK AND GODT 1989).

Allozyme surveys of conifers have generally revealed only slight differentiation of allelic frequencies among populations. When values of F_{st} have been used to estimate the number of migrants among populations, Nm, or the average number of migrants moving among populations per generation, is typically in the range of 5-20, indicating very high gene flow via wind borne pollen (SCHUSTER AND MITTON 2000; SCHUSTER et al., 1989; HAMRICK et al., 1994; LATTA et al., 1994).

2.1.2. Microsatellites

Microsatellites are tandem repeat motifs that occur in units of two to six nucleotides ($AAAA...A_n$ or $CAGCAGCAGCAG...CAG_n$). Because of these tandem repeat motifs, microsatellites are mutational hot spots, so they segregate for many alleles and have high heterozygosity. Little is known about the organization or function of microsatellites in the large conifer genome although they are abundant (SCOTTI et al., 2000). Microsatellites are especially useful for outcrossing species because they are codominant, easy to score and have multiple alleles per marker locus. A set of 200+ nuclear microsatellites have been developed for *Pinus taeda* L. and many are polymorphic in other hard pines (AUCKLAND et al., 2002).

Microsatellite markers are developed from short sequences of DNA enriched for the tandemly repeated units or repeat motifs. The short DNA sequences with repeat motifs are sequenced. If the sequences on either side of the repeat motif are long enough then a forward and a reverse primer can be designed for amplification via the polymerase chain reaction (PCR). In general a microsatellite marker is composed of five parts: 1) a region for binding the forward primer, 2) a flanking region, 3) the repeat motif, 4) another flanking region and 5) a region for binding the reverse primer. The DNA sequence is used to design a forward and reverse primer. Using the primers, each specific microsatellite sequence is amplified using the polymerase chain reaction and then run on a gel to observe the banding pattern. Scoring microsatellites is a measure of length polymorphisms.

Microsatellite alleles are measured by their band width using numbers of base pairs (bp). A homozygote (two copies of the same band) is viewed on a gel as a single band. A heterozygote is seen as two different bands, i.e. a difference in band length. What mutations cause changes in band width? A change in band width of an allele can be caused by a single base pair insertion or deletion or a change in the number of repeats. Single base pair changes can occur in the flanking or repeat regions. If a single base change occurs in the either of the two primer binding regions then null (absent) alleles occur. Primer binding is the essential step in obtaining a single PCR product.

Sequencing a sample of alleles is necessary to directly observe the type of nucleotide changes causing changes in allele band widths. Size homoplasy occurs when there are two alleles which have different sequence composition yet share the same band width (identical by state). For example, size homoplasy for one microsatellite allele has been observed between *P. palustris* (146 bp, repeat motif A_{14}) and *Picea rubens* (146 bp, motif $A_{14}CAA_6$) for locus PtTX3020. Similarly, size homoplasy has been observed between *P. radiata* (158 bp with a repeat motif CAG_7) and *P. strobus* (158 bp with repeat motif $CAG_2CAA_2CAG_2$) for locus PtTX2123 (KUTIL AND WILLIAMS 2001). In both cases, size homoplasy occurs at the phylogenetic extremes for microsatellite transfer so this does not appear to be a problem for intraspecific studies of mating systems. Because of size homoplasy, allele sequences should be checked when testing trans-specific markers and especially when microsatellites are used for determining phylogenetic relationships within a genus.

Polymorphism levels of these microsatellites vary within a species (ECHT et al., 1996), especially if the microsatellites were developed from the low-copy regions of the conifer genome (ELSIK et al., 2000). For the undermethylated (UM) microsatellites, allele numbers are lower and fewer are considered rare alleles (ZHOU et al., 2001). Data collection for microsatellites has become automated thus reducing both cost and time to conduct large marker studies (ZHOU et al., 2002). The high number of alleles per locus offers a method of measuring allelic diversity with higher resolution. For gene flow studies, the diagnostic or unique alleles within stands or subpopulations are useful.

To date, triplet repeat microsatellites developed in one species of hard pines can be polymorphic in other hard pines. DNA sequences appear highly orthologous and no ascertainment bias has been detected to date (SOKOL AND WILLIAMS 2005). Cross-amplification has been found in soft pines but not in hard pines. Contrary to earlier conjecture, use of microsatellites with a perfect repeat motif does not increase transfer rates within Pinaceae (KUTIL AND WILLIAMS 2001; SHEPHERD et al., 2001).

2.2. Organellar markers

The organellar genomes of pines are ideal for measuring gene flow, for mtDNA has maternal inheritance and cpDNA has paternal inheritance in pines (NEALE et al., 1986; STRAUSS et al., 1989). These different modes of inheritance allow us to explicitly identify gene flow mediated by pollen and by seeds. In addition, pollen and seeds have disparate potentials for dispersal. The wind-borne pollen has the potential to travel great distances, but in contrast, the seeds of pines usually fall within a circle that has a radius equal to the height of the tree. Interesting exceptions to this general pattern of organellar inheritance are found in coast redwood, *Sequoia sempervirens* (NEALE et al., 1989), and incense-cedar *Calocedrus decurrens*, (NEALE et al., 1991) in which both organelles are paternally inherited.

Wright (1951) described the relationship between gene flow and the variance in allelic frequencies among populations at evolutionary equilibrium:

$$F_{st} = \frac{1}{4Nm+1}.$$

F_{st} is the standardized variance in allelic frequencies, N is the population size, and m is the migration rate. This relationship is slightly modified for haploid genomes in monecious species, such as the organellar genomes of pines:

$$F_{st} = \frac{1}{2Nm+1}$$

Gene flow of nuclear genes is influenced by both pollen and seed dispersal, which are the paternal and maternal components of gene flow. As shown below, Ennos (1994) described the contributions of maternal and paternal gene flow to biparental gene flow in terms of F_{st}:

$$(\frac{1}{F_{st(b)}}) = (\frac{1}{F_{st(m)}}) + (\frac{1}{F_{st(p)}}).$$

This relationship indicates that the gene flow and population structure of biparentally inherited genes will be most similar to the component with the highest gene flow. That is, the gene flow of nuclear genes in conifers will reflect predominantly the gene flow mediated by pollen. This expectation is intuitive and it is consistent with empirical data.

2.2.1. Chloroplast (Cp) DNA

CpDNA, like mtDNA in plants, has relatively low levels of genetic variation within populations. For example, only three minor variants were found in a study of 100 individuals of the lupine, *Lupinus texensis* (BANKS AND BIRKEY 1985). A survey of 384 trees from 19 populations of Monterey pine, Bishop Pine, and knobcone pine revealed little or no cpDNA variation within or among populations of knobcone pine and Monterey pine (HONG *et al.,* 1993). Bishop pine had some variation within populations, but more variation among populations. A total of 20 forms or haplotypes of cpDNA were detected in the 384 trees. A summary of the percent nucleotide changes per site in 46 studies of cpDNA yielded a mean of 0.07% at the intraspecific level, 0.8% at the interspecific level, and 3.4% at the intergeneric level (Schaal *et al.,* 1991). In contrast, values of 1% to 8% were reported within species for mtDNA sequences from a diverse set of animals studied in the southeastern U. S. (AVISE 1992). Although variation in cpDNA is relatively low, it is sufficient to support studies of geographic variation in some conifer species (MATOS 1992; WAGNER *et al.,* 1987; WAGNER 1992).

2.2.2. Mitochondrial (Mt) DNA

Low mutation rate is probably the reason for the paucity of useful variation in plant mtDNA. In most animals, the mutation rate of mtDNA is substantially higher than the mutation rate of nuclear DNA. However, in plants, the ranking of mutation rates is much different; the mutation rates are highest in nuclear DNA, intermediate in cpDNA, and lowest in mtDNA (PALMER 1985, 1990). For example, Lynch (1997) lists the synonymous substitution rates (substitutions/site/billion years) for plant nuclear DNA, cpDNA, and plant mtDNA, to be 7.31 ± 0.58, 1.70 ± 0.09 and 0.51 ± 0.05, respectively. That is, nuclear DNA and cpDNA evolve at 14 times and three times, respectively, the rate of plant mtDNA. For a reference, the evolutionary rate of mammalian mtDNA is 7.75 ± 0.43, or 15 times the rate of plant mtDNA.

To date, most studies of population structure of mtDNA in pines have been limited to RFLPs revealed by Southern blots (DONG AND WAGNER 1993, 1994; HONG et al., 1993; STRAUSS et al., 1993) or to size variants of introns amplified by the polymerase chain reaction and separated on sequencing gels (LATTA AND MITTON 1997, 1999; LATTA et al. 1998; MITTON et al., 1999; SINCLAIR et al., 1999; SPERISEN et al., 2001; JARAMILLO-CORREA et al., 2004). More recently, DNA sequence data have been used to detect nucleotide substitutions that either produce or erase palindromes. When polymorphic sequences are digested with the appropriate restriction enzymes, the cleaved amplified polymorphic sequences (CAPS) can be genotyped on agarose gels by simply recording which sequences cut (had palindromes) and which did not. This efficient technique for surveying organellar genetic variation has been used to describe geographic patterns in organellar genomes of whitebark pine, *Pinus albicaulis* (RICHARDSON et al., 2002) and silver fir, *Abies alba* (LIEPELT et al., 2003).

3. GENE FLOW ESTIMATED WITH MOLECULAR MARKERS

The *average* dispersal distances of wind-dispersed seed are typically small, usually a matter of meters or even a few kilometers but using averages provides a closer measure of local neighbourhood dispersal (LND) rather than long-distance dispersal (LDD) (WILLIAMS et al., 2006). In contrast, pollen might be transported more than 100 km (see KATUL et al., 2006) or more especially during turbulent storms. Regardless of the distance method, one would expect a disparity between seed and pollen dispersal distances in conifers with biparental organelle inheritance. In short, comparing seed-dispersed mtDNA versus pollen-dispersed cpDNA would give contrasting patterns of geographic variation, both within and among populations.

3.1. Variation within a stand

A long-term study of ponderosa pine provided the demographic and genetic data to study the distribution of genotypes within a stand. The population is on a south-facing slope at the lower entrance to Boulder Canyon, CO, at an elevation of 1,738. Within the population, which covers approximately 2 hectares, 217 trees were permanently marked, and their genotypes are known for seven polymorphic allozyme loci (LINHART et. al., 1981), one polymorphic mtDNA marker and one polymorphic cpDNA marker (LATTA et al., 1998). Studies of old photos indicate that the site has filled in since the time since European settlers arrived. Analyses of mtDNA revealed a significant degree of patch structure at the site. Analyses of the ages of trees, their mtDNA haplotypes, and their allozyme genotypes revealed that the six clusters of trees on the site were half-sibs clustered around the maternal parent. In contrast the to pattern of mtDNA, no structure was detected in the pattern of cpDNA haplotypes.

3.2. Gene flow along an elevational transect

Potentials for dispersal of pollen and seed lead biologists to expect high gene flow in genes dispersed by pollen (nuclear genes, cpDNA) and low gene flow for genes dispersed solely by seed. This hypothesis was tested with a study of gene flow among populations of limber pine in the Front Range of Colorado (LATTA AND MITTON 1997). The populations were distributed from tree line at the Continental Divide to an isolated stand of trees 150 km to the east, on an escarpment on the Great Plains. Haplotype frequencies were used to calculate F_{st} for both cpDNA and mtDNA, and gene flow was inferred from $F_{st.}$ The F_{st} values were .02 and .68 for cpDNA and mtDNA, respectively, suggesting that the number of migrants among populations per year are 12.25 for pollen and 0.12 for seeds. The gene flow of cpDNA is high, and should tend to homogenize the frequencies of cpDNA haplotypes and nuclear genes among populations within distances of approximately 150 km. The data were consistent with this expectation; the value for nuclear allozymes was $F_{st} = 0.02$. In contrast, the gene flow of mtDNA is below the threshold at which the influence of genetic drift predominates. So mtDNA is expected to vary more among populations than nuclear genes and cpDNA, and genetic drift will cause populations to diverge with respect to mtDNA haplotypes.

3.3. Gene flow across the ponderosa pine transition zone

During the Wisconsin glaciation, ponderosa pine retracted to refugia in two general areas, northern Mexico and the Pacific Coast (BETANCOURT et al., 1990; RICHARDSON 1998; BONNICKSEN 2000 and references therein). At the end of the ice age, ponderosa pines began to spread north from Mexico, reaching the San Andres Mountains of southern New Mexico 14,920 ya, the Santa Catalina Mountains of southern Arizona a few centuries later, Chaco Canyon, NM, 9,500 ya, and the Grand Canyon about 10,000 ya. They arrived in eastern Nevada 6,100 ya, and in northern Colorado 5,090 ya. Ponderosa pines reached northeastern Wyoming about 4,000 ya, and continued north around the northern edge of the Great Basin where they formed a narrow transition zone in eastern Montana with the ponderosa pines spreading east from their Pacific refugia. The western subspecies, *Pinus ponderosa ponderosa*, met the eastern subspecies, *P. p. scopulorum*, as recently as 1,000 years ago.

The subspecies are adapted to different moisture regimes, can be distinguished with morphological characters, monoterpene profiles and allozymes. Their divergence is sufficient to be reflected in other members of their community. The western pine beetle, *Dendroctonus brevicomis*, utilizes ponderosa pine as one of its hosts. *D. brevicomis* contains cryptic species with mtDNA sequence divergence of 9% (KELLEY AND MITTON 1998). The cryptic species are segregated on the two subspecies of ponderosa pine.

A clear contrast between gene flow for mtDNA and cpDNA was reported in the transitional zone of ponderosa pine (LATTA AND MITTON 1999). Both cpDNA and mtDNA markers are diagnostic for these subspecies, allowing the tracking of gene flow by both pollen and seeds across the transition zone. The mtDNA markers reveal a sharp boundary between the subspecies, approximately 7 km wide, with no intermixing of eastern and western haplotypes (JOHANSEN AND LATTA 2003). The cpDNA had a gentle cline between the subspecies, with gene flow 100 km into the western subspecies and 150 km into the eastern subspecies.

3.4. Gene flow across the range of silver fir

A range-wide survey of 100 populations of silver fir, *Abies alba*, revealed sharply contrasting patterns of geographic variation for mtDNA and cpDNA markers (LIEPELT et al., 2000). An 80 bp deletion in the fourth intron of mitochondrial NAD5 produced two haplotypes that could be scored by amplifying the intron and running the PCR products on an agarose gel. A substitution in the chloroplast psbC gene produced a presence/absence polymorphism for a palindrome. Digestion of the psbC PCR products with HAEIII produced either three or four fragments that could be distinguished in agarose gels. More than 1000 trees from 100 populations were typed by these efficient methods.

The mitochondrial haplotypes showed two major regions in the range of silver fir, with relatively little mixing or introgression. Populations in the eastern portion of the range were fixed for one allele, and populations in the western portion of the range were fixed for the other allele. This pattern was consistent with earlier allozyme studies that postulated an eastern Mediterranean refugium in the Balkan Peninsula and a western Mediterranean refugium in the Apennines (KONNERT AND BERGMANN 1995). Mixed populations were found in Croatia, Slovenia, and northeastern Italy, but the majority of populations were fixed for a single allele, and therefore clearly descendant from either the eastern or western refugium.

The cpDNA haplotypes revealed a steep, continuous cline that spanned the entire range of the species. The refugia were probably fixed for alternate alleles during one of the last several glacial maxima, but gene flow among populations during the interglacials, when populations were in close proximity, produced a cline. This cline formed in the present interglacial, it would have formed about 7,500 years ago in southern Europe and between 1,000 and 4,000 years ago in the Carpathians (LIEPELT et al., 2000). However, it is likely that the cline was generated several interglacials ago (HEWITT 2000), and has been smoothed in the last several interglacials.

4. CONCLUSIONS

To date, DNA-based polymorphisms for tracking gene flow are plentiful for conifers but these methods have not yet been applied to the question of transgenic escape in forest trees. Knowledge of population genetics and of the conifer mating system in particular will be important to determining the degree of gene flow between transgenic conifer plantations and surrounding forests. The potential for using these methods to track transgenic escapes is great for three reasons: 1) genomics resources is providing better molecular marker systems, 2) better analytical and computational methods are coming available which are quite powerful for studying gene flow on either contemporary or evolutionary time scales and 3) conifer species mostly likely to be developed for transgenic plantations have biparental organelle inheritance. The latter point means that mitochondrial (mt) DNA and chloroplast (cp) DNA polymorphisms will provide two independent methods for tracking maternal and paternal genetic contributions, respectively. These markers can dissect gene flow and hybrid zones with high precision.

5. REFERENCES

ADAMS, W. T. AND R. J. JOLY, 1980 Allozyme studies in loblolly pine seed orchards: clonal variation and frequency of progeny due to self-fertilization. Silvae Genet. **29**: 1-4.

ALLARD, R. W., S. K. JAIN, AND P. L. WORKMAN, 1968 The genetics of inbreeding populations. Advances in Genetics **14**: 55-131.

AUCKLAND, L.D., T. BUI, Y. ZHOU, M. SHEPHERD AND C. G. WILLIAMS, 2002. *Conifer Microsatellite Handbook*. Corporate Press, Raleigh NC. 57 p.

AVISE JC., 1992 Molecular population structure and the biogeographic history of a regional fauna: a case history with lessons for conservation biology. Oikos **63**: 62-76.

BANKS J, BIRKEY C JR., 1985 Chloroplast DNA diversity is low in a wild plant, *Lupinus texensis*. Proc. Natl. Acad. Sci. USA. **82**: 6950-6954.

BERGMANN, F. AND W. RUETZ, 1991 Isozyme genetic variation and heterozygosity in random tree samples and selected orchard clones from the same Norway spruce poplations. Forest Ecology and Management **46**: 39-47.

BETANCOURT, J. L., W. S. SCHUSTER, J. B. MITTON, AND R. S. ANDERSON, 1991 Fossil and genetic history of a pinyon pine (*Pinus edulis*) isolate. Ecology **72**: 1685-1697.

BETANCOURT, J. L., T. R. VAN DEVENDER, AND P. S. MARTIN, 1990 Packrat Middens: The last 40,000 Years of Biotic Change. The University of Arizona Press, Tucson.

BONNICKSEN, T. M., 2000 America's Ancient Forests: From the Ice Age to the Age of Discovery. John Wiley and Sons, Inc. New York.

BOYLE, T. J. B. AND E. K. MORGENSTERN, 1986 Estimates of outcrossing rates in six populations of black spruce in central New Brunswick. Silvae Genet. **35**: 102-106.

BRAMLETT, D. AND T. POPHAM, 1971 Model relating unsound seed and embryonic lethals in self-pollinated pines. Silvae Genetica **20**: 192-193.

BROWN, A. H. D., 1990 Genetic characterization of plant mating systems. pp. 145-162 in A. H. D. Brown, M. T. Clegg, A. L. Kahler, and B. S. Weir. (eds.) Plant Population Genetics, Breeding, and Genetic Resources. Sinauer Associates, Inc., Sunderland, Ma.

BUCHHOLZ, J., 1918 Suspensor and early embryo of *Pinus*. Botanical Gazette **66**: 185-228.

BUCHHOLZ, J., 1920 Embryo development and polyembryony in relation to the phylogeny of conifers. American Journal of Botany **7**: 125-145.

BUCHOLZ, J. T., 1922 Developmental selection in vascular plants. Botanical Gazette **73**: 249-286.

BUSH, R. M. AND P. E. SMOUSE, 1991 The impact of electrophoretic genotype on life history traits in *Pinus taeda*. Evolution **45**: 481-498.
BUSH, R. M. AND P. E. SMOUSE, 1992 Evidence for the adaptive significance of allozymes in forest trees. New Forests **6**: 179-196.
CHAISURISRI, K. AND Y. EL-KASSABY, 1994 Genic diversity in a seed production population versus natural populations. Biodiversity and Conservation **3**: 512-523.
CHARLESWORTH, D. AND B. CHARLESWORTH, 1987 Inbreeding depression and its evolutionary consequences. Ann. Rev. Ecol. Syst. **18**: 237- 268.
CHARLESWORTH, B., M. MORGAN AND D. CHARLESWORTH. 1991 Multilocus models of inbreeding depression with synergistic selection and partial self-fertilization. Genetical Research **57**: 177-194.
CHELIAK, W. M., B. P. DANCIK, K. MORGAN, F. C. H. YEH, AND C. STROBECK, 1985 Temporal variation of the mating system in a natural population of jack pine. Genetics **109**: 569-584.
CLEGG, M. T. 1980, Measuring plant mating systems. BioScience **30**: 814-818.
CROW, J. F. AND M. J. SIMMONS, 1983 The mutation load in Drosophila. pp. 1-35 in The Genetics and Biology of *Drosophila*. Volume 3c, M. Ashburner, H. L. Carson, J. N. Thompson. Academic Press, London.
CRUMPACKER, D. W., 1967 Genetic loads in maize (Zea mays L. and other cross-fertilized plants and animals. Evol. Biol. **1**: 306-423.
DENTI, D. AND D. J. SCHOEN, 1988 Self-fertilization rates in white spruce: effect of pollen and seed production. J. Hered. **79**: 284-288.
DENG, H. AND Y.-X. FU, 1998 Conditions for positive and negative correlations between fitness and heterozygosity in equilibrium populations. Genetics **148**: 1333-1340.
DONG, J. AND D. B. WAGNER, 1993 Taxonomic and population differentiation of mitochondrial DNA diversity in *Pinus banksiana* and *Pinus contorta*. Theor. Appl. Genet. **86**: 573-8.
DONG, J. AND D. B. WAGNER, 1994 Paternally inherited chloroplast polymorphism in *Pinus*: estimation of diversity and population subdivision, and tests of disequilibrium with a maternally inherited mitochondrial polymorphism. Genetics **136**: 1187-94.
DYER, R. AND V. SORK, 2001 Pollen pool heterogeneity in shortleaf pine, *Pinus echinata* Mill. Mol. Ecol. **10**: 859-866.
EANES, W. F., 1999 Analysis of selection on enzyme polymorphisms. Ann. Rev. Ecol. Syst. **30**: 301-326.
ECHT C.S., G. G. VENDRAMIN, C. D. NELSON AND P. MARQUARDT, 1999 Microsatellite DNA as shared genetic markers among conifer species. Can. J. For. Res. **29**: 365-371.
EL-KASSABY, Y. AND K. RITLAND, 1996 Impact of selection and breeding on the genetic diversity of Douglas-fir. Biodiversity and Conservation **5**: 795-813.
EL-KASSABY, Y. A., K. RITLAND, A. M. K. FASHLER AND W. J. B. DEVITT, 1988 The role of reproductive phenology upon the mating system of a Douglas-fir seed orchard. Silvae Genet. **37**: 76-82.
EL-KASSABY, Y. A., F. C. YEH, AND O. SZIKLAI, 1981 Estimation of the outcrossing rate of Douglas-fir (*Pseudotsuga menziesii* (Mirb.) Franco) using allozyme polymorphisms. Silvae Genet. **30**: 182-184.
Elsik, C.G., V. T. Minihan, A. M. Scarpa, S. E. Hall. and C. G. Williams, 2000. Low-copy microsatellite markers for *Pinus taeda* L.Genome **43**: 550-555.
ENNOS, R. A., 1994 Estimating the relative rates of pollen and seed migration among plant populations. Heredity **72**: 250-259.
EPPERSON, B. K. AND R. W. ALLARD, 1984 Allozyme analysis of the mating system in lodgepole pine populations. J. Hered. **75**: 212-214.
ERICKSON, V. J. AND W. T. ADAMS, 1989 Mating success in a coastal Douglas-fir seed orchard as affected by distance and floral phenology. Can. J. For. Res. **19**: 1248-1255.
ERICKSON, V. J. AND W. T. ADAMS 1990 Mating system variation among individual ramets in a Douglas-fir seed orchard. Can. J. For. Res. **20**: 1672-1675.
FARRIS, M. A. AND J. B. MITTON, 1984 Population density, outcrossing rate, and heterozygote superiority in ponderosa pine. Evolution **38**: 1151-1154.
FOWLER, D. P., 1965 Effects of inbreeding in red pine, *Pinus resinosa* Ait. III. Factors affecting natural selfing. Silvae Genet. **14**: 36-46.

FRANKEL, R., 1983 Heterosis: *Reappraisal of Theory and Practice*. Springer-Verlag, Berlin. 290 pp.
FRANKLIN, E., 1969 Inbreeding depression in metrical traits of loblolly pine (*Pinus taeda* L.) as a result of self-pollination. Raleigh NC., School of Forest Resources, North Carolina State University.
FRANKLIN, E. C., 1970 Survey of mutant forms and inbreeding depression in species of the family Pinaceae. USDA For. Serv. Res. Pap. SE-61.
FRANKLIN, E. C., 1972 Genetic load in loblolly pine. Am. Nat. **106**: 262-265.
FRIEDMAN, S. T. AND W. T. ADAMS, 1985a Levels of outcrossing in two loblolly pine seed orchards. Silvae Genetica **34**: 157-162.
FRIEDMAN, S. T. AND W. T. ADAMS, 1985b Estimation of gene flow into two seed orchards of loblolly pine (*Pinus taeda* L.). Theor. Appl. Genet. **69**: 609-615.
FURNIER, G. R. AND W. T. ADAMS, 1986 Mating system in natural populations of Jeffrey pine. Amer. J. Bot. **73**: 1009-1015.
GIBSON, J. P. AND J. L. HAMRICK, 1991 Heterogeneity in pollen allele frequencies among cones, whorls, and trees of table mountain pine (*Pinus pungens*). Am. J. Bot. **78**: 1244-1251.
GRIFFIN, R. AND D. LINDGREN, 1985 Effect of inbreeding on production of filled seed in *Pinus radiata* - experimental results and a model of gene action. Theor. Appl. Genet. **71**: 334-343.
HAMRICK J. L. AND M. J. W. GODT, 1989 Allozyme diversity in plant species. pp. 43-63 In A. H. D. Brown, M. T. Clegg, A. L. Kahler, and B. S. Weir. (eds.) *Plant Population Genetics. Breeding and Genetic Resources*. Sinauer Press, Sunderland, MA.
HAMRICK, J. L., Y. B. LINHART, AND J. B. MITTON, 1979 Relationship between life history parameters and electrophoretically-detectable genetic variability in plants. Ann. Rev. Ecol. Syst. **10**: 173-200.
HAMRICK, J. L., J. B. MITTON, AND Y. B. LINHART, 1981 Levels of genetic variation in trees: influence of life history characteristics. In M. T. Conkle (ed.) *Isozymes of North American Forest Trees and Forest Insects*. USDA Gen. Tech. Rep. PSW-48.
HAMRICK, J. L., A. F. SCHNABEL, AND P. V. WELLS, 1994 Distribution of genetic diversity within and among populations of Great Basin conifers. pp. 147-161 in K. T. Harper, L. L. St. Clair, K. H. Thorne, and W. M. Hess (eds.) *Natural History of the Colorado Plateau and Great Basin* University Press of Colorado, Niwot, CO.
HEWITT G. M., 2000 The genetic legacy of the Quaternary ice ages. Nature **405**: 907-913.
HONG, Y.-P., V. D. HIPKINS AND S. H. STRAUSS, 1993 Chloroplast DNA diversity among trees, populations and species in the California closed-cone pines (*Pinus radiata, Pinus muricata,* and *Pinus attenuata*). Genetics **135**: 1187-96.
INNES, D. AND G. RINGIUS, 1990 Mating system and genetic structure of two populations of white spruce (*Picea glauca*) in eastern Newfoundland. Canadian Journal Botany **68**: 1661-1666.
JARAMILLO-CORREA, J. P., J. BEAULIEU AND J. BOUSQUET, 2004 Variation in mitochondrial DNA reveals multiple distant glacial refugia in black spruce (*Pice mariana*), a transcontinental North American conifer. Mol. Ecol. **13**: 2735-2747.
JOHANSEN, A. D. AND R. G. LATTA, 2003 Mitochondrial haplotype distribution, seed dispersal and patterns of postglacial expansion of ponderosa pine. Mol. Ecol. **12**: 293-298.
KATUL G., C. G. WILLIAMS, M. SIQUEIRA, D. POGGI, A. PORPORATO, H. MCCARTHY AND R. OREN. 2006. Dispersal of transgenic conifer pollen. Chapter 8. Editor: C.G. Williams. In: *Landscapes, Genomics and Transgenic Conifers*. Springer Publishers.
KELLEY, S. T. AND J. B. MITTON, 1998 Strong differentiation in mitochondrial DNA of *Dendroctonus brevicomis* (Coleoptera: Scolytidae) populations on different subspecies of ponderosa pine. Ann. Ent. Soc. Am. **92**: 193-197.
KING, J. N., B. P. DANCIK, AND N. K. DHIR, 1984 Genetic structure and mating system of white spruce (*Picea glauca*) in a seed production area. Can. J. For. Res. **14**: 639-643.
KLEKOWSKI, E. AND N. KAZARINOVA-FUKSHANSKY, 1984 Shoot apical meristems and mutation: selective loss of disadvantageous cell genotypes. Am. J. Bot. **71**: 28-34.
KNOWLES, P., G. R. FURNIER, M. A. ALEKSIUK, AND D. J. PERRY, 1987 Significant levels of self-fertilization in natural populations of tamarack. Can. J. Bot. **65**: 1087-1091.
KOSKI, V., 1971 Embryonic lethals of *Picea abies* and *Pinus sylvestris*. Commun. Institute of Forestalia Fennica **75**: 1-30.
KUITTINEN, H. AND O. SAVOLAINEN 1992 *Picea omorika* is self-fertile but outcrossing conifer. Heredity **68**: 183-187.
KUTIL, B.L. AND CLAIRE G. WILLIAMS, 2001 Triplet repeat microsatellites shared among hard and soft pines. Journal of Heredity **92**: 327-332.

KONNERT, M., AND F. BERGMANN, 1995 The geographical distribution of genetic variation of silver fir (*Abies alba*, Pinaceae) in relation to its migration history. Plant Syst. Evol. **196**: 19-30.

LATTA, R. AND K. RITLAND, 1994 The relationship between inbreeding depression and prior inbreeding among populations of four *Mimulus* taxa. Evolution **48**: 806-817.

LANDE, R. AND D. W. SCHEMSKE, 1985 The evolution of self fertilization and inbreeding depression in plants. I. Genetic models. Evolution **39**: 24-40.

LATTA, R. G., Y. B. LINHART, D. FLECK, AND M. ELLIOT, 1998 Direct and indirect estimates of seed versus pollen movement within a population of ponderosa pine. Evolution **52**: 61-67.

LATTA, R. G. AND J. B. MITTON. 1997. A comparison of population differentiation across four classes of gene marker in limber pine (*Pinus flexilis* James). Genetics **146**: 1153-1163.

LATTA R.G. AND J. B. MITTON, 1999 Historical separation and present gene flow through a zone of secondary contact in ponderosa pine. Evolution **53**: 769-776.

LEDIG, F., V. JACOB-CERVANTES, ET AL.,1997 Recent evolution and divergence among populations of a rare Mexican endemic, Chihuahua spruce, following Holocene climatic warming. Evolution **51**: 1815-1827.

LERNER, I. M., 1954 *Genetic Homeostasis*. Oliver and Boyd, Edinburgh. 134 pp.

LINHART, Y. B., J. B. MITTON, K. B. STURGEON, AND M. L. DAVIS, 1981 Genetic variation in space and time in a population of ponderosa pine. Heredity **46**: 407-426.

LYNCH, M., 1988 The rate of polygenic mutation. Genet. Res. Camb. **51**:137-148.

LYNCH, M., 1997 Mutation accumulation in nuclear, organelle, and prokaryotic transfer RNA genes. Mol. Biol. Evol. **14**: 914-925.

LIEPELT, S., R. BIALOZYT AND B. ZIEGENHAGEN, 2002 Wind-dispersed pollen mediates postglacial gene flow among refugia. Proc. Natl. Acad. Sci. USA **99**: 14590-14594.

LINHART, Y. B., J. B. MITTON, K. B. STURGEON, AND M. L. DAVIS, 1981 Genetic variation in space and time in a population of ponderosa pine. Heredity **46**: 407-426.

LINHART, Y. B. AND J. B. MITTON, 1985 Relationships among reproduction, growth rate, and protein heterozygosity in ponderosa pine Amer. J. Bot. **72**: 181-184.

LOVELESS, M. D. AND J. L. HAMRICK, 1984 Ecological determinants of genetic structure in plant populations. Annu. Rev. Ecol. Syst. **15**: 65-95.

LYNCH, M. 1988. The rate of polygenic mutation. Genet. Res. Camb. **51**: 137-148.

MATOS, J. 1992. Evolution within the *Pinus montezumae* complex of Mexico: population subdivision, hybridization, and taxonomy. Ph.D. Thesis. Washington University, St. Louis.

MITTON, J. B., 1983 Conifers. pp. 443-472 in S. Tanksley and T. Orton (eds) *Isozymes in Plant Genetics and Breeding*, Part B. Elsevier.

MITTON, J. B., 1992 The dynamic mating systems of conifers. New Forests **6**: 197-216.

MITTON, J. B., 1997 *Selection in Natural Populations*. Oxford University Press, New York. 272 pp.

MITTON, J. B. AND K. L. DURAN, 2004 Genetic variation in piñon pine, *Pinus edulis*, associated with summer precipitation. Mol. Ecol. **13**: 1259-1264.

MITTON, J. B. AND R. M. JEFFERS, 1989 The genetic consequences of mass selection for growth rate in Engelmann spruce. Silvae Genetica **38**: 6-12.

MITTON, J. B. AND M. C. GRANT, 1984 Associations among protein heterozygosity, growth rate, and developmental homeostasis. Ann. Rev. Ecol. Syst. **15**: 479-499.

MITTON, J. B. AND B. A. PIERCE, 1980 The distribution of individual heterozygosity in natural populations. Genetics **95**: 1043-1054.

MITTON, J. B., B. R. KREISER, AND R. G. LATTA, 2000a Glacial refugia of limber pine (*Pinus flexilis* James) inferred from the population structure of mitochondrial DNA. Mol. Ecol. **9**: 91-97.

MITTON, J. B., B. R. KREISER, AND G. E. REHFELDT, 2000b Primers designed to amplify a mitochondrial *nad1* intron in ponderosa pine, *Pinus ponderosa*, limber pine, *P. flexilis*, and Scots pine, *P. sylvestris*. Theor. Appl. Genet. **101**: 1269-1272.

MITTON, J. B., Y. B. LINHART, J. L. HAMRICK, AND J. S. BECKMAN, 1977 Observations on the genetic structure and mating system of ponderosa pine in the Colorado Front Range. Theor. Appl. Genet. **7**: 5-13.

MORAN, G. F., J. C. BELL, AND A. C. MATHESON 1980 The genetic structure and levels of inbreeding in a *Pinus radiata* D. Don seed orchard. Silvae Genet. **29**:190-193.

MORGANTE, M., G. VENDRAMIN, et al., 1991 Effects of stand density on outcrossing rate in Norway spruce (*Picea abies*) populations. Can. J. Bot. **69**: 2704-2708.
MUONA, O. AND A. HARJU, 1989 Effective population sizes, genetic variability and mating system in a natural stands and seed orchards of *Pinus sylvestris*. Silvae Genet. **38**: 221-228.
MULLER, G., 1977a Cross fertilization in a conifer stand inferred from gene-markers in seeds. Silvae Genet. **26**: 223-226.
MULLER, G., 1977b Investigations of the degree of natural self- fertilization in stands of Norway spruce (*Picea abies* (L.) Darst.) and Scots pine (*Pinus sylvestris* L.). Silvae Genetica **26**: 207-217.
NEALE, D. B. AND W. T. ADAMS, 1985a Allozyme and mating system variation in balsam fir (*Abies balsamea*) across a continuous elevational transect. Can. J. Bot. **63**: 2448-2453.
NEALE, D. B. AND W. T. ADAMS, 1985b The mating system in natural and shelterwood stands of Douglas-fir. Theor. Appl. Genet. **71**: 201-207.
NEALE, D. B., K. A. MARSHALL, AND D. E. HARRY. 1991. Inheritance of chloroplast and mitochondrial DNA in incense-cedar (*Calocedrus decurrens*). Can. J. For. Res. **21**: 717-720.
NEALE, D. B., K. A. MARSHALL, AND R. R. SEDEROFF, 1989 Chloroplast and mitochondrial DNA are paternally inherited in *Sequoia sempervirens*. D. Don Endl. Proc. Natl. Acad. Sci. USA **86**: 9347-9349.
NEALE, D. B., N. C. WHEELER, AND R. W. ALLARD, 1986 Paternal inheritance of chloroplast DNA in Douglas-fir. Can. J. For. Res. **16**: 1152-1154.
O'CONNELL, L. (2003). The evolution of inbreeding in western red cedar (*Thuja plicata*: Cupressaceae). Department of Forest Sciences, Faculty of Forestry. Vancouver, BC, University of British Columbia: 162 p.
PALMER, J. D., 1985 Evolution of chloroplast and mitochondrial DNA in plants and algae. In *Molecular Evolutionary Genetics*, R. J. MacIntyre (ed.) New York: Plenum Press, pp. 131-240.
PALMER, J. D., 1990 Contrasting modes and tempos of genome evolution in land plant organelles.Trends Genet. **6**: 115-120.
PALMER, J. D., 1992 Mitochondrial DNA in plant systematics: Applications and limitations. In *Molecular Systematics of Plants*, P. S. Soltis, J. E. Soltis, and J. J. Doyle (eds). New York: Chapman & Hall, pp. 36-48.
PERRY, D. J. AND P. KNOWLES, 1990 Evidence of high self- fertilization in natural populations of eastern white cedar (*Thuja occidentalis*). Can. J. Bot. **68**: 663-668.
RICHARDS, A., 1997 *Plant breeding systems*. London, Chapman & Hall.
RICHARDSON, B. A., J. BRUNSFELD AND N. B. KLOPFENSTEIN, 2002 DNA from bird-dispersed seed and wind-disseminated pollen provides insights into postglacial colonization and population geneti structure of whitebark pine (*Pinus albicaulis*). Mol. Ecol. **11**: 215-227.
SAVOLAINEN, O. AND P. HEDRICK, 1995 Heterozygosity and fitness: no associations in Scots pine. Genetics **140**: 755-766.
SARVAS, R., 1962 Investigations on the flowering and seed crop of *Pinus silvestris*. Institute Forestalis Fennica Comm. **53**: 1-198.
SCHAAL, B. A., S. L. O'KANE AND S. H. ROGSTAD, 1991 DNA variation in plant populations. Trends Ecol. Evol. **6**: 329-33.
SCHMIDT, P. S., D. D. DUVERNELL AND W. F. EANES, 2000 Adaptive evolution of a candidate gene fro aging in *Drosophila*. Proc. Natl. Acad. Sci. USA **97**: 10861-10865.
SCHOEDER, S., 1989 Outcrossing rates and seed characteristics in damaged natural populations of *Abies alba*. Silvae Genet. **38**: 185-189.
SCHUSTER, W. S., D. L. ALLES, AND J. B. MITTON. 1989. Gene flow in limber pine: evidence from pollination phenology and genetic differentiation along an elevational transect. Amer. J. Bot. **76**: 1395-1403.
SCHUSTER, W. S. F. AND J. B. MITTON, 2000 Paternity and gene dispersal in limber pine (*Pinus flexilis* James). Heredity **84**: 348-361.
SCOTTI, I., F. MAGNI, R. FINK, W. POWELL, G. BINELLI AND P. E. HEDLEY, 2000 Microsatellite repeats are not randomly distributed within Norway spruce (*Picea abies* K.) expressed sequences. Genome **43**: 41-46.
SHAW, D. V. AND R. W. ALLARD, 1981 Analysis of mating system parameters and population structure in Douglas-fir using single- locus and multilocus methods. pp 18-22 in M. T. Conkle (ed.) *Isozymes of North American Forest Trees and Forest Insects*. USDA For. Serv. Gen. Tech. Rep. PSW-48.

SHAW, D. V. AND R. W. ALLARD, 1982 Estimation of outcrossing rates in Douglas fir using allozyme markers. Theor. Appl. Genet. **62**: 113-120.

SHEA, K. L., 1987 Effects of population structure and cone production on outcrossing rates in Engelmann spruce and subalpine fir. Evolution **41**: 124-136.

SHEN, H. -H., D. RUDIN AND D. LINDGREN, 1981 Study of pollination pattern in a scots pine seed orchard by means of isozyme analysis. Silvae Genet: **30**: 7-15.

SHEPHERD, M., M. CROSS, T.L. MAGUIRE, M.J. DIETERS, C. G. WILLIAMS AND R.J. HENRY, 2002 Transpecific microsatellites for hard pines. Theor. Appl. Genet. **104**: 819-827.

SIEGISMUND, H. AND E. KJAER, 1997 Outcrossing rates in two stands of noble fir (*Abies procera* Rehd.) in Denmark. Silvae Genet. **46**: 144-146.

SIMMONS, M. J. AND J. F. CROW, 1977 Mutations affecting fitness in Drosophila populations. Ann. Rev. Genet. **11**: 49-78.

SINCLAIR, W. T., J. D. MORMAN AND R. A. ENNOS, 1999 The postglacial history of Scots pine (*Pinus sylvestris* L.) in western Europe: evidence from mitochondrial DNA variation. Mol. Ecol. **8**: 83-88.

SKINNER, D., 1992 Ovule and embryo development, seed production and germination in orchard grown control pollinated loblolly pine (*Pinus taeda* L.) from coastal South Carolina. Department of Biology. Victoria, BC, University of Victoria: 88.

SNYDER, T. P., D. A. STEWARD, AND A. F. STRICKLER, 1985 Temporal analysis of breeding structure in jack pine (*Pinus banksiana* Lamb.). Canadian J. For. Res. **15**: 1159-1166.

SOKOL, K.A. AND C. G. WILLIAMS, 2005 Evolution of a triplet repeat in a conifer. Genome **48**: 417-426.

SORENSEN, F. C., 1969 Embryonic genetic load in coastal Douglas fir, *Pseudotsuga menziessii var. menziessii*. Am. Nat. **103**: 389-398.

SORENSEN, F. C., 1971 Estimate of self-fertility in coastal Douglas-fir from inbreeding studies. Silvae. Genet. **20**: 115-120.

SORENSEN, F. C., 1982 The role of polyembryonal vitality in the genetic system of conifers. Evolution **36**: 725-733.

SORENSEN, F. AND R. S. MILES, 1982 Inbreeding depression in height, height growth, and survival of Douglas-fir, ponderosa pine, and noble fir to 10 years of age. For. Sci. **28**: 283-292.

SPERISEN, C., U. BÜCHLER, F. GUGERLI, G. MÁTYÁS, T. GEBUREK AND G. G. VENDRAMIN, 2001 Tandem repeats in plant mitochondrial genomes: application to the analysis of population differentiation in the conifer Norway spruce. Mol. Ecol. **10**: 257-263.

STAUFFER, A. AND W. ADAMS,1993 Allozyme variation and mating system of three Douglas-fir stands in Switzerland. Silvae Genet. **42**: 254-258.

STRAUSS, S. 1986 Heterosis at allozyme loci under inbreeding and crossbreeding in *Pinus attentuata*. Genetics **113**: 115-134.

STRAUSS, S. AND W. LIBBY 1987 Allozyme heterosis in radiata pine is poorly explained by overdominance. Amer. Nat. **130**: 879-890.

STRAUSS, S., H., Y.-P. HONG AND V. D. HIPKINS, 1993 High levels of population differentiation for mitochondrial DNA haplotypes in *Pinus radiata*, *muricata*, and *attenuata*. Theor. Appl. Genet. **85**: 6065-71.

STRAUSS, S. H., D. B. NEALE, AND D. B. WAGNER, 1989. Genetics of the chloroplast in conifers. Journal of Forestry **87**: 11-7.

WAGNER DB., 1992 Nuclear, chloroplast, and mitochondrial DNA polymorphisms as biochemical markers in population genetic analyses of forest trees. New Forests **6**: 373-390.

WAGNER D. B., G. R. FURNIER, M. A. SAGHAI-MAROOF, S. M. WILLIAMS, B. P. DANCIK BP AND R. W. ALLARD, 1987 Chloroplast DNA polymorphisms in lodgepole and jack pines and their hybrids. Proc. Natl. Acad. Sci. USA **84**: 2097-200.

WILLIAMS, C.G. AND O. SAVOLAINEN, 1996 Inbreeding depression in conifers: implications for breeding strategy. Forest Science **42**: 102-117.

WILLIAMS, C.G., S. L. LADEAU, R.A. OREN AND G.G. KATUL. 2006. Modeling seed dispersal distances: implications for transgenic *Pinus taeda*. *Ecological Applications* (in press).

WRIGHT, J. W., 1976 *Introduction to forest genetics*. Academic Press, New York.

WRIGHT S. 1931., Evolution in Mendelian populations. Genetics **16**: 97-159.

WRIGHT, S., 1951 The genetical structure of populations. Ann. Eugen. **15**: 323-354.

WRIGHT, S., 1969. *Evolution and the Genetics of Populations. Volume 2. The Theory of Gene Frequencies*. University of Chicago Press, Chicago. 511 pp.

WRIGHT, S., 1977 *Evolution and Genetics of Populations. Vol. 3. Experimental Results and Evolutionary Deductions*. University of Chicago Press, Chicago. 613 pp.

ZHOU, Y., T. BUI, L. D. AUCKLAND AND C. G. WILLIAMS, 2001 Undermethylation as a source of microsatellites in large plant genomes. Genome **46:** 809-816

ZHOU Y, T. BUI, L.D. AUCKLAND AND C. G. WILLIAMS, 2002 Direct fluorescent primers are superior to M13-tailed primers for automated genotyping of *Pinus taeda* microsatellites. *Biotechniques* **32:** 46-52.

CHAPTER 10

PINES AS INVASIVE ALIENS: OUTLOOK ON TRANSGENIC PINE PLANTATIONS IN THE SOUTHERN HEMISPHERE

DAVID M. RICHARDSON

Centre for Invasion Biology, Department of Botany and Zoology, University of Stellenbosch, Stellenbosch, South Africa

REMY J. PETIT

INRA, UMR Biodiversité, Gènes & Ecosystèmes, Equipe de Génétique, Cestas Cedex, France

Abstract. Several species of pines (*Pinus* spp.; Pinaceae) are highly invasive in parts of the Southern Hemisphere, where pines are widely planted in commercial forestry plantations. Problems associated with the spread of pines from plantations have increased substantially over the past few decades. We review the current extent of the problem and the research that has been undertaken to explain different facets of these invasions, including the factors contributing to species invasiveness and the susceptibility of ecosystems to invasion. Recent interest in producing transgenic pines for wood production raises important issues when considering future scenarios for pine invasions and for sustainable commercial forestry. This chapter considers the genetic diversity in introduced versus native pine populations and then examines the potential for transgene escape from pine plantations in the Southern Hemisphere. Propagule pressure appears to play a major role in these invasions. Commercial plantations have typically introduced a large share of the species' existing genetic diversity, resulting in rapid adaptation to local conditions and favoring the spread of feral pine populations. The extent to which inherent invasiveness of transgenic pines will differ from non-transgenic pines will depend on the properties conferred by the transgenes, but differences could be substantial. Even subtle changes in species-environment interactions could affect the dynamics of pine invasions. Genetic engineering for reproductive sterility could potentially reduce invasiveness, but criteria for forest certification current prohibit the use of any genetically-modified planting, thus blocking a potentially useful avenue of intervention. Integrated programs for managing pine invasions in the Southern Hemisphere will need to give serious attention to transgenic plantation forestry.

1. INTRODUCTION

Biological invasions have emerged as one of the most pressing threats to biodiversity worldwide in the past few decades (MACK *et al.,* 2000). All types of organisms are invasive and cause a wide range of threats to biodiversity. Invasive alien plants are among the most damaging of invaders. All growth forms and most major taxonomic groupings are represented in the growing list of invasive alien plants worldwide (REJMÁNEK et al., 2005). Certain plant groups are, however, overrepresented, both because of inherent features of their biology, but also because of features that make them more likely to be widely translocated by humans. Among the most widespread and damaging invaders are species that are intentionally moved around and used in various ways by humans. With rapid developments in biotechnology, many plant species are being genetically manipulated to enhance their usefulness. Concerns have been raised regarding the implications of genetic engineering for invasiveness (WILLIAMSON 1993; DALE 1997).

Many of the world's worst plant invaders are herbs and grasses, but there are a growing number of woody plants that cause major problems as invaders, including many species of trees. Trees have many uses for humans and hundreds of species are grown outside their natural range, for forestry, shelter, as ornamentals, and for many other uses. Rapid advances in the application of biotechnology to trees in the past decade provide important prospects for enhanced productivity (WALTER et al., 1998; CAMPBELL et al., 2003; WALTER 2004). Simultaneously, important concerns have emerged, some of which are related to the relatively long rotations of forestry plantings compared to other crops (OWUSU 1999; WALTER 2001). In particular, there are fears that the aptitude conferred by transgenes might not be very durable. So far, transgenes have been limited to a few characters under the control of a major, dominant gene. As a consequence, if the gene confers some resistance to a pest or pathogen, there is a considerable risk that it will eventually be overcome by new races of the parasite or pathogen, with particularly detrimental economical consequences for long-lived species. Another possibility, which is not specific to trees, is that the transgenes could be transferred to nearby weedy populations, compromising efforts made to control them.

The phenomenon of introduced trees as important invasive alien species has emerged as a major environmental issue in the past century, especially in the last few decades. Alien tree invasions pose special problems. Many alien trees are highly beneficial in parts of the landscape, but troublesome invaders outside of areas set aside for their cultivation. The benefits of some invasive species are so widely accepted that conflicts of interest arise when control measures are advocated (DE WIT et al., 2001). Among the most widespread and damaging of invasive alien trees worldwide are taxa in the genera *Acacia* (Fabaceae), *Acer* (Aceraceae), *Ailanthus* (Simaroubaceae), *Elaeagnus* (Oleaceae), *Fraxinus* (Oleaceae), *Maesopsis* (Rhamnaceae), *Pinus* (Pinaceae), *Leptospermum* (Myrtaceae), *Melaleuca* (Myrtaceae), *Miconia* (Melastomataceae), *Mimosa* (Fabaceae), *Myrica* (Myricaceae), *Paulownia* (Fabaceae), *Pittosporum* (Pittosporaceae), *Prosopis* (Fabaceae), *Pseudotsuga* (Pinaceae), *Robinia* (Fabaceae), *Sapium* (Euphorbiaceae), *Salix* (Salicaceae), *Schinus* (Anacardiaceae) and *Tamarix* (Tamaricaeae). Most species belonging to this taxonomically disparate assemblage are useful to humans and have therefore been moved around the world to varying degrees, e.g. for erosion control, drift sand stabilization and reclamation (*Acacia, Paulownia, Tamarix*), fodder for livestock (*Prosopis*), fuel wood (*Acacia*), ornamental use (*Acer, Fraxinus, Elaeagnus,*

Leptospermum) and timber (*Pinus*). Transgenic trees have been regenerated in the past decade in at least two of the important invasive tree genera listed above – *Pinus* and *Robinia* (VAN FRANKENHUYZEN & BEARDMORE 2004). This adds further complexity to understanding and managing invasions in these genera. We need to derive principles and guidelines for the effective and objective management of alien tree invasions in the face of changing environments and forestry practices, including increased emphasis on biotechnology for Southern Hemisphere forests.

This chapter focuses on the phenomenon of pines (*Pinus* spp.) as invasive alien species in the Southern Hemisphere. We provide a short overview of the history and current dimensions of the problem and some recent attempts to understand how these invasions have arisen. We are particularly interested in exploring the implications of developments in forestry, notably the use of transgenic trees, for invasions and sustainable management of exotic plantations. We show that biotechnology has the potential to exacerbate or ameliorate problems caused by alien tree invasions.

2. BACKGROUND ON *PINUS*

The genus *Pinus* (with roughly 111 species) is arguably the most ecologically significant and economically important group of trees in the world (RICHARDSON 1998). Within their natural range in the Northern Hemisphere (only one species, *P. merkusii*, has a natural range extending marginally into the Southern Hemisphere), pines play a major role in net primary production, forest structure, biogeochemical processes, hydrologic flow, fire regimes, and in supplying habitat and food for animals. Huge areas of natural forest in the Northern Hemisphere are dominated by pines, including taiga, temperate and tropical montane forests, and tropical and subtropical savannas. Natural pine forests have been used by humans for a wide range of uses over millennia. Large-scale planting of pines started in the second half of the 19th century in Europe, but sustained, large-scale forestry only became widespread in Europe in the early 20th century, and expanded to other parts of the world in the second half of the 20th century. Some pines have proved highly successful for use in plantations well outside their natural ranges where coniferous species suitable for large-scale production of fibers and solid wood products are scarce.

The main pine species planted in the tropics and subtropics are *P. caribaea*, *P. elliottii*, *P. kesiya*, *P. oocarpa*, *P. patula*, *P. pinaster*, *P. radiata* and *P. taeda*. Reasons for the widespread use of pines in exotic forestry plantations include their simple structural design with straight trunks and an almost geometrical branching habitat that makes them ideal for timber production. Pines are also very suitable for cultivation in commercial plantations because their leaf area index (LAI) can be doubled or trebled through silviculture, greatly enhancing their productivity. They grow faster than many other potential species, and are easier to manage in plantations. Their seeds are easy to collect, store and germinate; and they are ideally suited for planting in grasslands or scrublands (marginal forest lands) where most afforestation take place. For these reasons, pines form a very substantial part of the exotic forestry enterprises in Southern Hemisphere countries, usually alongside eucalypts (SOHNGEN *et al.,*1997).

Several features of pines make them an intriguing genus to explore theoretical and practical aspects of invasion ecology (REJMÁNEK & RICHARDSON 1996; RICHARDSON & HIGGINS 1998; RICHARDSON & REJMÁNEK 2004, and references therein). Firstly, many pine species have features that favor their dissemination by humans. This has ensured the sustained widespread translocation of many species to nearly all parts of the world, affording many species ample opportunities to sample a wide range of habitats. Secondly, the diversity of life-history traits, especially those relating to reproduction and dispersal, allows us to explore links between life-history syndromes and invasiveness. Third, genetic transformation clearly has the potential to alter key facets of pine biology that are implicated in their invasiveness. Fourth, the paleoecology of pines has been well studied, providing us with a valuable historical template for understanding the ecological determinants of range shifts.

Not surprisingly, pines have been an important focus for genetic transformation. The genus *Pinus* is second only to *Populus* (a model for transgenic tree research) in the number of permit applications for confined field releases (VAN FRANKENHUYZEN AND BEARDMORE 2004). The targeted species include *P. pinaster*, *P. radiata*, *P. strobus*, *P. sylvestris*, and *P. taeda*. The list will grow as techniques to produce transgenic conifers improve. Although there are still no commercial plantations, industry permits are accounting for an increased fraction of the total number of applications (about half of permits at the end of 2003; VAN FRANKENHUYZEN AND BEARDMORE 2004). The company GenFor hopes to have transgenic *P. radiata* ready for commercial plantation in Chile by the year 2008 (http://www.wrm.org.uy/bulletin/88/south.html, 2004). According to its proponents, recent achievements such as the production of *P. radiata* resistant to insects have the potential to mitigate serious risks to forestry, using an environmentally sustainable technology (GRACE *et al.,* 2005). However, studies of risk assessment are still in their infancy. In particular, the question of the release of transgenic pines into commercial plantations has not yet been discussed in connection with the existence of invasive pines. The possibility that transgenic pine plantations could trigger new invasions through the escape of seeds or could mate with nearby preexisting weedy pine populations, conferring them new genetically engineered attributes, represent important ecological issues that need to be addressed.

3. PINE INVASIONS IN THE SOUTHERN HEMISPHERE – SPECIES, REGIONS, HABITATS

Table 1 summarizes information on species known to be naturalized and invasive in the Southern Hemisphere. Of the 23 species listed, 18 are known to be invasive; five species are only known to be naturalized. Regions where pines are known to be invasive and/or naturalized are: Argentina (4/4); Australia (8/3); Brazil (3/0); Chile (5/0); Madagascar (1/0); Malawi (1/0); New Caledonia (1/0); New Zealand (12/3); Reunion (0/1); South Africa (8/2); and Uruguay (1/0) (numbers in brackets show

Table 1. List of *Pinus* taxa known to be naturalized or invasive (bold) in the Southern Hemisphere (updated from RICHARDSON & REJMÁNEK 2004). Definitions for naturalized and invasive follow RICHARDSON et al. (2000) and PYŠEK et al. (2004).

Species	Country/region where the species is known to be invasive (bold) or naturalized
Pinus banksiana	**New Zealand**
P. brutia	Australia (WA)
P. canariensis	Australia (WA; SA); **South Africa**
P. caribaea	**Australia** (WA; SA); **New Caledonia**
P. contorta	Argentina; **Australia** (NSW); **Chile**; **New Zealand**
P. elliottii	**Argentina**; **Australia** (NSW); **Brazil**; New Zealand; South Africa
P. halepensis	**Australia** (SA; Vic); **New Zealand**; **South Africa**
P. jeffreyi	Australia
P. kesiya	**Brazil**; South Africa
P. monticola	Argentina
P. mugo	**New Zealand**
P. muricata	**New Zealand**
P. nigra	**Australia** (NSW; Vic; SA); **New Zealand**
P. patula	**Madagascar**; **Malawi**; **New Zealand**; **South Africa**
P. pinaster	**Australia** (SA; Vic; NSW; TAS); **Chile**; **New Zealand**; Reunion; **South Africa**; **Uruguay**
P. pinea	Australia (NSW); **South Africa**
P. ponderosa	**Argentina**; **Australia** (SA); **Chile**; **New Zealand**
P. radiata	**Australia** (WA; SA; Qld; NSW; Vic; TAS); **Chile**; **New Zealand**; **South Africa**
P. roxburghii	South Africa
P. strobus	**New Zealand**
P. sylvestris	Argentina; **Chile**; **New Zealand**
P. taeda	**Argentina**; **Australia** (NSW, Qld); **Brazil**; New Zealand; **South Africa**
P. uncinata	New Zealand

numbers of invasive and naturalized species; Table 1; see also Richardson & Rejmánek 2004). The species that are invasive/ naturalized in the most countries/regions are *P. pinaster* (6 countries), *P. elliottii*, *P. patula*, and *P. taeda* (5), *P. halepensis* and *P. radiata* (4), and *P. contorta* and *P. ponderosa*. This list contains the most widely planted species. Among the major plantations species, only *P. caribaea* (2 countries), *P. kesiya* (2) and *P. oocarpa* (0) are not among the most widespread invaders. This is at least partly due to the limited extent and/or shorter history of plantings of these species. *Pinus pinaster* and *P. halepensis*, two Old World Mediterranean species, are both over-represented as naturalized/invasive species compared to their current plantings, but both have been very widely planted in the Southern Hemisphere in the past two centuries and have been afforded many opportunities to invade. RICHARDSON AND HIGGINS (1998) summarize the relative importance of different species as invaders in the countries where this has been assessed in most detail, namely Australia, New Zealand and South Africa.

4. DETERMINANTS OF PINE INVASIONS

We focus on a few critical aspects that seem particularly relevant to the issue of pine invasiveness. Commercial forestry in the Southern Hemisphere can no longer ignore the risks of invasions incurred by the deployment of large plantations, even more so when based on transgenic material. First, the deployment of transgenic pine populations could in principle contribute to new invasions in the Southern Hemisphere, with consequences potentially as severe as those resulting from the use of conventional plantations in the past. Second, transgenic pine populations could in principle ruin efforts under way to control already existing weedy pine populations, by conferring them with new aptitudes favoring invasiveness.

Figure 1. The naturalization-invasion continuum conceptualizing the various barriers that an introduced plant (whether transgenic or not) must overcome in a new environment (adapted from RICHARDSON et al., 2000).

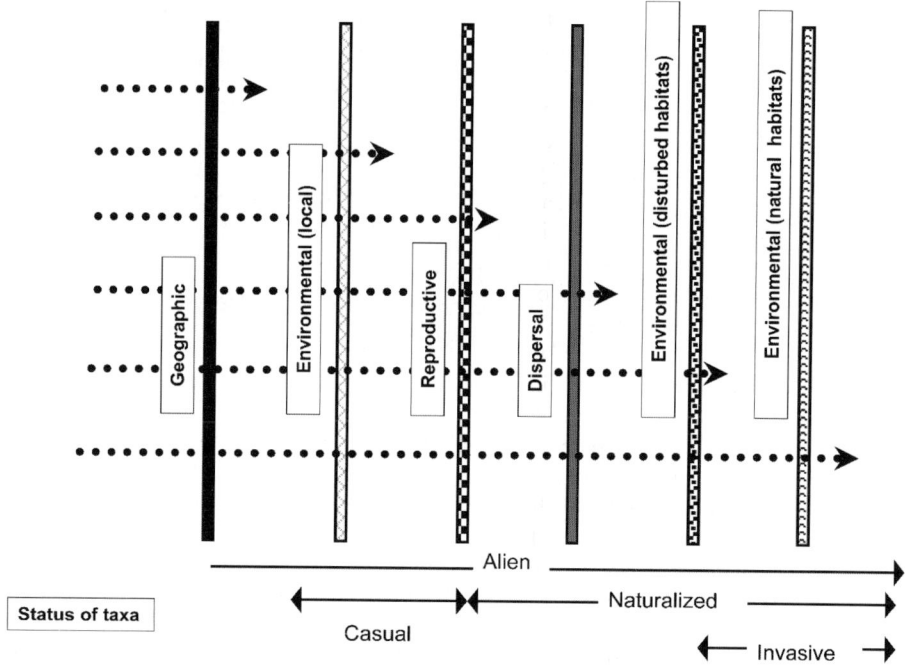

The conceptualization of the "naturalization-invasion continuum" (Figure 1) can serve as a basis for discussion. We consider the main barriers to invasion and phases through which species must pass to become a casual alien, naturalized, and invasive - first in disturbed habitats, then in natural habitats.

Almost all pine species have enjoyed widespread dissemination and planting as aliens, but a few species now dominate in exotic forestry (see above). For the Southern Hemisphere, there is a positive correlation between the extent of use in forestry and the extent of naturalization and invasion (RICHARDSON et al., 2004), pointing to the important role of propagule pressure as a driving force in pine invasions. This effect is observed not only at the global and national scales, but also at the regional scale (ROUGET AND RICHARDSON 2003). Clearly, the greater the number of propagules of a pine species that are available (a function of the number of adult trees [or total plantation size] and the time they have been present), the better the chance of establishment, naturalization, and invasion. Prolific seed production is a pivotal issue in pine invasion ecology.

In the early years of exotic forestry in the Southern Hemisphere, the absence of appropriate mycorrhizal symbionts was a major barrier to establishment and therefore obviously naturalization and invasion. This barrier was rapidly overcome through the introduction and dissemination of fungi.

Traits associated with reproduction and dispersal separate highly invasive from less invasive and non-invasive pines. REJMÁNEK AND RICHARDSON (1996) found that small seed mass (<50 mg), short juvenile period (<10 yrs), and short intervals between large seed crops (1 to 4 yrs) consistently separated invasive from non-invasive pines. A discriminant function using these three traits provides one of the most robust screening criteria for assessing invasiveness in plants. The syndrome of life-history traits associated with invasiveness (small seeds, short juvenile periods and short intervals between large seed crops) is also associated with the fastest-growing species, many of them among the most important species for commercial forestry (and hence the targets of genetic engineering efforts). For pines and indeed all woody plants, these traits seem important for allowing the species to disperse passively over long distances (inherent vagility and innate capacity for rapid population growth; HIGGINS AND RICHARDSON 1999), win in competition against other plants, and to survive (and often proliferate) under local disturbance regimes. The roles of two of the traits are clear. Short juvenile period and short interval between large seed crops translate into early and consistent reproduction. Both factors are clearly advantageous in frequently disturbed habitats. Small mean seed mass is associated with several potentially important phenomena: larger numbers of seeds produced, better wind dispersal, high initial germinability, and shorter chilling period needed to overcome dormancy (references in RICHARDSON AND REJMÁNEK 2004). Short juvenile period may be related to fast growth in general, which may be related to high leaf-area ratio (LAR = leaf area/total biomass) exhibited by at least some of the invasive pine species exhibit (see RICHARDSON AND REJMÁNEK 2004).

The above factors equip many widely planted pines, whether transgenic or not, with the inherent ability to invade in a wide range of sites. Several studies have demonstrated the importance of biotic resistance as a factor limiting seedling establishment and thus invasion potential. In a global review of 53 cases within and outside the natural range of *Pinus* where pines had extended their ranges and/or increased their densities, interactions between pine seedlings and resident biota was found to be vital determinant of invasibility (RICHARDSON AND BOND 1991). Numerous studies have documented invasion events at different scales. For example in South African fynbos, RICHARDSON (1988) described the role of fire in structuring invasive populations of *P. halepensis* in South African fynbos, and RICHARDSON AND BROWN (1986) documented fire and other landscape-scale processes as drivers of *P. radiata* invasion. RICHARDSON AND COWLING (1992) conceptualized the fire-driven dynamics of fynbos ecosystems that define the invasion window for pines. Working in a highly fragmented system in fynbos, ROUGET *et al.* (2001) explored the complex interplay of disturbance and environmental factors in structuring *P. halepensis* invasions. Pine invasions in Southern Hemisphere systems without regular fire are usually mediated by interplays between biotic and abiotic factors, as shown for *P. contorta* invasions in New Zealand (LEDGARD 2001) and several conifer species in Patagonia (SIMBERLOFF *et al.*, 2002).

Recent modeling studies have revealed many opportunities for improving our understanding of the complex processes and interactions that mediate pine invasions. For example, HIGGINS AND RICHARDSON (1998) modeled interactions between life-history traits, environmental features, and disturbance levels for pines. They showed that interactions between these factors were at least as important as the main effects. The implication is that determinants of invasion at scales smaller than landscapes are hugely complex. Outcomes (e.g., invade or not invade) can switch with subtle changes to any of the implicated factors. In general, a genetically-modified pine species will possess the same general attributes that make the species more or less prone to invasions. However, the transgene(s) will generally have altered aspects of its life history and reproductive biology (e.g. seed production, performance under local conditions) in more or less subtle ways. These should be examined on a case-by-case basis for their effects on invasiveness. Conceptually, however, exactly the same framework as for a non-transgenic population (see above) can be used.

5. GENETIC DIVERSITY IN INTRODUCED VERSUS NATIVE PINE POPULATIONS

Although typically not listed as a determinant of invasions, genetic diversity could condition local adaptation under new conditions and hence invasiveness. Alternatively, genetic diversity could be preserved in most invasions as a mere byproduct of propagule pressure (cf. the positive correlation between the extent of use of a species in forestry and the extent of naturalization and invasion discussed previously). Regardless of whether it is a cause or a consequence of invasions, it seems appropriate to monitor genetic diversity in exotic plantations (including, in the future, in transgenic ones) and in feral populations. Such information is relevant to programs aiming at controlling invasive pine populations, in particular programs of biological control relying on introduced pests and diseases. Resistances or tolerances are more likely to be already present or to rapidly evolve in host tree populations that have a broad genetic basis, thereby rendering eradication or control measures more difficult as naturalized populations adapt new countermeasures.

A recent review concluded that biological invasions of trees, even in remote and isolated places, have resulted in surprisingly little loss of genetic diversity (PETIT et al., 2004). Similar conclusions have been reached for other organisms (Wares et al., 2005). Unfortunately, although *Pinus* is probably genetically the most thoroughly investigated tree genus (e.g. LEDIG 1998), we know of no population genetic studies of feral populations of invasive pines in the Southern Hemisphere. However, there is some work on the genetic differences between native and introduced pine populations based in the latter case on material sampled in plantations. Three examples are provided here. A fourth example involving an associated pathogenic fungus is provided to illustrate that pests too can travel without experiencing dramatic bottlenecks.

The first example is based on a comparison between an elite domesticated population of *Pinus taeda* in Zimbabwe and five indigenous populations from the U.S., using a set of 18 microsatellite loci (WILLIAMS et al., 2000). No loss of genetic diversity is apparent (Figure 2). Further analyses indicated that the Zimbabwe introduced population was an admixture of eastern and western U.S. populations. Historical records from early 20th century of multiple introductions of U.S. germplasm and extensive plantings throughout southern Africa further support this finding. Introgression with *Pinus elliottii*, also introduced in the region and closely related to *P. taeda*, might have further accentuated this trend, given the unusual reproductive phenology of the two species in their new African environment.

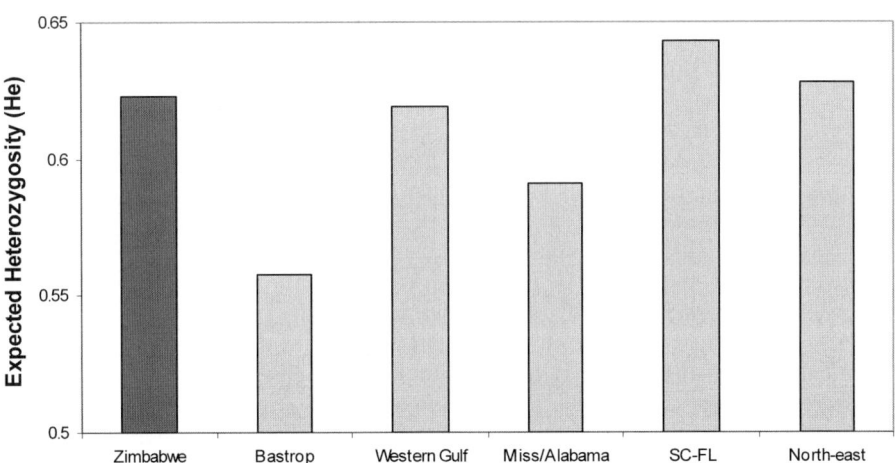

Figure 2. Comparison of gene diversity H between an elite domesticated population of Pinus taeda in Zimbabwe and five indigenous populations from the U.S., using a set of 18 microsatellite loci.[source of data: WILLIAMS et al. 2000].

The second example is based on a study of allozyme diversity of *P. radiata* in Australia by MORAN AND BELL (1987). These authors have shown that a substantial portion of genetic diversity had been captured in local breeding programs of this pine: within-population diversity is $H = 0.091$, compared to 0.098 in the (restricted) natural range in North America. Admixture was also inferred, with two of the five native *P. radiata* populations (Monterey and Año Nuevo) representing the major sources of the original introductions.

A third example is provided by an allozyme study of *Pinus caribaea* in its native range and in two exotic plantations growing in Australia and China (ZHENG AND ENNOS 1999). The Australian population had average gene diversity comparable to that of other populations from the same variety (var. *bahamensis*), whereas the Chinese population had 30% more diversity than other populations of the *caribaea* variety to which it belongs. Since the species, introduced in China in the 1970s, had significant heterozygote deficit, it is likely that it represents a mixture of germplasm from different islands in the native range.

The same trends could be observed in a pine pathogen inadvertently introduced in the Southern Hemisphere. The asexual fungus *Sphaeropsis sapinea*, an endophyte and latent pathogen of pines, is assumed to have been introduced to the Southern Hemisphere along with its host. BURGESS *et al.* (2001) compared the genetic diversity of native and introduced population of this fungus. Some of the introduced populations are genetically highly variable, often more so than in the native part of the range of its host tree.

The findings of maintenance, if not increased, of genetic diversity during the early stage of pine invasions is not surprising in view of the patterns observed during natural spread (e.g. postglacial migration) of pines. For instance, the Canary Island pine (*Pinus canariensis*) retains levels of genetic diversity equivalent to those found in Mediterranean continental species, *P. pinaster* and *P. halepensis*, even though it is considered to result from colonization by a single continental source (GÓMEZ *et al.*, 2003). On the contrary, *Pinus resinosa*, a widespread pine species in northeastern North America, is one of the most genetically depauperate tree species. However, during its postglacial colonization, diversity increased rather than decreased. This was interpreted as a consequence of the admixture of populations originating from different refugia, as indicated by a study of cpDNA variation (WALTER AND EPPERSON 2001). Finally, an extensive allozyme survey of lodgepole pine, *Pinus contorta*, across its range in western North America by WHEELER AND GURIES (1982) was reanalyzed by CWYNAR AND MACDONALD (1987) in the historical context of postglacial colonization of the species. There was no evidence for decreased gene diversity H, but the number of alleles did decreased somewhat during the extensive northward spread, whereas dispersal ability increased through the decrease of seed mass and the concomitant increase of wing loading. The reduction in allelic richness might be related to the relatively narrow range of the species along the Rocky Mountains, precluding admixture of populations. This study illustrates that even when successive founder events took place along a narrow latitudinal transect, much diversity can be preserved. Furthermore, it suggests that invasive populations might be selected for increased dispersal ability.

It appears from this list of case studies that, in many respects, human-mediated invasions do not differ much from natural ones and that the widespread expectation of reduced diversity following translocation into places remote from the native range is not borne out. The causes of this maintenance of diversity probably include admixture and the high propagule pressure associated with most pine invasions (see above). Whatever the causes, high genetic diversity should facilitate rapid adaptation of naturalized pines to their new habitats. In fact, pines are considered among the most genetically variable of all species, as revealed by measures of quantitative genetic variation (CORNELIUS 1994). As a consequence, local adaptation is to be expected in most cases, even though only few generations have elapsed since the first introduction. Results from provenance tests in the introduced range confirm this expectation whenever local material is contrasted with material sampled directly in the native range. Such studies clearly suggest that rapid local adaptation has taken place, with populations introduced a few generations ago generally outcompeting newly introduced provenances from the native range. This is the case for *Pinus radiata* in northern Spain (ESPINEL et al., 1995) and in New Zealand (BURDON AND BANNISTER 1973). These findings imply that biological control of weedy pines in the Southern Hemisphere could be difficult, as the abundant store of diversity should facilitate the evolution of tolerance or resistance in these populations. However, the first few generations of evolution in the new environment could have resulted in a decrease of allocation of resources to defense in pines, due to relaxed selection in the absence of at least some of their natural enemies, as shown for other introduced tree species (e.g. SIEMANN AND ROGERS 2001). This could have contributed to the initial outstanding productivity of pine plantations in the Southern Hemisphere. Nevertheless, the longevity of trees implies that little of the original diversity would be completely lost, so that sources of resistances and tolerances are likely to be still present, even if at reduced frequency. Hence, naturalized populations will probably represent a moving (i.e. evolving) target for most type of control measures. Introduction of additional genetic material from the native range should only exacerbate the problem.

Interestingly, with current genetic engineering techniques, the use of somatic embryogenesis means that few genotypes are used for genetic transformation. Hence, future transgenic plantations might have a reduced genetic basis, compared to past commercial plantations. This might be interpreted as an advantage of this form of forestry, by reducing the risk to introduce long-term viable (i.e. adaptable) invasive pine populations. However, in many cases, feral populations already exist near plantation areas so the introduction of genetically modified trees will only add to the existing and substantial genetic diversity.

6. POTENTIAL FOR TRANSGENE ESCAPE FROM PINE PLANTATIONS

The escape of transgenes from commercial plantations could further broaden the genetic basis of feral naturalized populations. For instance, herbicides or pest resistance genes could benefit not only target commercial plantations but also weedy populations of the same species or of a related species through hybridization and introgression. Even in the absence of invasive populations in the vicinity of a commercial transgenic plantation, an invasion could be initiated *de novo*, through seed dispersal rather than through pollen dispersal. Such new invasive populations would then include the transgene(s) present in the plantations. To fully evaluate the risks of dissemination of transgenes, information on gene flow through seeds and through pollen is needed. Below, we briefly review the relevant literature on gene flow for *Pinus* (see also MITTON AND WILLIAMS, this volume, and KATUL et al., this volume).

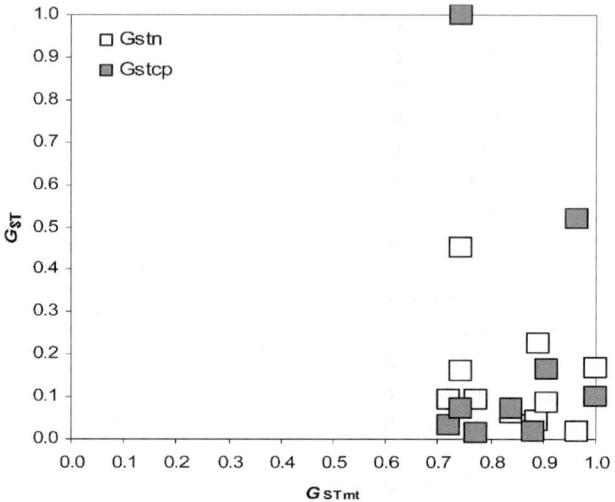

Figure 3. Comparison of genetic subdivision of diversity at maternally-inherited markers (mtDNA) and paternally- (cpDNA) or biparentally- (nuclear) inherited markers in Pinus. The species involved are the following: Pinus albicaulis, P. chiapensis, P. contorta, P. densata, P. flexilis, P. muricata, P. parviflora, P. pinaster, P. ponderosa, P. radiata, P. sylvestris. The much greater differentiation at mtDNA (abscissa) than at nuclear DNA markers (ordinate) indicates that pollen flow is more efficient than seed flow in pines, although both can play important roles for the escape of transgenes. Data extracted from PETIT et al., (2005).

Pines have as much allozyme diversity within their populations as other long-lived woody perennials (H_{es} = 0.14), and much more than plants with other life forms (HAMRICK et al., 1992). The proportion of genetic diversity distributed among populations is particularly low (G_{ST} = 0.065; N = 93 studies), pointing to efficient gene flow in most pine species. To examine whether gene flow takes place mostly by seeds or by pollen, measures of subdivision of diversity based on differentially inherited markers are particularly useful. In *Pinus*, chloroplast DNA (cpDNA) is typically paternally inherited and mitochondrial DNA (mtDNA) maternally inherited (WAGNER et al., 1989, 1991). Cases of paternal leakage of mitochondria have however been reported in the genus (WAGNER et al., 1991). Maternal inheritance implies dispersal through seeds only, whereas paternal (and biparental) inheritance implies dispersal through both pollen and seeds. Contrasting the organization of diversity based on maternally-inherited markers (mtDNA) and paternally-inherited (cpDNA) or biparentally-inherited ones (nuclear markers such as allozymes) should give an idea of the relative importance of gene flow through seeds and pollen. We extracted the information concerning *Pinus* from a recent review of the population genetic literature (PETIT *et al.*, 2005). Only species for which data on mtDNA variation was available were included (11; see Figure 3). Average G_{ST} is 0.85 for mtDNA, 0.13 for nuclear markers and 0.22 for cpDNA. By relying on equilibrium models of gene flow, these mean estimates suggest that the long-term historical ratio of pollen to seed gene flow is ~36 in pines (i.e. pollen flow would be between one and two orders of magnitude greater than seed flow). In plants, this ratio amounts to ~17, with great variation across species (PETIT *et al.*, 2005).

Hence, in principle, risks of transgene escapes in the wild, from pine plantations into nearby feral populations, should mainly take place through pollen movement. The high degree of fixation at maternally-inherited (mtDNA) markers suggests that seed flow between established populations is much slower and comparatively negligible. However, in the absence of pre-existing feral pine populations, dispersal by seeds could still play a role, as invasion largely depends on rare long-distance seed dispersal events (see PETIT *et al.*, 2004 for a review) and is largely uncoupled from average levels of gene flow. As a matter of fact, WILLIAMS *et al.* (2006) have recently shown that in *P. taeda*, the probability of long-distance dispersal of transgenic seeds at distances exceeding 1 km approaches 100%, and mention distances beyond 30 km. Notwithstanding the possibility of transgene escape through seeds, it is clear that pines have already become invasive in many regions of the Southern Hemisphere, so dissemination by pollen is the greatest concern. There is an important literature on long-distance pollen flow in pines as a consequence of the concern for genetic pollination of seed orchards established in the frame of the numerous pine breeding programs (e.g. BRIDGWATER *et al.*, 1998; PLOMION *et al.*, 2001; see also KATUL *et al.*, in *this* volume). The general finding is that, despite the strong efforts to limit contamination of seed orchards (which directly compromise the breeders' efforts), such as the establishment of pollen dilution zones or supplemental pollination, pollen flow from outside remains important, typically representing 20-60% of the mating events.

In a landscape comprising both commercial (transgenic) pine populations and feral (naturalized) ones, the risk of transgene escape will depend on the relative size of the two populations and the fitness provided by the transgene. Simulations indicate that increased size of the source population will facilitate transgene escape and that rare long-distance successful pollen flow events should play a major role in spreading transgenes (reviewed in KATUL *et al.*, this volume).

Hence, evaluating maximum or long-distance dispersal distances is probably more important than estimating the extent of local neighborhood dispersal. A growing literature indicates that (wind-pollinated) forest trees can pollinate relatives distant from several kilometers and probably even hundreds of kilometers (see KATUL *et al.*, this volume). Indeed, a portion of the pollen cloud can be caught in updrafts, thereby potentially traveling to considerable distances, without necessarily losing its viability (DI GIOVANNI *et al.*, 1996). Moreover, compared to neutral expectations, positive selection of the transgene could further boost the spread of a favorable allele by one or two orders of magnitude, even when selection coefficients are quite low (MORJAN AND RIESEBERG 2004; RIESEBERG AND BURKE 2001; SLATKIN 1976, 1987), although this effect could be limited in time. Finally, commercialized transgenes have been so far universally dominant rather than recessive (ELLSTRAND 2003). Dominance implies that first generation hybrids between plantation and feral trees will express the gene, enabling even faster spread of advantageous transgenes, although it should also equally likely slow the spread of genes that lower fitness in wild populations (e.g. modified, weakened lignin) (STRAUSS AND DIFAZIO 2004 for poplar; WILLIAMS AND DAVIS 2005 for *P. taeda*).

In view of the much lower level of gene flow through seeds and of the maternal inheritance of cpDNA in most angiosperms, there has been much excitement about genetically modified cpDNA (e.g. DANIELL 2002). This is clearly not an interesting option in conifers, given that cpDNA is paternally-inherited in this group. Moreover, we have seen that long distance seed dispersal is also taking place in pines, so even the maternally-inherited mtDNA is not an option for biocontainment of transgenes. More generally, LU (2003) has argued that allocating more resources to develop sophisticated biotechnology tools for transgenes containment is not the best strategy. Instead, he suggests that the ecological impact of transgenes be assessed more extensively, given that they will inevitably spread in the environment.

7. MANAGEMENT ISSUES

Systematic control programs for invasive pines have evolved over the past few decades in South Africa and New Zealand. In South Africa, most work has been done in fire-prone fynbos vegetation, where control operations involve the integration of large-scale felling of trees and the application of prescribed fire. Given the regular occurrence of intense fires in these systems, the crucial role that fire plays in the regeneration dynamics and invasion ecology of serotinous pines in fynbos, and the large areas and inaccessible nature of the invaded areas, this is the only viable option for large-scale control. Despite the considerable expense of such operations, the major impacts of the invading pines, notably on scarce water resources (GÖRGENS AND VAN WILGEN 2004), makes these measures cost-effective (see also VAN WILGEN *et al.*, 1998). Similar, though less intensive, approaches are followed in other parts of South Africa. In the other pine invasion hotspot of the world, New Zealand, fire is not as widely used as a major part of control operations, and the integrated control strategy involves mechanical control and herbicide application (LEDGARD 2001). Approaches such as the above tackle the symptoms of the problem and are necessary to mitigate the harmful effects of the invasions in the short term.

In both regions some progress has been made towards implementing policies and approaches aimed at reducing the problem in the long term. For example, in South Africa recent legislation allows for commercially important pines and other trees to be placed in a special category of invaders ("category 2" in the "CARA list" – a list published as part of the regulations of the Conservation of Agricultural Resources Act; Act 43 of 1983 – http://www.nda.agric.za/docs/Act43/Eng.htm). Species in this category may be grown only in demarcated areas, and landowners must control spread beyond such demarcated areas (the 'polluter pays' principle). There has been widespread buy-in to this approach by forestry companies and other landowners, although some issues relating to seed pollution from plantations remain contentious. Some progress has been made towards spatially-explicit risk assessment at a national scale as a basis for objective demarcation of areas suitable for plantations (ROUGET et al., 2002). In New Zealand, guidelines based on simple landscape-ecology principles are provided to forestry companies and other landowners to reduce the incidence of pine invasions. These include recommendations relating to plantation design (e.g. orientation in relation to prevailing wind), species composition, and optimum land management around plantations to reduce the incidence of invasions (LEDGARD AND LANGER 1999).

Prolific seed production is a key driver of pine invasions, and the need for reducing the output of seeds in plantations is widely recognized. Biological control using seed-attacking insects has been considered as a promising long-term control option. Introduced seed-attacking insects have been successfully applied against other invasive alien trees in South Africa (notably for several Australian *Acacia* species; ZIMMERMANN et al., 2004). Exploratory work conducted in South Africa has revealed that the most promising seed attackers for invasive pines are possibly vectors for diseases such as pine pitch canker (MORAN et al., 2000). Risks of devastating outbreaks of pitch canker in commercial plantations are prohibitive, and research on bio-control options for invasive pines in South Africa has been suspended. Similar concerns have been raised in New Zealand (BROCKERHOFF AND KAY 1998). For these reasons and for others discussed above it appears that bio-control using imported insects holds little promise for playing a major role in dealing with the problem of pine invasions.

What are the implications of potential use of transgenic pine plantations for the management of pine invasions in the Southern Hemisphere? Firstly, research on pine invasions in the past few decades has shown that subtle changes in species-environment interactions can result in altered invasion dynamics. Various facets of global change, notably changing land use and altered climates, are likely to have marked effects on pine invasions in the next few decades. It is impossible to predict the extent to which transgenic pines will differ from non-transgenic conspecifics in their ability to negotiate the various barriers to invasion as conceptualized in Figure 1. The potentially lower level of diversity of transgenic plantations is unlikely to reduce invasiveness since transgenic pines will usually be planted in regions with large non-transgenic plantations. Differences in inherent invasiveness will depend on the property (or properties) conferred by the transgene(s), and these can be quite subtle, as pointed it out by WILLIAMS AND DAVIS (2005). Were the transgenic trees markedly more pest- or herbicide resistant, this could lead to substantial changes, but this is probably unlikely. More effort is likely to be directed towards producing trees with superior wood properties, and the spin-offs of such manipulations for invasiveness are difficult to predict. On the other hand, genetic engineering of sterility has important implications for pine invasiveness. Research on genetically engineered reproductive sterility is driven by the need to contain transgenes inserted into modified trees before permission will be granted for their commercial use (STRAUSS et al., 1995). The advantages of such a strategy could be to decrease transgene escape through pollen and/or seeds. Propagule pressure is hugely important in driving biological invasions (see above). If feral populations are already present, then reducing seed and especially pollen production could strongly limit transgene escape into nearby feral populations. The use of sterility/fitness reduction genes flanking the gene that provides a positive selection benefit effect could further facilitate transgene containment. However, biological invasions are largely controlled by rare events (TRAKHTENBROT et al., 2005), and total sterility of every single individual in large plantations is highly unlikely even if the stability of transgenes is generally very high (STRAUSS et al.,2004). Hence, whatever new containment techniques, working on the ecological impact of the transgenes will remain a necessity, as pointed out above, since even a single case of transgene escape could trigger a new invasion or could infiltrate already present invasive pines populations, thereby rendering their control potentially more difficult (KAREIVA et et al.,1996). Another important stumbling block in the path of genetically-engineered sterile trees in commercial forestry lies with organizations such as the FORESTRY STEWARDSHIP COUNCIL (FSC). FSC certification of commercial forests aims to promote sustainable forest practices. Although the FSC-certification criteria include stipulations relating to invasive species (see STRAUSS et al., 2001), the FSC expressly forbids any use of genetically-modified species. This is blocking a potentially valuable avenue of intervention.

It should be noted that invasion-related risks associated with transgenic pines are different for those pines that are already highly invasive and those that have not shown invasiveness yet (see Table 1 for a description of the current situation). For species without widespread invasive populations, the combined use of sterile material and the low genetic diversity due to the utilization of a few transformed genotypes could reduce the risks of invasiveness through reduced propagule pressure and reduced adaptability. Avoiding future invasions is easier than fighting against existing ones and this solution is certainly an improvement over the current situation (i.e. the use of genetically highly diverse and fertile material in commercial plantations). On the other hand, for those species that are already invasive, and thus with a large receptive population, it is more important to ensure that transgenes do not confer enhanced fitness (and ideally have reduced fitness), although reduced risk of escape, through sterility genes, would be a useful complementary precaution.

Clearly, the positive and negative effects of genetically engineering pines on their invasiveness need to be brought to the table when the issues surrounding GMOs and forest certification are debated (COVENTRY 2001). The presence of pine production plantations raises just as many environmental problems in the Southern Hemisphere (where feral weedy pine populations have become increasingly abundant) as in the Northern Hemisphere (where native pine germplasm exists that ought to be protected). The effective management of weedy pines is set to remain a challenge to foresters.

Acknowledgements: D. Richardson acknowledges financial support from the DST/NRF Centre of Excellence for Invasion Biology. R. Petit was supported by the Bureau des Ressources Génétiques. Helpful comments from Steve Strauss on an earlier version of the manuscript are gratefully acknowledged.

8. REFERENCES

ARONEN, T. S., T. O. NIKKANEN, and H. M. HÄGGMAN, 2003 The production of transgenic Scots pine (*Pinus sylvestris* L.) via the application of transformed pollen in controlled crossings. Trans. Res. **12**: 375-378.

BRIDGWATER, F. E., D. L. BRAMLETT, T. D. BYRAM, and W. J. LOWE, 1998 Controlled mass pollination in loblolly pine to increase genetic gains. The For. Chron. **74**: 185-189.

BURDON, R. D., and M. H. BANNISTER, 1973 Provenances of *Pinus radiata*: their early performance and silvicultural potential. NZ J. For. **18**: 217-232.

BROCKERHOFF, E.G., and M. KAY, 1998 Prospects and risks of biological control of wilding *Pinus contorta* in New Zealand. Proceedings of the 51st New Zealand Plant Protection Conference, pp. 216-223. New Zealand Plant Protection Society Incorporated.

BURGESS, T., B.D. WINGFIELD, and M.J. WINGFIELD, 2001 Comparison of genotypic diversity in native and introduced population of *Sphaeropsis sapinea* isolated from *Pinus radiata*. Mycol. Res. **105**: 1331-1339.

CAMPBELL, M.M., A.M. BRUNNER, H.M. JONES, and S.H. STRAUSS, 2003 Forestry's fertile crescent: the application of biotechnology to forest trees. Plant Biotech. J. **1**: 141-154.

CORNELIUS, J., 1994 Heritabilities and additive genetic coefficients of variation in forest trees. Can. J. For. Res. **24**: 372-379.

COVENTRY, P., 2001 *Forest certification and genetically engineered trees: Will the two ever be compatible?* OFI Occasional Papers No. 53. Oxford Forestry Institute, Oxford University.

CWYNAR, L.C., and G.M. MACDONALD, 1987 Geographical variation of lodgpole pine in relation to population history. Am. Nat. **129**: 463-469.

DALE, P.J, 1997 Potential impacts from the release of transgenic plants into the environment. Acta Physiol. Plant. **19**: 595-600.

DANIELL, H., 2002 Molecular strategies for gene containment in transgenic crops. Nature Biotech. **20**: 581-586.

DE WIT, M., P. CROOKES, and B.W. VAN WILGEN, 2001 Conflicts of interest in environmental management: estimating the costs and benefits of a tree invasion. Biol. Inv. **3**: 167-178.

DI GIOVANNI, F., P.G. KEVAN, and J. ARNOLD, 1996 Lower planetary boundary layer profiles of atmospheric conifer pollen above a seed orchard in northern Ontario, Canada. For. Ecol. Mgmt. **83**: 87-97.

DIFAZIO, S.P., G.T. SLAVOV, J. BURCZYK, S. LEONARDI, and S.H. STRAUSS, 2004 Gene flow from tree plantations and implications for transgenic risk assessment. In WALTER, C., and M. CARSON (eds.) Plantation Forest Biotechnology for the 21st Century. Research Signpost. ISBN: 81-7736-228-3, 405-422.

ELLSTRAND, N.C., 2003 *Dangerous Liaisons: When Cultivated Plants Mate with Their Wild Relatives.* Baltimore, Johns Hopkins University Press.

ESPINEL, S., A. ARAGONÉS, and E. RITTER, 1995 Performance of different provenances and of the local population of the Monterey pine (*Pinus radiata* D. Don) in northern Spain. Annales des Sciences Forestières **52**: 515-519.

GÓMEZ, A., S.C. GONZALEZ-MARTINEZ, C. COLLADA, J. CLIMENT, and L. GIL, 2003 Complex population genetic structure in the endemic Canary Island pine revealed using chloroplast microsatellite markers. Theor. Appl. Gen. **107**: 1123-1131.

GÖRGENS, A.H.M., and B.W .VAN WILGEN, 2004 Invasive alien plants and water resources in South Africa: current understanding, predictive ability and research challenges. SA J. Sci. **100**: 27-33.

GRACE, L.J., J.A. CHARITY, B. GRESHAM, N. KAY, and C. WALTER, 2005 Insect-resistant transgenic Pinus radiata. Plant Cell Reports. DOI: 10.1007/s00299-004-0912-x

HAMRICK, J.L., M.J.W. GODT, and S.L. SHERMAN-BROYLES, 1992 Factors influencing levels of genetic diversity in woody plant species. New Forests **6**: 95-124.

HIGGINS, S.I., and D.M RICHARDSON, 1998 Pine invasions in the southern hemisphere: modelling interactions between organism, environment and disturbance. Plant Ecol. **135:** 79-93.

HIGGINS, S.I., and D.M. RICHARDSON, 1999 Predicting plant migration rates in a changing world: The role of long-distance dispersal. Am. Nat. **153:** 464-475.

KAREIVA, P., I.M. PARKER, and M. PASCUAL, 1996 Can we use experiments and models in predicting the invasiveness of genhetically engineered organisms? Ecology **77:** 1670-1675.

KAUFMAN, S.R., P.E. SMOUSE, and E.R. ALVAREZ-BUYLLA, 1998 Pollen-mediated gene flow and differential male reproductive success in a tropical pioneer tree, *Cecropia obtusifolia* Bertol, Moraceae): a paternity analysis. Heredity **81:** 164-173.

KOSKI, V., 1970 A study of pollen dispersal a mechanism of gene flow in conifers. Communicationes Instituti Forestalis Fenniae **70:** 1-78.

LEDIG, F.T., 1998 Genetic variation in Pinus. In: RICHARDSON, D.M, (Editor) *Ecology and Biogeography of Pinus*. Cambridge University Press, Cambridge, pp. 251-280.

LEDGARD, N., 2001 The spread of lodgepole pine (*Pinus contorta*, Dougl.) in New Zealand. For. Ecol. Manage. **141:** 43-57.

LEDGARD, N.J., and E.R. LANGER, 1999 *Wilding prevention. Guidelines for minimising the risk of unwanted wilding spread from new plantings of introduced conifers*. New Zealand Forest Research Institute Limited.

LU, B-R., 2003 Transgene containment by molecular means — is it possible and cost effective? Environ. Biosafety Res. **2:** 3-8.

MACK, R.N., D. SIMBERLOFF, W.M. LONSDALE, H. EVANS, M. CLOUT, *et al.*, 2000 Biotic invasions: Causes, epidemiology, global consequences, and control. Ecol. Appl. **10:** 689-710.

MORAN, G.F., and J.C. BELL, 1987 The origin and genetic diversity of *Pinus radiata* in Australia. Theor. Appl. Gen. **73:** 616-622.

MORAN, V.C., J.H. HOFFMANN, D. DONNELLY, B.W. VAN WILGEN, and H.G. ZIMMERMANN, 2000 Biological control of alien, invasive pine trees (*Pinus* species) in South Africa. Proceedings of the X international symposium on biological control of weeds, 941-953.

MORJAN, C.L. and L.H. RIESEBERG, 2004 How species evolve collectively: implications of gene flow and selection for the spread of advantageous alleles. Mol. Ecol.**13:** 1341-1356.

OWUSU, R.A, 1999 *GM technology in the forest sector*. WWF-UK, Surrey, U.K.

PETIT, R.J., R. BIALOZYT, P. GARNIER-GÉRÉ, and A. HAMPE, 2004 Ecology and genetics of tree invasions: from recent introductions to Quaternary migrations. For. Ecol. Manage. **197:** 117-137.

PETIT, R.J., J. DUMINIL, S. FINESCHI, A. HAMPE, and D. SALVINI, *et al.,* 2005 Comparative organization of chloroplast, mitochondrial and nuclear diversity in plant populations. Mol. Ecol. **14:** 689-701.

PLOMION, C., G. LEPROVOST, D. POT, G.G. VENDRAMIN, and S. GERBER, 2001 Pollen contamination in a maritime pine polycross seed orchard and certification of improved seeds using chloroplast microsatellites. Can. J. For. Res. **31:** 1816-1825.

PYŠEK, P., D.M. RICHARDSON, M. REJMÁNEK, G.L. WEBSTER, and M. WILLIAMSON, 2004 Alien plants in checklists and floras: towards better communication between taxonomists and ecologists. Taxon **53:** 131-143.

REJMÁNEK, M., and D.M. RICHARDSON, 1996 What attributes make some plant species more invasive? Ecology **77:** 1655-1661.

RICHARDSON, D.M, 1988 Age structure and regeneration after fire in a self-sown *Pinus halepensis* forest on the Cape Peninsula, South Africa. SA J. Bot. **54:** 140-144.

RICHARDSON, D.M. (Editor), 1998a *Ecology and Biogeography of Pinus*. Cambridge University Press, Cambridge.

RICHARDSON, D.M, 1998b Forestry trees as invasive aliens. Conserv. Biol. **12:** 18-26.

RICHARDSON, D.M, 1999 Commercial forestry and agroforestry as sources of invasive alien trees and shrubs, pp. 237-257 in *Invasive species and biodiversity management,* edited by O. SANDERLUND, P. SCHEI, and A. VIKEN. Kluwer Academic Publishers, Dordrecht, The Netherlands.

RICHARDSON, D.M., and W.J. BOND, 1991 Determinants of plant-distribution - evidence from pine invasions. Am. Nat. **137:** 639-668.

RICHARDSON, D.M., and P.J. BROWN, 1986 Invasion of mesic mountain fynbos by *Pinus radiata*. SA J. Bot. **52:** 529-536.

RICHARDSON, D.M., and R.M. COWLING, 1992 Why is mountain fynbos invasible and which species invade? pp. 161-181 in *Fire in South African mountain fynbos* edited by B.W. VAN WILGEN, D.M. RICHARDSON, F.J. KRUGER, and H.J. VAN HENSBERGEN. Springer-Verlag, Berlin.

RICHARDSON, D.M., R.M. COWLING, and D.C. LE MAITRE, 1990 Assessing the risk of invasive success in *Pinus* and *Banksia* in South African mountain fynbos. J. Veg. Sci. **1**: 629-642.

RICHARDSON, D.M., and W.J. BOND, 1991 Determinants of plant-distribution - evidence from pine invasions. Am. Nat. **137**: 639-668.

RICHARDSON, D.M., P. PYŠEK, M. REJMÁNEK, M.G. BARBOUR, and F.D. PANETTA et al., 2000 Naturalization and invasion of alien plants: concepts and definitions. Divers. Distrib. **6**: 93-107.

RICHARDSON, D.M., P.A. WILLIAMS, and R.J. HOBBS, 1994 Pine invasions in the Southern Hemisphere: determinants of spread and invadability. J. Biogeog. **21**: 511-527.

RICHARDSON, D.M., and M. REJMÁNEK, 2004 Conifers as invasive aliens: a global survey and predictive framework. Divers. Distrib. **10**: 321-331.

RIESEBERG, L.H., and J. M. BURKE, 2001 The biological reality of species: gene flow, selection, and collective evolution. Taxon **50**: 47-67.

ROUGET, M., D.M. RICHARDSON, S.J. MILTON, and D. POLAKOW, 2001 Predicting invasion dynamics of four alien *Pinus* species in a highly fragmented semi-arid shrubland in South Africa. Plant Ecol. **152**: 79-92.

ROUGET, M., D.M. RICHARDSON, J.L. NEL, and B.W. VAN WILGEN, 2002 Commercially important trees as invasive aliens – towards spatially explicit risk assessment at a national scale. Biol. Inv. **4**: 397-412.

ROUGET, M., and D.M. RICHARDSON, 2003 Inferring process from pattern in plant invasions: A semimechanistic model incorporating propagule pressure and environmental factors. Am. Nat. **162**: 713-724.

SIEMANN, E., and W.E. ROGERS, 2001 Genetic differences in growth of an invasive tree species. Ecol. Letters **4**: 514-518.

SIMBERLOFF, D., M.A. RELVA, and M. NUNEZ, 2002 Gringos en el bosque: introduced tree invasion in a native *Nothofagus/Austrocedrus* forest. Biol. Inv. **4**: 35-53.

SIMONS, A.M., 2003 Invasive aliens and sampling bias. Ecol. Letters **6**: 278-280.

SLATKIN, M., 1976 The rate of spread of an advantageous allele in a subdivided population, pp. 767-780 in *Population Genetics and Ecology,* edited by S. KARLIN, and E. NEVO. Academic Press Inc., New York.

SLATKIN, M., 1987 Gene flow and the geographic structure of natural populations. Science **236**: 787-792.

SOHNGEN, B., R. MENDELSOHN, R. SEDJO, R., and K. LYON, 1997 An analysis of global timber markets. Discussion Paper 97-37. Resources for the Future, Washington, D.C. http://www.rff.org/Documents/RFF-DP-97-37.pdf.

STRAUSS, S.H. and S.P. DIFAZIO, 2004 Hybrids abounding. Nature Biotech. **22**: 29-30.

STRAUSS, S.H., W.H. ROTTMANN, A.M. BRUNNER, and L.A SHEPPARD, 1995 Genetic engineering of reproductive sterility in forest trees. Mol. Eng. **1**: 5-26.

STRAUSS, S.H., A.M. BRUNNER, V.B. BUSOV, C. MA, and R. MEILAN, 2004 Ten lessons from 15 years of transgenic *Populus* research. Forestry **77**: 455-465.

STRAUSS, S.H., P. COVENTRY, M.M. CAMPBELL, S.N. PRYOR, and J. BURLEY, 2001 Certification of genetically modified forest plantation. Int. For. Rev. **3**: 85-102.

TRAKHTENBROT, A., R. NATHAN, G. PERRY, and D.M. RICHARDSON, 2005 The importance of long-distance dispersal in biodiversity conservation. Divers.Distrib. **11**: 173-181.

VAN FRANKENHUYZEN, K. and T. BEARDMORE, 2004 Current status and environmental impact of transgenic forest trees. Can. J. For. Res. **34**: 1163-1180.

VAN WILGEN, B.W., D.C. LE MAITRE, and R.M. COWLING, 1998 Ecosystem services, efficiency, sustainability and equity: South Africa's Working for Water programme. TREE **13**: 378.

WAGNER, D.B., J. DONG, M.R. CARLSON, and A.D. YANCHUK, 1991 Paternal leakage of mitochondrial DNA in *Pinus*. Theor. Appl. Gen. **82**: 510-514.

WAGNER, D.B., D.R. GOVINDARAJU, C.W. YEATMAN, and J.A. PITEL, 1989 Paternal chloroplast DNA inheritance in a diallel cross of jack pine (*Pinus banksiana* Lamb. Heredity **80**: 483-485.

WALTER, C., 2001 Genetic engineering in conifer forestry: Technical and social considerations. In Vitro Cellular & Develop. Biol. **40**: 434-441.

WALTER, C., 2004 Genetic engineering in conifer forestry: technical and social considerations. In Vitro Celluar & Develop. Biol. **40**: 434-441.

WALTER, R., and B.K. EPPERSON, 2001 Geographic pattern of genetic variation in *Pinus resinosa*: area of greatest diversity is not the origin of postglacial populations. Mol. Ecol. **10**: 103-111.

WALTER, C., L.J. GRACE, A. WAGNER, D.W.R. WHITE, and A.R. WALDEN, et al., 1998 Stable transformation and regeneration of transgenic plants of Pinus radiata D. Don. Plant Cell Reports **17**: 460-468.

WARES, J.P., A.R. HUGHES, and R.K. GROSBERG, 2005 Species introductions and invasions: insights into the mechanisms that drive evolutionary change. Submitted.

WHEELER, N.C., and A.R. GURIES, 1982 Biogeography of lodgepole pine. Can. J. Bot. **60**: 1805-1814.

WILLIAMS, C.G., and B. H. DAVIS, 2005. Rate of transgene spread via long-distance seed dispersal in *Pinus taeda*. For. Ecol. Mgmt. **217**: 95-102.

WILLIAMS, C., S. L. LADEAU, R. A. OREN, and G. G. KATUL, 2006. Modeling seed dispersal distances: implications for transgenic *Pinus taeda*. Ecological Applications. In press.

WILLIAMS, C.G., C.G. ELSIK, and R.D. BARNES, 2000 Microsatellite analysis of Pinus taeda in Zimbabwe. Heredity **84**: 261-268.

WILLIAMSON, M., 1993 Invaders, weeds and the risk from genetically manipulated organisms. Experientia **49**: 219-224.

ZHENG, Y.Q., and R.A. ENNOS, 1999 Genetic variability and structure of natural and domesticated populations of Caribbean pine (*Pinus caribaea* Morelet) Theor. Appl. Gen. **98**: 765-771.

ZIMMERMANN, H., V. MORAN, and J. HOFFMANN, 2004 Biological control in the management of invasive alien plants in South Africa, and the role of the Working for Water programme. SA J. Sci.**100**: 34-40.

SECTION-IV

ECONOMICS OF TRANSGENIC TECHNOLOGY ADOPTION

CHAPTER 11

ECONOMIC PROSPECTS AND POLICY FRAMEWORK OF FOREST BIOTECHNOLOGY IN THE SOUTHERN U.S.A. AND SOUTH AMERICA

FREDERICK W. CUBBAGE
Department of Forestry and Environmental Resources,
North Carolina State University, Raleigh, North Carolina, USA

DAVID N. WEAR
USDA Forest Service, Research Triangle Park, North Carolina, USA

ZOHRA BENNADJI
Forestry Research Department,
Instituto Nacional de Investigación Agropecuaria, Tacuarembó, Uruguay

Abstract. An economic framework is presented for analyzing forest biotechnology with a focus on the case of transgenic forest trees in the southeastern U.S., Uruguay, and South America. Prospective economic benefits of forest biotechnology could reach hundreds of millions of dollars per year, but greatly increased research expenditures will also be required to achieve this potential. Commercial use of transgenic forests also must overcome biological, social, and policy issues related to social values as well as risk and control of dispersion that are unique with forest species. Benefits are likely to be realized earlier in South America than in the U.S., where timber growth rates and financial returns are much higher and clonal technology more prevalent, especially with *Eucalyptus* species. All major South American countries have ratified the Protocol of Cartegena on Biosafety, which requires risk assessments for the use of biotechnology of agricultural and, by extension, forestry. More detailed research can assess benefits, costs, and risks of transgenic forest trees and other biotechnology innovations using the framework presented here.

1. INTRODUCTION

Forest biotechnology research promises substantial returns through the development of genetically modified or transgenic forest trees. Benefits identified to date have focused on enhancing timber productivity and the quality of timber products, but other benefits such as enhanced conservation of biological diversity and bioremediation are also possible. Other applications of forest biotechnology also offer the substantial benefits of tree improvement and breeding (LI et al., 1999, MCKEAND et al.,2003) at a rapidly accelerated rate and with greater improvements in desirable tree characteristics. Research and development of transgenic forest trees also will prompt social and economic issues, ranging from evaluation of economic feasibility and economic benefit, ecological and social impacts, and the promulgation of regulatory institutions to ensure safe deployment and acceptance of transgenic trees.

We begin by outlining the opportunities and discussing a framework for evaluating the direct and secondary economic effects of these advances. We then examine general policy issues with an emphasis on the role and perspectives of forest certification programs. An examination of policy responses in Uruguay provides insights into the potential for technology deployment in South America and the potential for shifts in production and trade patterns.

2. OPPORTUNITIES FOR FOREST BIOTECHNOLOGY

In theory, transgenic forest trees have DNA constructs (transgenes) which makes them a form of genetically modified organisms (GMOs). This technology is a promising opportunity in plantation forestry but also has substantial drawbacks. It could provide higher yields, better wood quality, lower risk associated with pests and pest management, as well as offer engineered genetic diversity at the cell, stand, ecosystem level, all much faster than waiting for generations of tree breeding and testing and plantation establishment. However, the research and development costs and risks are much higher, and seedling costs will be higher at least in the short run. There also are major technical challenges, social issues, environmental risks, and market acceptance questions.

In the European Union, forest trees consisted of 17 out of 1,649 agricultural GMO trials, or about 1% of the total as of 2001. The EU licenses for environmental releases of forestry GMOs have ranged from one to three per year (LINDGREN 2003). In the United States, there have been 150 field trials for GMOs. These have focused on poplar, pines, walnut, and cottonwood (http://:www.isb.vt.edu/cfdocs/fieldtests1/cfm.

LUCIER et al.,(2002) suggested five scenarios that illustrate potential future impacts of biotechnology in the forest sector. Transgenic trees are the most immediate applications of forest biotechnology, but many other promising opportunities exist. For silviculture, transgenic trees can enable significant improvements in timber yields and resource use efficiency. We also could move toward more landscape management, with short-rotation wood production concentrated on sites that are well-suited to intensive management. Genetic mapping and identification, and gene conservation and banks could protect intra- and inter-species diversity.

For wood quality, biotechnology could enable improvements in fiber length, angle strength, and density. Better wood quality enables improvements in product quality and energy efficiency. Biotechnology also could enable new pulping technology based on selective enzymatic cleavage of lignin polymers. New biopulping technology might greatly reduce capital costs and improve pulp yields, product quality, and environmental performance (LUCIER *et al.,* 2002).

Forest biotechnology could enhance environmental management by more efficient bio-treatment of wastewater and conversion of solid residuals into bio-energy. Trees could be modified to adapt to contaminated sites, for sequestering carbon on marginal lands, or for ecological restoration. We could achieve enhanced biodiversity based on gene mapping. New bio-processing technologies could enable conversion of wood and wood residuals into valuable chemical feedstocks. New trees could produce endogenous chemicals or bio-pharmaceuticals that are extracted from wood prior to conversion into traditional products (LUCIER *et al.,* 2002). There might be many other market or nonmarket benefits of the applications of forest biotechnology.

3. ECONOMIC ANALYSES

Based on the prospective benefits listed above, one could take various approaches to analyzing the economic benefits and costs of transgenic forest trees. This section discusses the economic theory and an analytical framework that could be used to make such assessments in detail, and illustrates some of these potential economic benefits based on currently existing data or prior research literature. A thorough and complete analysis of such research benefits and costs will take more effort, but we will at least provide some concepts for discussion and future research.

3.1. Economic theory

An initial set of economic considerations focuses on the decision of producers to adopt biotechnology alternatives in timber production. Ultimately the actions of producers reveal the outcome of these considerations in determining whether or not deployment of the technology is feasible. Components of their decision-making calculus would include expected costs versus benefits and risk profiles, including an understanding of who bears these risk burdens. Risks include the probability of crop failure, damage by natural disturbance regimes, environmental damages spilling over from the plantings, the possibility of a backlash in public opinion, and simple market risk—i.e., demand and supply factors that determine prices in the future. Risks and risk burdens are especially important considerations in the deployment of any GMO.

Enhanced financial returns would derive from two qualitatively distinct changes in the timber production function. One is a simple outward shift in the timber volume produced over time, which could reduce the maturity age for a timber stand or increase the volume harvested at the maturity age. The other is a change in the quality of the timber volume produced that could be valued in the market (i.e., either a more uniform timber crop or a physical/chemical property that enhances the utility of the fiber). More uniform crops would reduce harvesting costs while altered fiber properties could reduce costs in a downstream production process such as pulping.

Productivity gains can be viewed as a continuation of a long line of productivity enhancing research in the southeastern United States. Tree breeding and fertilization programs have resulted in a more than doubling of timber productivity in the region over the past 50 years. Research that reduces the losses due to insects or diseases likewise enhances supply. These types of productivity gains (using transgenic trees or other technologies) are readily evaluated by the landowner because the gains are realized on site. That is they do not depend on the development of a new market for a qualitatively distinct product.

In contrast, economic returns to changes in physical/chemical properties of fiber depend on developments in downstream technologies and demands for the properties. For example, changes in pulping quality must be identified by consumers of pulpwood. Another key element for financial success is some means of branding or identifying the enhanced product. Clear product demarcation can be guaranteed by secure production chains, for example where the consumer of the fiber also grows the timber. However, recent divestitures of forest management from wood products divisions of forest products firms raise additional challenges in this regard.

In addition to the questions that producers face when evaluating the decision to adopt and deploy a biotechnology alternative, there are economic issues surrounding the potential social welfare impacts of new technology. Adoption may provide a competitive advantage for individual producers, but the impacts on consumers and total economic welfare may or may not be substantial. The total economic impact of the new technology depends on how the technology affects supply or demand relationships for timber. Enhanced productivity typically shifts supply outward— i.e., more timber is supplied for the same price—and causes prices to fall. Lower prices increase consumer welfare by allowing increased consumption with the same budget.

3.2. Economic welfare analyses

Several national studies have shown that shifts in the supply of or the demand for timber products causes discernable economic welfare impacts in the U.S. Demand for timber products from the U.S. was influenced by strong restrictions on lumber imports from Canada, its largest trading partner in this sector between 1986 and 1991. The largest shift in timber supplies resulted from the approximately 85 percent reduction in timber production from federal forests in the western United States starting in 1993 (harvest reductions of about 10 billion board feet).

These two episodes provide useful insights into the relative magnitude of welfare impacts from additional changes in timber productivity. WEAR AND LEE (1993) estimated that the Memorandum of Understanding on Softwood Lumber Imports resulted in a significant impact on production in the U.S. The shift yielded about $0.4 billion (15%) in additional benefits to the lumber-producing sector and reduced consumer welfare (prices increased) by about $2.00 per household per year. In the case of the public harvest reductions since 1992, the supply contraction resulted in gains of up to $1 billion for southern producers and a loss of about $10.00 per household per year in consumer benefits. While these episodes applied to the lumber sector it is hard to imagine that biotechnology would have a more substantial influence on the structure of supply. Social benefits from technology changes in the forest sector are likely to be relatively small (not inconsistent with this range of $2 to $10 per household per year).

Changes in the timber production technology not only alter how products flow from land but also the uses to which land is dedicated. Changes in timber prices relative to the competitive uses of forest land will give rise to land use change in a privately held landscape. In addition, increased productivity allows more production to be concentrated on a smaller land base.

Differences in the regulatory environment between countries could amplify the distributive effects of the modified organism. Returns to the comparative advantage of a superior product would accrue in countries that allow planting of the GMO, as long as no market restrictions were imposed on GMO products in importing countries. Production and trade patterns would adjust accordingly. While we lack any detailed estimates of how transgenic trees could alter productivity or wood quality, it is likely that the latter type of change is more likely to alter distributions of benefits and trade patterns because of the resulting changes on the demand for products from native trees and the exclusivity of the production technology.

3.3. Potential southern U.S.A. impacts

Timber market projections conducted to support the Southern Forest Resource Assessment (PRESTEMON AND ABT 2002; WEAR AND GREIS 2002) provide some insights into how timber markets could potentially be affected by productivity gains. Baseline projections suggest that roughly 12 million acres of forest losses associated with urbanization forecast to 2010 would be offset by price induced afforestation activities (e.g., timber prices increasing relative to agriculture prices leads to tree planting). Productivity gains give rise to timber price declines (outward shifts in supply), thereby decreasing the area in forest cover and increasing agricultural uses relative to this base case.

While additional productivity gains would give rise to a loss of forest area, the composition of forests would also be affected. PRESTEMON AND ABT (2002) show that productivity gains would concentrate intensive or plantation management on a smaller land base and leave a higher proportion of forests in a naturally regenerated condition. Net changes from their projections indicate that increased productivity would gives rise to an overall decrease in areas of both plantation and naturally regenerated forests, suggesting the possibility of social costs related to other benefits of forests not included in the timber market estimates.

Changes in the quality of fiber—in essence, developing markets for a new products—are not as easily evaluated because they involve structural change in timber markets rather than a shift in an existing supply or demand relationship. The introduction of a superior pulpwood product would cause the demand for the standard pulpwood product to retract. Prices would fall for the standard product and rise for superior pulpwood. Higher rents would accrue to those holding the patent for the new modified organism and lower returns would flow to those producing pulpwood using unmodified native organisms.

LUCIER et al. (2002) made a simple estimate of the potential economic impacts of forest biotechnology in general that creates better investment returns, which will in turn increase timber supply in the U.S. South. Given the southern pine 1995 harvest of 5.5 billion cubic feet and a blended stumpage price of $0.80/cubic foot, the approximate annual softwood total stumpage cost was $4.4 billion. The potential cost reduction for same volume, with 5% supply shift and 0.4 supply and demand elasticities, would be $220 million per year. There also would be increased regional output of about 5% for same total cost, leading to greater revenue, profits, and regional competitiveness. Of course, forest biotechnology is not likely to be applied just in one region, which may alter comparative advantages among the timber supply regions in the world. But overall, it is apt to lower marginal costs of producing timber, shift the supply curve for timber out, and lower equilibrium supply and demand price levels.

Another promise of forest biotechnology is that it will greatly reduce or eliminate major forest diseases, and perhaps pests. CUBBAGE et al. (2000) calculated the potential South-wide impacts of fusiform rust elimination. The historical fusiform rust research program has provided $40 to $60 million per year of benefits. There is a potential for practical elimination of fusiform rust, with many molecular markers for fusiform rust resistance already identified. The annual benefit of complete elimination would range from $120 million to $920 million, with median of about $200 million. The current forest biotechnology research costs are likely less than $10 to $20 million per year in the United States, although this level of investment surely will need large increases to realize the predicted benefits of forest biotechnology. The identification of fusiform rust resistance genes using biotechnology has been one of the salient results in the short run, illustrating that such significant gains are possible.

3.4. World timber investment returns

Successful adoption of transgenic forest trees could significantly alter the financial returns to forest investments throughout the world. Preliminary research by CUBBAGE et al. (2005) estimated the investment returns for plantations in South America. Eucalypts (*Eucalyptus* spp.) have the greatest internal rates of return, ranging from 13% to 25% per year for typical industrial applications in Uruguay, Argentina, and Brazil. *Pinus radiata* in Chile had average internal rates of return of 17%, and *Pinus taeda* in Uruguay, Argentina, and Brazil had rates of return ranging from 10% to 17%. Average internal rates of returns for typical plantations in the United States were closer to 9% to 12%.

The excellent returns in South America are based on excellent growth rates coupled with moderate costs. This combination is apt to attract the most investment in transgenic forest trees to the southern hemisphere as well. The calculated rates of return did not include land purchase costs, which have increased rapidly in South America due to demand for soybean production, especially in Brazil. Inclusion of land costs would decrease the preceding timber investment internal rates of returns by 5 to 10 percentage points per annum, and have the most adverse effects on returns in Brazil. This would make average internal rates of return for new land and timber investments more comparable among the southern U.S. and the Southern Cone. Higher land costs will accelerate the trend toward more intensive forestry, including transgenic trees, in order to minimize the factor costs for scarce land.

Increases in yield or wood quality would generate significant economic benefits, and have substantial impacts on comparative advantages among countries and forest products trade. These impacts of increased growth or quality or changes in trade could then be examined for their economic welfare impacts and for the benefits and costs of the research itself.

4. POLICY ISSUES AND RESPONSES

4.1. Scientific and public concerns

Biotechnology, including for forestry, surely has significant hazards as well, commensurate with its large potential. Environmental press comments, and a host of comments and critiques on the Internet illustrate some of these issues.

Technical and financial issues abound with forestry applications. Biotechnology is easiest to use with a set of common clones, but there are few in *Pinus taeda*. The desire for a fixed-clonal technology for GMOs may impede breeding advances. Scientists will want to stick with old clones, and thus overuse the best clones. Transgenic technology is most likely to be practical for short rotation exotics, and may never reach *Pinus taeda* stage, at least in the U.S.A. Given the comparative levels of investments between agriculture and forest trees, it may be decades before transgenic forest tree benefits are realized. Medical biotechnology, for example, has probably received 1000 times more funding than forest biotechnology, and just beginning to see the first commercial successes after more than 30 years of research and development (LINDGREN 2003).

Transgenic forest trees need long field testing before deployment to test sterility and to ensure viability outdoors, e.g., making sure low lignin trees do not break apart in the field. Perhaps field tests should be longer for transgenic trees than other types of trials given higher risk and uncertainty. Transgenic forest trees are likely to spread given the fertile breeding habits of trees, such as with exotic pines in South Africa and Argentina. Contamination of native maize in Mexico and normal grains in Canada with transgenes has been reported already. Sterile forest trees, yet to be developed, are difficult and expensive. Use of terminator technology may fail, or succeed too well and infect native populations and lead to extinction. Also, the regulatory process is long and cumbersome (LINDGREN 2003). The number of transgenic escapes in systems other than trees has increased rapidly, and trees will surely be more difficult to contain than annual crops.

The biological uncertainties of GMOs beget social issues. Forests are perceived as the epitome of nature, with large spiritual values, and thus should be protected in a natural state. Playing God with (tree) genetics may be perceived as even less desirable than with human medical advances, by both green groups and religious conservatives. Transgenic forests will have neither the glamor of curing human ills or the pedestrian nature of agriculture, which domesticated most wild species decades ago. Domestication of forest trees is seldom a goal of anybody except forest geneticists, and the fear of kudzu-like Frankentrees is far more pervasive. More sabotage such as that claimed by the Earth Liberation Front (ELF) at Rhinelander, Wisconsin and the University of Washington seems likely. Many green environmental groups have already protested against GMOs, plantations, and timber harvests. All these issues will increase costs of transgenic forest trees. Indirect costs of transgenic forest trees include drawing scarce funding away from traditional, successful breeding and forest management programs with more immediate payoffs and much less risk.

GMOs may have market and government problems stemming from the social issues. As noted below, operational transgenic forest trees are proscribed in FSC certification standards, and may be addressed in other certification standards later. This may limit European market acceptance. Several U.S. retail lumber and office products firms have adopted or are considering a non-GMO policy. Government officials, legislators, and Congress members will be extremely careful and require many regulatory steps with GMOs in order to protect the public and their reputations.

LINDGREN (2003) concluded that forestry GMOs may be worth considering within decades if the rotation length is less than 10 years (reasonable testing time); exotics are planted (limited escape risk); clonal forestry is established (easy mass propagation available); there is an ongoing breeding program (adapted materials and competence for field testing available); and there is a good interface with science. Use of transgenic forest trees will require the will to make uncertain investment. If applied, such criteria would constrain applications in slow growing plantation species and in lower return regions of the world.

4.2. Forest certification

Some of these broader public concerns are reflected very specifically in the current forest certification programs, which govern forest operations on the brunt of the industrial forest land in the United States and Canada, and a small but increasing share of the industrial and other forest land in South America. These guidelines are particularly important for transgenic forest trees on forest industry lands in the short run.

Forest certification is designed to measure, monitor, audit, and improve forest management processes and practices at the forest level. In the United States, the Sustainable Forestry Initiative (SFI), which was initiated by the forest industry, is the dominant certification system. The Forest Stewardship Council (FSC), which was initiated by environmental nongovernment organizations, is less prevalent in the U.S., but remains a benchmark for green certification, and is more common in applications to industrial forest lands in South America. In partnership with other groups and the government, the forest industry in Brazil and Chile also have started forest certification programs termed Cerflor and Certfor, respectively, which have had significant enrollment by forestry companies in their countries.

THE U.S. SUSTAINABLE FORESTRY INITIATIVE (2002) generally accepts forest biotechnology when applied with due diligence to avoid the problems listed above, at least to the extent possible. Cerflor and Certfor also accept forest biotechnology, prudently applied. At this time, however, only FSC has explicit proscriptions against the use of genetically transformed trees for certified forest lands.

Key FSC standards related to tree improvement and forest biotechnology (FSC 2001; Smartwood 2001) include section 6.3.b. genetic, species, and ecosystem diversity. This includes 6.3.b.1, select trees for harvest, retention, and planting to maintain genetic diversity and species diversity of residual stand. Standard 6.3.b.2 requires that diverse native habitats be maintained; and 6.3.b.3 requires use of locally adapted seed of known provenance be used for artificial regeneration.

Standard 6.6 requires development and adoption of environmentally friendly non-chemical methods of pest management. Standard 6.8 requires that use of biological control agents shall be documented, minimized, monitored, and strictly controlled. Furthermore, it states that use of genetically modified organisms shall be prohibited. This includes a statement of: "Applicability Note: Genetically improved mechanisms (e.g., …Mendelian crossed) are not considered to be GMOs and may be used. The prohibition of GMOs applies to all organisms including trees." In addition, standard 6.8.a states that exotic predators used only as part of IPM strategy if other methods ineffective.

The FSC standards for tree improvement and forest biotechnology also limit plantations, especially of exotic species. They state that: 6.9 The use of exotic species shall be carefully controlled and actively monitored to avoid adverse ecological impacts; 6.9.a. that they should be contingent on peer-reviewed scientific evidence that any species in question is non-invasive and does not diminish biodiversity…use must be actively monitored; 6.9.b. owners must use control measures for invasive plants. Furthermore, they mandate that: 6.10 Forest conversion to plantations or non-forest uses shall not occur, except for (a) when it occurs as a limited portion of FMU; (b) does not occur in high conservation value forests; and (c) provides long term conservation benefits.

Plantations converted from natural forests after November 1994 normally shall not qualify for certification, unless the manager/owner is not responsible directly or indirectly for conversion. However, typical southern forests regenerated from old farm fields are not considered natural, so this may not be daunting as it appears. This characteristic flexibility also may carry over to FSC proscriptions on the use of GMOs as well. For example, one firm in Latin America had to destroy a multi-million dollar research program in transgenic eucalypts and burned all it transgenic plants in order to receive FSC certification for their forest lands. On the other hand, another American firm has reportedly received FSC certification for its forest lands, <u>excluding</u> some experimental trials that it currently has of transgenic trees. These are just anecdotes, but suggest that FSC will continue to wrestle with the treatment of GMOs. FSC has needed to be flexible to attract members into its programs, and must continue to be so given the recent development of new competing programs in Latin America.

5. REGULATORY RESPONSES: URUGUAY AND SOUTH AMERICA

Governmental policy responses clearly will affect development and deployment of transgenic forest trees and other applications of forest biotechnology. While responses are developing in the U.S., it is instructive to examine how South America is coping with this issue. This is especially useful, because transgenic forest trees are likely to be applied first in the southern hemisphere, where the forest rotations are much shorter and the financial returns much greater. Uruguay is particularly useful as a case study in this regard because it has many international forest products firms that have obtained FSC forest certification. All of these export the brunt of their raw material (roundwood or wood chips) or their manufactured products (lumber and plywood) to other countries. Due to its central geography and size, Uruguay also could serve as a scientific and business model in many respects for all of South America.

5.1. Level of forest biotechnology adoption in Uruguay

Research in forest biotechnology in Uruguay is currently underway in both private and public sectors, focused exclusively on tree improvement for fast growing species. Most research focuses on *Eucalyptus grandis* and *E. globulus*, two short rotations exotics widely planted in Uruguay since the promulgation of a forestry law in 1987 with a set of incentives and fiscal exoneration. To date, more than 700,000 ha of *Eucalyptus* and *Pinus* plantations have been established in the four regions of the country that qualify as forest priorities zones dues to their soil and climate conditions (DIRECCIÓN GENERAL FORESTAL 2005). International forest companies are well represented in these forest and plantation research programs (e.g., Botnia, Ence, Weyerhaeuser).

Table 1. Major Eucalyptus spp. and Pinus spp. tree improvement results and forest biotechnologies used in the INIA-Forestry Department in Uruguary.

Species	Prospecting of genetic variability		Variety release	More than one generation cycle of improvement	Use of molecular markers
	Local	Introduced			
Eucalyptus grandis	Yes	Yes	Yes	Yes	Yes
Eucalyptus globulus	Yes	Yes	Yes	Yes	No
Eucalyptus maidenii	Yes	Yes	Yes	No	No
Eucalyptus saligna	No	Yes	No	No	No
Eucalyptus dunnii	No	Yes	No	No	No
Pinus taeda	Yes	Yes	No	No	No
Pinus elliottii	No	Yes	No	No	No

Source: INIA-National Forestry Program data bases

Conventional tree breeding programs for *E. grandis* and *E. globulus* have long been pursued by both private and public institutions in Uruguay. The selection traits studied are growth and disease resilience. Two improvement cycles have been accomplished for both *E. grandis* and *E. globulus* and, since 2000, INIA's Forestry Department (Instituto Nacional de Investigación de Agropecuria) has released three varieties and nine clonal lines. The use of INIA *Eucalyptus* varieties could allow a growth increase of almost 15 to 30%, compared to mean growth currently obtained in commercial plantations. The annual growth gains expected with the use of clones compared to the use of unimproved seed are estimated to be more than 50% (BENNADJI 2004a).

Clonal forestry for *E. grandis* has reached a commercial scale in the northern and western regions of the country. Almost 4,000,000 clonal seedling were produced in 2003 and 4,000 ha of clonal plantations were established in the same period. In addition, INIA reports tree improvement and biotechnology progress in several species of *Eucalyptus* and *Pinus* (Table 1) (BENNADJI 2004b).

Forest biotechnology research in Uruguay at present currently falls into two broad categories: tissue culture and molecular markers. It does not include transgenic trees at this time. Tissue culture is used to enhance clonal propagation and as a tool to support large scale production of uniform materials. Micropropagation is used for the establishment of clonal gardens and for gene conservation. Molecular markers are used at a small scale to quantify genetic diversity between breeding populations and individual trees and to establish variety identification (genetic fingerprinting) (BENNADJI 2002). High costs currently prevent the application of micropropagation techniques to plantation establishment at the commercial level.

Field trials of a transgenic forest tree were reported for the first time in 1997 in the western region of the country, as an initiative of FOSA (Forestal Oriental), the Uruguayan branch of Shell. The transgenic trees were glyphosate-resistant *Eucalyptus grandis*. These field trials were stopped in 2000, when FOSA sought FSC certification.

5.2. Principal restrictions to the use of transgenic forest trees in Uruguay

As described above, there are considerable economic and socio-political challenges to adoption and deployment of transgenic forest trees. Production costs currently restrict widespread application. Technology application also requires a costly infrastructure of research, development, and deployment and building the human and financial capacity requires considerable resources. In addition, environmental NGOs stand in strong opposition to transgenic forest trees in Uruguay. Most major international pulp and paper firms in Uruguay have obtained FSC forest certification. These companies currently export all of their pulpwood roundwood or chips to Europe or Japan. These export markets and the FSC certifications have limited enthusiasm for GMOs in Uruguay.

Until 2000, the Ministry of Livestock, Agriculture and Fishery was in charge of the risk analysis of GMOs introduction in Uruguay. A scientific advisory commission supported the tasks of this Ministry. The commission was integrated by (i) the Public Health Ministry, (ii) the Environment Ministry, (iii) the National Institute of Seeds and (vi) the National Agricultural Research Institute. In August 2000, a Risk Evaluation Commission was officially created by decree and in 2001, Uruguay ratified the Protocol of Cartagena on Biosafety (PCBS).

Uruguay has a good legal framework for protecting intellectual property rights. As a member of the UPOV (Union pour la Protection des Obtencions Végétales), the country applies a set of legal rules for the protection of improved vegetative material of reproduction. On the other hand, Uruguay is also a signatory on the Convention on Biodiversity, which prohibits patents of living material.

5.3. Status of Transgenic Forest Trees in South America

In terms of adoption levels for transgenic forest trees, the situation among South American countries is quite heterogeneous. Argentina, Brazil, Chile, Mexico and Uruguay could be classified as advanced countries, compared to Andean and Central American Countries. There is a lack of information on transgenic forest tree field trials testing and only three trials of transgenic tree species *Eucalyptus globulus*, *Pinus radiata* and *Eucalyptus grandis* are reported in Chile and Uruguay; all three are directly supervised by the private sector. The transgenic traits are lignin modification and herbicide tolerance. However, most of the South American countries have committees or commissions in charge of transgenic risk analysis, usually related to agricultural species.

In Table 2, the status of the regulation and policy frameworks of some countries is presented. These regulatory frameworks put in place to oversee the experimental and commercial release of transgenic forest trees are broadly similar. However, the instruments were established for agricultural species and therefore do not fully account for environmental considerations surrounding genetically modified trees.

All of these Latin American countries have responded to concerns regarding genetically modified trees to one degree or another. All have now ratified the Protocol of Cartagena on Biosafety (PCBS) but not all have established institutional infrastructure for evaluating the risks and potential effects of these transgenic forest trees. In most cases, these institutions were established to evaluate genetically modified (GM) or transgenic crops and have yet to adopt protocols necessary for evaluating perennial plants such as forest trees.

Regardless of the investment in GM research and institutions to evaluate risks, the establishment of transgenic tree plantations may be trumped by the demand for FSC certification by forest products firms. Prohibition of transgenic forest trees by this certification body has and will likely have strong influence over the adoption of transgenic trees, reflecting the sensitivity of firms to public concerns regarding the environmental quality of their land management.

Table 2. Regulation and policies for GMOs in selected Latin-American countries.

Country	Institution in Charge of the GMO Risk Analysis	Protocol of Cartegena on Biosafety (PCBS)
Argentina	CONABIA (Comisión Nacional Asesora de Bioseguridad Agropecuaria)	Ratified May 2000
Bolivia	CNB (Comité Nacional de Bioseguridad)	Ratified April 2003
Brazil	CNTBio (Comisión Técnica Nacional de Bioseguridad)	Ratified November 2003
Chile	CALT (Comité Asesor para la Liberación de Organismos Transgénicos)	Ratified May 2000
Colombia	CTN (Comité Técnico National)	Ratified May 2003
Ecuador	-	Ratified January 2003
Mexico	-	Ratified august 2002
Paraguay	C.B (Comisión de Bioseguridad)	Ratified March 2004
Peru	CONABID (Comisión Nacional de Diversidad Biológica)	Ratified April 2004
Uruguay	CERVGM (Comisión de Evaluación de Riesgos de Vegetales Genéticamente Modificados)	Ratified June 2001
Venezuela	CNBio (Comisión Nacional de Bioseguridad)	Ratified May 2002

6. CONCLUSIONS

Forest plantations and forest biotechnology have been promoted with the promise of providing more of the world's wood fiber needs on a smaller land base, and as a means of sparing timber harvests on natural forests (SEDJO AND BOTKIN 1997; WORLD WILDLIFE FUND 2003). We have about 204 million ha of planted stands in total (5.3% of world forests), and 100 million ha of fast grown industrial plantations (2.0% of world forests) (FAO 2003; SIRY et al., 2005). Plantations provide about 25% of world industrial wood fiber, and are projected to increase this share to about 40%. The U.S. South has about 12 million ha of plantations (SMITH et al., 2001), and Brazil, Argentina, Uruguay, and Chile have almost 10 million ha of plantations, which grow much faster than in the U.S. Plantations provide the basis for and target of most of our genetic improvement efforts. Genetically improved trees and wood, achieved through traditional tree improvement programs, are generally accepted for private lands, both by forest industry and NIPFs.

Genetic improvement has been widely adopted throughout the world and implemented at moderate cost. Forest biotechnology promises to take us to another level technological innovation. It also can provide tools to identify and protect biological diversity at the molecular level, and perhaps provide means to achieve environmental remediation of industrial wastes. Forest biotechnology research is proceeding at a rapid pace, albeit with significant costs. However, transgenic forest trees and forest biotechnology will face far greater public opposition and regulatory challenges than any other forestry technology to date.

The decision to deploy transgenic trees will rest not only on estimates of their costs and returns but also on perceptions regarding risks. Returns will either increase productivity or provide qualitative changes in wood fiber. Our review suggests that annual benefits from transgenic forest trees and even forest biotechnology in general could be worth hundreds of millions of dollars per year, but only with successful research and implementation. Research costs would need to increase by tens of millions of dollars per year to realize these potential benefits—a level far beyond current expenditures. Risks of transgenic forest trees include standard concerns for production and market risk but also hazards associated with perceived or real environmental damages and backlash in public opinion.

Types of economic impacts depend on the nature of the effects of GMOs. For productivity enhancements, shifts in timber supply not unlike previous returns to research would shift production to smaller land bases and reduce timber prices. The proportion of intensively managed forest land would likely fall but so would the total area of forests in the U.S. South. Consumer benefits would accrue but, as previous studies have shown, because wood products are a small portion of the U.S. economy, total benefits are likely to be relatively small.

The economic impacts of a change in fiber quality could be more substantial, because it would reduce demands for pulpwood generated from native organisms. High prices for transgenic forest trees would lead to lowered prices for existing pulpwood products. In addition, because of differences in the timber growth rates and regulation of transgenic forests trees between countries, it is likely that altered comparative advantage would result in greater changes in production and trade patterns. Research and development in transgenic forest trees is more likely to be beneficial for the fast-growing, high-return plantations in South America.

Forest certification, which mandates and audits standards of forestry practice at the stand or ownership level, has potential for a much larger impact of forest management, tree improvement, and forest biotechnology. The Sustainable Forestry Initiative specifically requires that program participants demonstrate that they conduct or support forestry research in health and productivity, water quality, and wildlife and biodiversity. SFI encourages use of plantations, tree improvement, and forest management, and infers that forest biotechnology would be acceptable. With appropriate safeguards, exotics are legitimate under SFI, although there are not many exotic timber species being planted in the U.S. yet. SFI might offer specific opportunities in tree, stand, or ecosystem biodiversity for applying the science of tree improvement or forest biotechnology.

Forest certification by FSC requires that managers favor natural stands and biodiversity. FSC allows plantations and tree improvement with fairly extensive strictures to protect natural stands and ecosystems. It explicitly prohibits use of transgenic forest trees. FSC has been very flexible in decisions, allowing a large number of forests with exotic plantations to be certified if they have a large natural stand/reserve component as well. It does require refereed science to justify the use exotics and ensure that they do not cause any environmental harm. Its rigor has varied to a total ban of all transgenic plants ands trees for firms that are FSC-certified to a partial exclusion of the experimental lands that include transgenic trees. Debates over transgenic forest trees and GMOs will increase within the FSC as the prospects for operational use become more likely.

The case study of GMOs and transgenic forest trees in Uruguay and related regulations in South America is informative. Uruguay just began its significant forest plantation program in 1987 with a new forest law providing subsidies and favorable tax status for forest plantations of exotic species. Since then, it has begun prospecting for genetic variability, released varieties of several species, begun first and second generation tree breeding programs, and begun to use molecular markers for one species, *Eucalyptus grandis*. Uruguay has significant government oversight of agricultural and forest biotechnology applications, and does seek to employ the most modern technology to increase productivity and enhance export opportunities. Since 2001, Uruguay and every other major South America country have ratified the Protocol of Cartegna on Biosafety, which addresses transgenic agricultural crops, and trees by extension of those policies. This protocol establishes a structure for evaluating risks and potential effects of genetically modified organisms. The U.S. approach to regulation is even more complex and uncertain.

Public perceptions influence forest certification programs and the potential use of transgenic forest trees as well. Transgenic forest trees offer promise, charisma, and financial support, and issues. A distant promise is that of designer trees perfect for specific wood, paper, or environmental remediation applications, with known genetic diversity at the tree, stand, or ecosystem level. Given that even well-supported agriculture applications have been limited to modest herbicide resistance, the promise of complex wood quality or growth improvements seems distant. The production economics and costs and returns for transgenic forest trees are significant, and the social acceptance may be more challenging. FSC forest certification prohibits use of transgenic forest trees or for IPM, and several wood and paper retailers have or are considering adopting this policy. On the other hand, perhaps some of the outstanding recent medical breakthroughs, such as RNAi, can be duplicated in forestry at a much lower cost using similar technology. Maybe medicine and agriculture will pave the way for much less expensive subsequent forestry applications.

Increased government research in forest biotechnology is necessary, but costly, and the payoff will be distant. We still need to map the genomes of model tree species, discover molecular controls of key processes, assess ecological issues and opportunities, and understand risk management. We must link our forest biotechnology programs with traditional tree improvement, silviculture, and forest management, and vice versa. One avenue for this could be to use forest biotechnology to identify desirable characteristics, as described before, and then use vegetative propagation and/or somatic embryogenesis to rapidly ramp up and develop container stock for planting of superior trees. Perhaps this would avoid the social and certification antipathy for transgenic trees or other GMOs and still allow rapid implementation of the best science at reasonable costs.

Benefits of any forest biotechnology breakthroughs are likely to be realized soonest in South America, where timber growth rates are much higher, and clonal technology more prevalent, especially with *Eucalypts* species. All major Latin American countries have ratified the Protocol of Cartegena on Biosafety, which requires risk assessments for the use of biotechnology of agricultural and, by extension, forestry. More detailed research can assess benefits, costs, and risks of forest biotechnology using the framework presented here.

For example, biotechnology in medical applications and in the stock market achieved their most dramatic gains ever in 2002 and 2003. If possible, we should capitalize on such advances in medicine and agriculture, apply them well in forestry, answer pressing social or market questions, and then integrate biotechnology with existing tree improvement and forest research and development programs. With such advances, forest biotechnology and genetic improvement might help us achieve the widely accepted paradigm of Sustainable Forest Management that promotes economic, ecological, and social benefits for this and future generations.

7. REFERENCES

BENNADJI, Z., 2002 Propagatión vegetativa de especies del género *Eucalyptus*. Serie Aftercare Forestal INIA-JICA. N°9. 37 p.

BENNADJI, Z., 2004a Elección y uso de semillas forestales de calidad: un proceso factible en el Uruguay. Revista INIA **1**: 23-25.

BENNADJI, Z., 2004b Programas de mejoramiento genético forestal en el INIA. Revista Seragro **10**(117):25-28.

CUBBAGE, F.W., J.M. PYE, T.P. HOLMES, AND J.E. WAGNER, 2000 An economic evaluation of fusiform rust protection research. Southern Journal of Applied Forestry **24**(2):77-85.

CUBBAGE, F.W., P. MACDONAGH, M. NOEMI BÁEZ, J. SIRY, J. SAWINSKI JR, A. FERREIRA, V. HOEFLICH, G. FERREIRA, V. MORLAES OLMOS, R. RUBILAR, J. ALAREZ, AND P. DONOSO, 2005 Timber investment returns for plantations and native forests in the Americas. Poster presentation at: Symposium on Tropical Forests, University of Florida, Gainesville, Florida, February 13-16.

DIRECCIÓN GENERAL FORESTAL, 2005 Boletín Estadístico 2004. Ministerio de Ganadería, Agricultura y Pesca. Uruguay. Accessed at : http://www.mgap.gub.uy, March 03, 2005.

FAO, 2003 State of the World's Forests, 2003. Food and Agriculture Organization of the United Nations. Rome. 243 p.

FOREST STEWARDSHIP COUNCIL, 2001 Forest Stewardship Council United States Principles and Criteria. Accessed at: http://www.fscus.org/standards_policies/principles_criteria/index.html. July 6, 2003.

LI, B., S. MCKEAND, AND R.WEIR, 1999 Tree improvement and sustainable forestry—impact of two cycles of loblolly pine breeding in the U.S.A. Forest Genetics **6**(4):229-234.

LINDGREN, D., 2003 Is there a future for genetically modified (GMO) trees in forestry in this century? Powerpoint talk presented at NC State University, Department of Forestry, March 3, 2003. Raleigh, NC. electronic file. Personal communication. Author contact at Swedish University of Agricultural Sciences, Umea, Sweden.

LUCIER, A, D. PARKS, AND F.W. CUBBAGE, 2001 Economic analysis of the potential of forest biotechnology. Presented at: Biotech Branches Out: A Look at the opportunities and impacts of Forest Biotechnology. Atlanta, GA. December 4-5. Invited presentation.

McKeand, S. T. Mullin, T. Byram, and T. White, 2003 Deployment of genetically improved loblolly and slash pines in the South. Journal of Forestry **101**(4/5):32-37.

Montreal Process, 2003 The Montreal Process. http://www.mpci.org. Accessed on July 6, 2003.

Prestemon, J.P., and R.C. Abt, 2002 Timber markets. In: Greis, J., and D.N. Wear (editors), The Southern Forest Resource Assessment. USDA-Forest Service, General Technical Report SRS-53. Asheville, North Carolina. p. 299-326.

Sedjo, R., and D. Botkin, 1997 Using forest plantations to spare natural forests. Environment **39**(10): 14-20.

Siry, J.P., F.W. Cubbage, and M.R. Ahmed, 2005 Sustainable Forest Management. Journal of Forest Policy and Economics. In press.

SmartWood, 2001 Generic guidelines for assessing natural forest management. Accessed at: www.smartwood.org/guidelines/forest-management-generic.html. March 20, 2001.

Smith, B., J. Vissage, R. Sheffield., and D. Darr, 2001 Forest Resources of the United States, 1997. General Technical Report NC-219. St. Paul, MN: USDA Forest Service North Central Forest Experiment Station. 190 p.

Sustainable Forestry Initiative, 2002 The Sustainable Forestry Initiative Standard, 2002-2004 SFI Standard. American Forest and Paper Association. Washington, D.C.

Wear, D.N., and J.G. Greis (editors), 2002 The Southern Forest Resource Assessment: Technical Report. USDA Forest Service, General Technical Report SRS-53, 635 pp.

Wear, D.N., and B.C. Murray, 2004 Federal timber restrictions, interregional spillovers, and the impact on U.S. softwood markets. Journal of Environmental Economics and Management **47**(2): 307-330.

Wear, D.N., and K.J. Lee, 1993 U.S. policy and Canadian lumber: effects of the 1986 Memorandum of Understanding. Forest Science **39**:799-815.

World Wildlife Fund, 2003 The forest industry in the 21st century. Accessed at: http://www.panda.org/forestandtrade/latest_news/publications/ on January 25, 2003

CHAPTER 12

PRIVATE FORESTS AND TRANSGENIC FOREST TREES

MARK A. MEGALOS

Forest Stewardship / Legacy Program, North Carolina Division of Forest Resources, Raleigh North Carolina, USA

Abstract. Molecular domestication of forest tree species is becoming available to corporate timberlands. To what extent other types of forest landowners have access to this technology in the seedling marketplace depends on perceived needs, available information and proof of its safety. Available information must be configured for an increasingly science-illiterate population. Adoption rates of genetically modified (GM) forest trees by U.S. private forest owners with acreages over 1,000 are predicted to be low. GM conifers or other forest tree species enhanced for wood quality, phytomediation, herbicide or pesticide resistance will be less attractive than clonal (non-GM) forest tree species planted for the purposes of wildlife, aesthetics and restoration of heritage trees. With or without GM technology adoption, private landowners must be part of the discussion and policy-making related to commercial-scale use of GM conifers by virtue of the vast forests they control.

1. INTRODUCTION

Genetically modified (GM) or transgenic trees have been promoted as a solution to the loss of biodiversity, preventing insect and disease infestations, remediating pollutants in the environment, reducing tree stress and the chemicals needed to produce paper, carbon sequestration, even a source of fuel and pharmaceuticals (CAMPBELL *et al.*, 2003; FENNING AND GERSHENSON 2002; HERSCHBACH AND KOPRIVA 2002). At the same time, GM trees are being lambasted as a threat to biodiversity via gene escape and introgression into wild gene pools (VAN FRANKENHUYSEN AND BEARDMORE 2004), or because of invasiveness, harm to insects, proprietary ownership by multi-national companies and perceived threat to individual family forests (SFN 2005; LANG 2004). GM trees at present are largely confined to the domain of laboratories and initial field trials where risks and benefits can be intensively studied. Research about the ecological risks of transgenic trees is ongoing yet is scant in comparison to similar studies on the deployment of transgenic crops (VAN FRANKENHUYSEN AND BEARDMORE 2004).

Genomics and the concomitant understanding of gene expression, metabolites and plant physiology are here to stay. GM trees, as research tools, are poised to yield significant insights about plant evolution, wood formation, growth, stress and metabolic functions within plants (HERSCHBACH AND KOPRIVA 2002). With at least 33 genetically transformed forest tree species having already been successfully produced, it appears that once regulatory hurdles are cleared, GM trees could be available in the marketplace within the decade. Following the trend in the horticultural tree crop arena, GM tree research could soon be shifting towards modification of traits that are more consumer oriented (DANDEKAR et al., 2002). How, when, and where genetically modified trees reach the marketplace will impact the private forests in the United States.

Private forests comprise 392 million acres in the United States. There are currently over 10.3 million private and family-owned forests nationwide. Owners of more than 100 acres of forestland number only 534,000, yet they control one third of all private forest land, a staggering 139 million acres (BUTLER AND LEATHERBERRY 2004). The importance of private forestlands, especially larger tracts, cannot be overstated, as they have traditionally been the source of public goods (e.g. water, wildlife habitat, and scenic beauty) and a consistent supply of wood fiber and stumpage for the forest products industry. Private forests, and the renewable resources they generate and protect, are critical to employment, sustainability and viability of many rural economies. Urbanization, globalization, corporate mergers and shifting overseas production threatened traditional domestic markets and, thus, the stability and viability of private forests. Concerns about the sustainability and viability of domestic private forests fuel both the development and the discouragement of the technology behind GM trees.

Understanding the needs and motivations of private forest landowners yields insights into their likelihood of adoption and potential support for GM trees. This chapter provides a review of the steps that precede innovation adoption as a precursor to a discussion on the involvement and education of private forest landowners about GM tree development and deployment. Predictions of GM tree products likely to be popular with private forest owners are advanced. Suggestions for additional opinion research are made in an effort to ascertain which GM solutions may provide the greatest public benefit and appeal for private forests and their owners.

2. WHERE DO PEOPLE STAND ON GM TREES?

Discussions about the promise or potential problems of genetically modified trees have largely taken place outside of the media. The GM tree debate has been confined to its proponents, researchers, regulators, academics and the organized community of opponents. The GM tree debate is a present-day example of what HOBAN (2000) describes as *social arena theory,* whereby interest groups try to sway public policy in directions that support their own self-interest. As with other controversies, the protagonists and antagonists involved in the in public debate reflect opposing sides of a continuum. Typically, as in this instance, somewhere in the middle exist the vast majority of consumers who exhibit little interest or concern about the technological risks until such time as they are directly impacted by them (e.g. food as it relates to GM crops).

The discussions about the potential merits and detriments of GM trees are still in their infancy, and reflect the fact that experimental plants exist only within laboratory and a few field test sites. The debate is largely academic and hypothetical by necessity, because the technology is too recent to truly study its impact. Many opponents believe that a moratorium is needed to allow for the study of all potential impacts (transgene spread, non-target impacts, ecosystem and gene stability). It is the very lack of knowledge about how transgenes interact with the environment that impedes public confidence and regulatory approval that can lead to commercial release (van FRANKENHUYSEN AND BEARDMORE 2004).

The phenomenal advances in genomics and new techniques (microarrays, automated sequencing, genetic transformation) have in many ways outpaced the ability of our current intellectual and regulatory systems to process them. Other chapters in this volume provide updated overviews of the present regulations of transgenic releases. Moreover, the science and techniques required of this emerging science is complex and requires a significant investment of time and resources simply to understand and be conversant in this emerging area of science. The potential impacts of GM trees on private lands and owners have yet to be fully explored, yet opinion research on agricultural biotechnology can yield insights about likely acceptance of GM trees.

The controversy surrounding the release and trade disputes about GM crops provides insights into the likely acceptance of transgenic trees. An insightful, historical account of the development GM crops, their release in the U.S. and opposition in Europe is detailed in a popular press book entitled *Dinner at the New Gene Cafe* (LAMBRECHT 2001). Clearly, a difference in acceptance exists within and across societies and nations. HOBAN (2000) noted that 15% of U.S. consumers listed biotechnology as a serious food risk over a three-year survey period. Conversely, three-quarters of those same consumers rated microbial contamination and nearly two-thirds rated pesticide residues as serious food hazards during the 1995-1997 time frame. HOBAN (2000) also noted a consistently higher perception (rating) of serious food hazard risk of GM foods by Europeans compared to U.S. consumers. Forty-four percent of Europeans respondents rated genetic engineering as a serious food hazard. HOBAN (2000) acknowledges that consumer attitude toward transgenic technology is related to general beliefs about science, technology and food and thus was rather negative about the prospects for food biotechnology in Europe.

Table 1. U.S. Total Forestland Acreage by Broad Ownership Type, Region, 2003.

REGION	PRIVATE	PUBLIC	TOTAL
North	128,317,000	41,368,000	169.685,000
South	188,845,000	25,758,000	214,603,000
West	75,569,000	160,448,000	236,017,000
United States	392,730,000	227,574,000	620,305,000

2.1. Extent, Size and Importance of Private Forests

United States forests cover 620 million acres, of which 392 million acres are in private ownership (SMITH et al., 2004; Table 1). More than one half of all private forest acreage, nearly 189,000,000 acres, is located in the South. Private forests located in the northern and central states comprise roughly a third or 128,000,000 acres of the U.S. private forestland. The remainder of the private forests are located west of the Mississippi River and account for more than 75,000,000 acres.

Family ownership accounts for 261 million acres, more than one in three of the nation's forested acres. Business interests control approximately 131,000,000 forest acres, one-third of all private forests (BUTLER AND LEATHERBERRY 2004; SMITH AND VISSAGE 2004). Private forestlands outnumber corporate holdings by 2 to 1 in the North and South, whereas in the Western half of the United States private forests approach parity between family and business ownership. Note the dominance of Southern forests in both business and family ownership.

Table 2. Private U.S. Forestland Acreage by Region, 2003 (Butler and Leatherberry 2004; Smith et al. 2004).

REGION	FAMILY	BUSINESS	TOTAL
North	93,866,000	34,451,000	128,317,000
South	127,559,000	61,286,000	188,845,000
West	40,215,000	35,354,000	75,569,000
United States	261,639,000	131,091,000	392,730,000

2.2. Large Private Forests

A strong relationship exists between the size of forestland holdings and observed probability of seeking management advice and harvest timber (BUTLER AND LEATHERBERRY 2004). Larger forest ownerships undoubtedly offer economies of scale not typically available to smaller tracts. Moreover, a business-like approach is required of larger ownerships simply because few entities can afford to amass or maintain larger forests without systematic timber receipts or other incomes flows derived from management. As we shall review later, large ownerships have been positively correlated with education, socio-economic class, education and other variables to forest management adoption. If GM trees are viewed simply as a new innovation for forest management, it is likely that larger ownerships will deploy them first and most successfully. The dominance of large private forest ownership and acreage within the Southern region suggest that GM trees or similar innovations may have greater acceptance there, as opposed to other regions within the United States.

Table 3. Size Distribution, Number and Acreage (in Millions) of U.S. Large Private Forest Ownerships By Region, 2003 (Source: BUTLER AND LEATHERBERRY 2004).

Acreage Class	Northern Owners (acres)	Southern Owners (acres)	West Owners (acres)	U.S. Total Owners (acres)
100–499	180,000 (29.9 M)	243,000 (43.9 M)	64,000 (11.7 M.)	487,000 (85.74 M)
500–999	6,000 (3.6 M.)	18,000 (11.1 M)	9,000 (5.6 M)	33,000 (20.3 M)
1000–999	2,000 (2.7 M)	8,000 (13.7 M)	4,000 (7.5 M)	14,000 (24.0 M)
5000 +	< 1,000 (1.5 M)	< 1,000 (5.5 M)	< 1,000 (2.3 M)	1,000 + (9.3 M)
Total	< 189,000 (37.8 M)	< 270,000 (74.4 M)	< 78,000 (27.2 M)	535,000 (139.3 M)

Note: Data may not add to totals because of rounding.

2.3. What Motivates Private Forest Landowners?

With a population of 10 million and rising, individual private forest landowners in the U.S. nearly defy categorization. Most private forest landowners have multiple reasons for owning their forests, yet four reasons predominate: 1) part of their home or farm; 2) maintaining the natural, aesthetic nature of their holdings; 3) recreational purposes; and 4) maintaining a family legacy (BUTLER AND LEATHERBERRY 2004). Interestingly, the order and primacy of these ownership objectives have not changed appreciable in the last decade (BIRCH 1996)

Private landowners in the American South differ from national ownership objectives on three key areas: land investment, family legacy and timber production. By viewing these differences on an acreage or area basis the objectives of the largest landowners gain prominence. Timber production as an ownership objective historically ranks near the bottom of most survey results. The southeastern United States follows this same trend when viewed in terms of ownership numbers, yet when interpreted on an area or acreage basis, timber production ranks third in reason for ownership (Figure 1). Regional differences are clear: Timber production is very important or important to landowners that control 4 in 10 forest acres in the South (BUTLER AND LEATHERBERRY 2004). The importance of timber as an ownership objective on larger tracts is well documented and has ramifications for technology transfer and deployment of GM trees as will be discussed in more depth below.

Figure 1. Comparison of Ownership Objectives by Area of U.S. versus Southern Family-Owned Forest Land (Source of Data: NWOS, 2003).

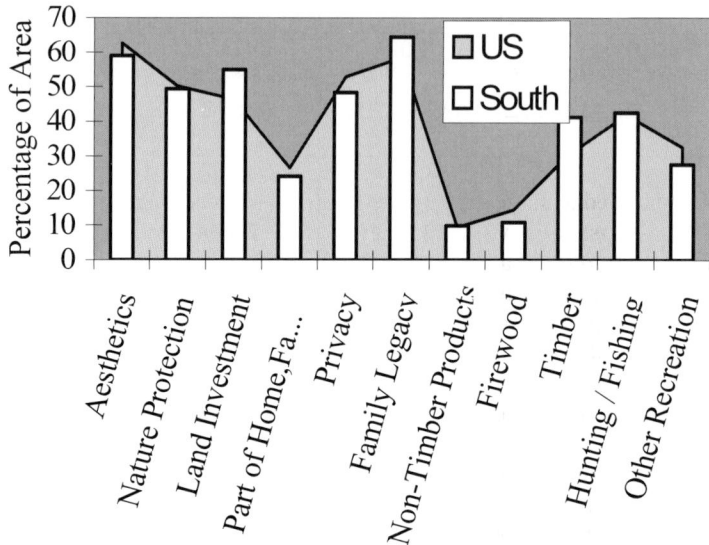

3. INNOVATION-DECISION PROCESS AND GM TREES

The innovation-decision process applies to the introduction of transgenic forest trees to private forest owners. It refers to a series of actions and choices by which an individual or entity evaluates a new idea and determines its suitability to their business or personal activities (ROGERS 2003). It begins once the awareness or knowledge of an innovation is received. A person, upon becoming aware of an innovation, forms an attitude about it, followed by a decision to adopt or reject it. If the decision is favorable, then the innovation can be implemented in its present form. If judged unfavorably, then the innovation can be modified to suit situation or need during the implementation stage or simply dismissed outright. A confirmation period typically follows innovation adoption, where reinforcement of the decision is sought. Should confirmation fail, individuals may reverse the decision to undertake an innovation. Reversals would likely follow the receipt of conflicting contradicting, or unfavorable messages about the innovation.

The innovation-decision process is theorized to include decision stages, as shown below in a linear, orderly progression (Figure 2). In actuality, the acceptance and adoption of innovations may be less systematic, returning to its beginning or a previous step as additional information or insights occur.

Figure 2. The Innovation-Decision Process (ROGERS 2003).

---

knowledge ⇨ persuasion ⇨ decision ⇨ implementation ⇨ confirmation

---

Knowledge – The knowledge stage constitutes the first exposure to an innovation or practice. Some debate exists over whether the exposure to an innovation is accidental or actively sought by an individual. ROGERS (2003) notes that individuals generally tend to expose themselves to ideas that are in accordance with their interests, needs or existing attitudes and avoid discordant messages. When an individual is unaware of an existing or potential need, a change agent may help to increase awareness of that need or how the innovation can address that need directly. In other cases, a change agent can advance the understanding of an innovation in an effort to increase its adoption.

An advice source or change agent for private forest landowners can be as varied as a professional forester to a fellow landowner. In the recent NATIONAL WOODLAND OWNER SURVEY (2003), more than one in five US southern family forest owners identified state forestry and private consultants as their advice source. Less than ten percent of those same landowners identified federal agency, industry or extension as their source of advice (BUTLER AND LEATHERBERRY 2004b). Information processed during the knowledge stage is cognitive in nature (e.g. how-to information or principles of how the innovation works). Should public and private foresters become *de facto* GM trees advice sources, then a wholesale continuing educational effort would be required before any wholesale adoption could be realized.

Persuasion – During the persuasion stage, an individual forms a favorable or unfavorable attitude about the innovation. Individuals actively seek information during the persuasion stage, in contrast to the passive way in which they received knowledge about the innovation itself. A general perception is formed of an innovation and the perceived attributes are judged based on their relative advantage, compatibility, and complexity.

Individuals will think hypothetically about the merits of an innovation. Persuasion can take place individually and might entail reading, discussions and assessment of the risks and uncertainties as they compare to anticipated benefits. If the envisioned benefits outweigh the anticipated risks the individual may be persuaded to adopt the innovation. Peers are the preferred source for much of the information used to decide whether to adopt or reject an innovation. Unlike the thinking style used during the knowledge stage, the affective or feeling style is more commonly used during the persuasion stage (ROGERS 2003).

Decision – The decision stage occurs when an individual or entity chooses to adopt or reject an innovation. Often the decision process can include a small scale trial or 'trial by others'. Trials can minimize the risk or uncertainties associated with an innovation. Successful demonstration or small scale testing of GM trees by private landowners would be an essential component of a technology transfer effort, especially at this critical decision stage.

Implementation – Putting an innovation into place marks the implementation stage. During this stage, time and resources are invested to seeking information that can help avoid or overcome problems typically encountered with innovation implementation. The length of the implementation stage varies with the complexity of the innovation, its availability, and the extent to which modifications (reinvention) are necessary (ROGERS 2003). The implementation phase is complete when an innovation becomes standard, or routine.

Confirmation - The confirmation stage marks an individual's attempt to reinforce their innovation decision after it has been made. The confirmation process is an attempt to quell or eliminate any dissonance that an adopter feels about their decision. This second-guessing approach can stem from the size of capital investment, fears about the future, and the negative messages and rumors about the drawbacks of an innovation.

4. CHARACTERISTICS INFLUENCING LANDOWNER USE OF NEW INNOVATIONS

The perception of the benefits and drawbacks of innovations can help explain the rate of adoption by individuals. Understanding the five key characteristics, by which individuals and entities evaluate innovations, can facilitate better prediction of the adoption of forestry practices or participation in incentive programs. More than half, and as much as 87% of the variance in rate of adoption of innovations can be explained by the perceived attributes described below (ROGERS 1995, 2003). Innovation characteristics collectively determine the manner in which an individual will perceive the 'total package' – a sure thing, on one hand, or a risky bet, on the other.

Figure 3. Perceived Attributes of Innovations (from ROGERS 2003).

Relative Advantage	the degree to which an innovation is perceived as better than the idea it supersedes.
Compatibility	the degree to which an innovation is perceived as consistent with existing values, past experience, and the needs of potential adopters.
Complexity	the degree to which an innovation is perceived as difficult to understand and use.
Trialability	the degree to which an innovation may be experimented on a limited basis.
Observability	the degree to which the results of an innovation are visible to others.

Relative advantage is synonymous with 'what's in the best interest of the potential adopter.' The extent to which GM trees reward or benefit landowners will ultimately impact his/her decision to adopt. Individuals weigh the potential benefits by measures familiar to them: profitability, low relative cost, and quick return. HAYMOND (1988b) found the importance of quick monetary return to be a significant predictor of silvicultural practice adoption by opinion leaders in South Carolina. Commercial thinning and timber stand improvement, both yielding income, were used by the vast majority of landowners in her study. Yet adoption isn't always about money *per se*. Relative advantage of forestry practices, like GM trees, relates to how well those practices match the non-timber desires of landowners. Improvement of other uses (other than timber) was found to be a significant attitudinal variable for predicting silvicultural practice use by opinion leaders in South Carolina (HAYMOND 1988b).

New ideas and forestry practices must be compatible with the values, needs, and experiences of potential adopters. Furthermore, the delivery or communication of innovative practices by foresters and resource professionals must be made with respect to the norms or beliefs of the potential adopters. ROGERS (2003) details the importance of homophily in the flow of communication of innovations, that is, individuals are more apt to act upon information received from people who are like them. Thus potential purveyors of GM forest tree technology must be viewed as of the same mind as potential adopters.

The degree to which an innovation is perceived as difficult to understand and use has been a persistent barrier to adoption. Simplicity in message of the benefits and rewards of an innovation are critical to its adoption. New ideas are slow to become accepted, as was the case with improved genetics of pine trees, because of the long time period required to demonstrate success (ZOBEL AND SPRAGUE 1993). GM trees will succeed or fail based on how well they are understood and presented to potential users.

The degree to which forestry management may be tested on a pilot scale is central to the missions of extension and technical assistance to private landowner.

Limited trials of GM technology via single trees or small plots with limited inputs, will likely receive greater acceptance (e.g. heritage tree restoration or clonal wildlife mast tree). Whereas, GM trees with improved growth, strength or quality traits will require intensive site preparation, fertilization and herbicides for controlling competing vegetation to fully succeed. Such trials will lag in adoption rates by all but the most intensive managers who have the necessary capital, expertise and resources.

Observations of most forest practices typically occurs at the macro scale and may take a decade or more to become apparent. When an open field is planted to pines the result is evident in a few years. Other cultural treatments (herbicides, bedding, thinning) may be more difficult to discern without a comparison treatment nearby. Companies or entities with a mission to promote GM trees and technology would advance their cause greatly by providing observations at the forefront of their promotional activities. Therein lies the dilemma: promotion of GM fields and even greenhouse trials can lead to their untimely destruction by radical opponents who may destroy the plant materials or the laboratory itself.

4.1. Innovation vs. Diffusion?

The value of an innovation-decision model is its ability to describe the component parts of the adoption process, and its near universal applicability across cultures (FEDER *et al.*, 1982). Diffusion refers to the process by which new ideas are communicated and is central to the innovation-decision process. Diffusion research explores the manner in which an innovation is communicated through certain channels, over time, among the members of a social system (ROGERS 2003). Alerting potential adopters of GM forest tree merits can succeed or fail largely on how information diffusion takes place and by whom.

The adoption of an innovation is primarily an outcome of the learning or communication process (BROWN 1981). Diffusion research seeks to identify and examine effective information flow, the characteristics of that flow, how the information is received, and resistance to adoption. Communication channels describe the manner in which messages are transmitted between individuals. The relationship between individuals impacts the nature of information exchange and ultimately, the receptivity or likelihood of innovation adoption.

Two types of communication channels are advanced by diffusion research: mass media, and interpersonal communication. Each channel type is further subdivided by its source: local or cosmopolitan. Mass media raises the awareness or knowledge level about an innovation. Mass media communication can rapidly reach large audiences, create knowledge and spread information. At best, mass media can change weakly-held attitudes. Mass media is important at the knowledge stage and can prompt innovators to try a new idea, especially in developed countries but to a much lesser extent in the developing world.

Interpersonal channels describe the information exchange between or among individuals. Interpersonal communication can form and change strongly held beliefs. Interpersonal communication is the dominant manner by which late adopters learn about innovations and persuaded to adopt them. Two-way exchange of information allows for the clarification and the breaching of social-psychological barriers that influence perception, retention, and exposure to the benefits of an innovation (ROGERS 2003). Late adopters are unlikely to be swayed by mass media communications because media lack persuasive power to motivate. Thus, late adopters rely on interpersonal channels nearly exclusively when deciding to adopt an innovation.

Cosmopolite communication channels originate from outside a social system are important to the knowledge phase of innovation adoption. These sources may include national and regional press, university research, farm shows and exhibitions. External communication is of lesser importance in the persuasion stage. Localite sources, in contrast, increase in importance during the persuasion stage, especially for later adopters. Should GM trees be made available to private forest landowners the use of local communications channels will be of paramount importance to late adopters, especially after early adoption has taken place.

MUTH AND HENDEE (1980) stress that new information flow in social systems is non-random; key individuals are sought out for information, opinions, and suggestions. Through key opinion leaders, innovations are diffused laterally to their peers. The basic tenet of diffusion strategy is that individuals may be persuaded to adopt through communications, both through a change agent and media communications (BROWN 1981). DOOLITTLE AND STRAKA (1987) confirmed the greater likelihood of changing client perception about forest regeneration using interpersonal communications, rather than by impersonal means. The authors note that late adopters were less likely to belong to highly interconnected social systems, were more clannish, and exhibited little contact among groups than did early adopters. They advise that personal contacts or individual mailings be employed to inform non-industrial private forestland owners.

Personal contact is more effective than targeted mailing or mass media in transmitting specific NIPF management advice to Michigan forest landowners (WEST et al., 1988). Their data showed that mass media had almost no effect on adoption of NIPF practices. They found that peer-based information and advice was received about as often as contact from professional foresters or from mass media. The importance of sound, effective management advice transmitted through carefully selected opinion leaders was stressed. The implications for GM tree advocates and detractors are clear: that opinion leaders are the key to reaching private forest owners.

4.2. Characteristics of Active Management Forest Landowners

The forestry research literature is replete with studies of the explanatory factors or characteristics of nonindustrial private forest landowners linked to propensity to reforest, harvest timber or manage for timber production (ROYER AND KAISER 1983; GREENE AND BLATNER 1986; ROYER 1987, ROYER AND MOULTON 1987; HYBERG AND HOLTHAUSEN 1989; HARDIE AND PARKS 1991; NAGUBADI et al., 1996; RATHKE AND BAUGHMAN 1996). Input from sociological theory, like innovation adoption behavior, has been scant in forestry research. This is particularly problematic because forest management activities are truly classified as innovations (GRAMANN 1984).

Early studies of the adoption characteristics of woodland owners (SOUTH et al., 1965; SOLLIE 1965) established positive relationships between forest management practice adoption and certain demographic or land characteristics (Table 4).

Table 4. Characteristics of landowners and relationship to active forest management.

Characteristic	Relationship / Correlation
Education	Positive
Acreage (Farm and Woodland)	Positive
Attitude Toward Woodland / Forest Conservation	Positive
Occupation of Type of Farmer / Owner	Varied
Use of Outside Information	Positive
Use of Numerous Information Sources	Positive
Socio-economic Status or Rank	Positive
Knowledge of Cost-share Assistance	Positive
Source of Family Income from Farm	Positive

Recent diffusion research on pine regeneration conducted in Alabama (DOOLITTLE AND STRAKA 1987) confirmed many of ROGERS (2003) generalizations about the characteristics of early adopters. Early adopters were found to have more education, higher social status, larger production units, a commercial - economic orientation, more specialized operations, favorable attitude toward credit, and higher level of importance placed on timber management. Several variables were found to be positive and significant in explaining landowner regeneration behavior following harvest on pine sites (Table 5).

Table 5. Explanatory Variables of Southern Landowners Linked to Pine Regeneration and Silviculture Practice Implementation (DOOLITTLE AND STRAKA 1987 and HAYMOND 1988a, respectively).

Explanatory Variable	Statistical Significance
PINE REGENERATION	
Venturesome Rating	$p < .05$
Innovative Rating	$p < .05$
High School Graduate	$p < .05$
Timber Management Important	$p < .01$
Income Above $15,000/ year	$p < .01$
Opinion Leadership (self-rated)	$p < .01$
Owned Over 100 Acres	$p < .01$
SILVICULTURAL PRACTICES	
Size of Forest Landholding	$p < .01$
Importance of Quick Monetary Return	$p < .01$
Importance of Improvement for Other Uses	$p < .02$
Importance of Increase Timber Production	$p = .09$

Adoption research of silvicultural practice use by South Carolina opinion leaders suggest that a timber orientation and costs of adoption may not fully explain adoption behavior to the extent that earlier studies have surmised. The importance of forest improvement for non-timber uses is clear and significant for the landowners in their study region. The apparent increased adoption behavior of multi-objective landowners was also found in a study of Finnish forest owners (KUULUVAINEN et al.,1996) and for Missouri private forest owners of mixed-objective typology as defined by KURTZ AND LEWIS (1981).

5. UNCERTAINTIES AND THE FUTURE OF GM TREES PLANTED ON PRIVATE FORESTS

Transgenic or GM forest trees and the promise that they may hold for private forestland are locked at a scientific and social crossroads. Before private forests can be a part of the restoration of American chestnut, carbon sequestration or other public benefits, broad-scale education and support must take place. More research is needed about the potential risks, benefits and ethics associated with GM tree deployment (GRESSEL AND ROTTEVEEL 2000; THOMPSON AND STRAUSS 2000). And from the author's perspective, that research and its subsequent dissemination must focus on the economic, ecological and sociological perspective both in the developed and undeveloped world. A wealth of resources has coalesced around forest tree biotechnology and its potential commercial uses in general.

Additional public support can be garnered and sustained as this technology portfolio is focused on environmental and energy concerns of greater society. The subject of forest trees galvanizes many political causes ranging from individual rights, economic viability to environmental justice. A serious and deliberate assessment of the risks associated with the deployment of GM trees may not quiet all opposition. To address that likely scenario, all efforts to ensure biosafety through regulation and agency oversight must be transparent and coordinated (MCLEAN AND CHAREST 2000; MULLIN AND BERTRAND 1998). In addition, the exploration of ethics for tree biotechnologists related to procedures, unintended consequences and intellectual property rights is warranted (THOMPSON AND STRAUSS 2000).

YANCHUK (2001) suggested that research and development conducted by forestry biotechnologists will largely be used in support of advancing clonal forestry. Initial deployment of GM trees will, by necessity, occur in the countries with the most experience with clonal and exotic forestry. GM trees are not likely to follow the rapid adoption observed in GM crops in the United States, at least not on national and private lands. That being said, it is important to note that clonal forestry has finally arrived on the North American continent. A commercial supply of clonal forest trees, produced via somatic embryogenesis (SE) from non-GM conifer stock, was sold to industrial and private landowners during the 2004 planting season (ATTREE 2004).

Currently, SE clones cost 8 to 10 times conventionally grown seedlings but potentially have captured a quantum leap of genetic gain over their fertilized brethren. As anticipated cost savings result from the increased production of clonal plantlets, a viable platform for DNA recombination will emerge. LIBBY (2004) speculates that clones will become the norm for intensive plantation management. Continued advances in propagation technology, supported by conventional tree-breeding programs will synergistically culminate in use of genetically engineered specific clones (LIBBY 2004). Assuming GM trees can successfully pass deregulation for an unconfined release, they may become a fixture in the world's forested landscape within the decade, if not sooner. Should GM trees become deregulated for open release in the United States, their theorized acceptance might follow the projected schedule (Table 6).

SALLEH (2001) offers a more robust prediction for GM trees in Asia and the Pacific, complete with finite milestones:

> " In 2050, a new transgenic tree that grows fast, absorbs carbon efficiently and produces edible shoots and fruits will be undergoing genetic modification to produce sap that can be used as fuel to drive automobiles without any processing. Scientists will aim to start economic production of this fuel by the year 2075."

Whether or not we will see a fuel derived directly from GM forest trees is anyone's guess. What is certain is that the molecular domestication of several commercial tree species is at hand. To what extent GM technology is to be brought to the marketplace depends on societal need, support and proof of its safety to an increasingly science-illiterate populace. The private landowners, by virtue of the forests they control, should be part of the discussion and ultimate decisions related to use of GM trees.

Table 6. Predictions for GM tree adoption by United States private forest owners with acreage over 1,000 (i.e. probable early adopters).

Innovation Type	Near term	Long term
Clones w/ Novel Traits for Beauty Enhancement or Wildlife (mast-fruit)	Medium	High[+]
Clones w/ Novel Traits for Restoration/ Heritage trees	Medium	High[+]
Clones w/ Novel Traits for Growth Enhancement	Low-Medium	Medium[*]
Phytoremediation	Low	Medium-High[*+]
IR – Insect Resistance	Low	Medium[*+]
HR- Herbicide Resistance	Low	Low-Medium[*]
LM- Lignin Modification	Low	Low-Medium[*]

[*] With favorable licensing, availability and awareness of successful trials or demonstrations
[+] Site Specific Locations

6. REFERENCES

ATTREE S.M. 2004. Developing a commercial somatic embryogenesis production platform for conifers. *Abstract from:* 2004 IUFRO Joint Conference: Forest Genetics and Tree Breeding in the Age of Genomics: Progress and Future. Nov. 1-5,2004. Charleston, S.C. USA.
BROWN A. L. 1981. *Innovation Diffusion: A New Perspective*. Methuen, New York. 345 p.
BUTLER B.J AND E.C. LEATHERBERRY. 2004. America's family forest owners. J. For. **102**: 4-9.
CANCIAN F. 1979. Innovator's Situation: Upper-Middle Class Conservatism in Agricultural Communities. Stanford Univ. Press, Stanford. 159 p.
CAMPBELL M.M., A.M. BRUNNER, H.M. JONES AND S.H. STRAUSS. 2003. Forestry's fertile crescent: the application of biotechnology to forest trees. Plant Biotechnology Journal **1**: 141-154.
CARLSON J.E. AND D. A. DILLMAN. 1988. The influence of farmers' mechanical skill on the development and adoption of a new agricultural practice. Rural. Sociol. **53**: 235-245.
DANDEKAR A.M. *et.al.*, 2002. Different genes for different folks in tree crops: what works and what does not. HortScience **37**: 281-286.
DOOLITTLE L., AND T.J. STRAKA. 1987. Regeneration following harvest on non-industrial private pine sites in the South: A diffusion of innovations perspective. South. J. Appl. For. **11**: 37-41.

ENGLISH B.C., C.D.BELL, G.R.WELLS, AND R.K. ROBERTS. 1997. Stewardship incentives in forestry: Participation Factors in Tennessee. South. J. Appl. For. **21**: 5-10.

FEDER G., R.E. JUST, AND D. ZILBERMA. 1982. Adoption of Agricultural Innovations in Developing Countries: A Survey. Staff Working Pap. #542., World Bank, Washington. 65 p.

FENNING T.M. AND GERSHENZON J. 2002.Where will the wood come from? Plantation forests and the role of biotechnology. Trends in Biotechnology **20**: 291-296.

GRAMANN J.H. 1984. Sociological issues in non-industrial private forestry. The Rural Soc. **4**: 364-368.

GREENE J.L. AND K.A. BLATNER. 1986. Identifying woodland owner characteristics associated with timber management. Forest Science **32**: 135-146.

HARDIE I.W., AND P.J. PARKS. 1991. Individual choice and regional acreage response to cost-sharing in the South, 1971-1981. Forest Science **37**: 175-190.

HAYMOND J.L. 1988a. NIPF opinion leaders: What do they want? Journal of Forestry **86**: 30-32, 34, 35.

HAYMOND J.L. 1988b. Adoption of silvicutural practices by opinion leaders who own nonindustrial private forestland. South. J. Applied Forestry **12**: 20-23.

HERSCHBACH C. AND KOPRIVA K. 2002. Transgenic trees as tools in tree and plant physiology. Trees **16**: 250-261.

HOBAN T.J. 2000. Social controversy and consumer acceptance of agricultural biotechnology. J. Biolaw & Bus. **3**: 3.

HODGES D.G. AND F.W CUBBAGE. 1990. Adoption behavior of technical assistance foresters in the Southern pine region. Forest Science **36**: 516-530.

HYBERG B.T. AND D. M. HOLTHAUSEN. 1989. The behavior of nonindustrial private forest landowners. Can. J. For. Res. **19**: 1014-1023.

KUULUVAINEN J., H. KARPPINEN, AND V. OVASKAINEN. 1996. Landowner objectives and nonindustrial timber supply. Forest Science **42**: 300-309.

LANG C. 2004. Genetically modified trees- the ultimate threat to forests. Rainforest Movement and Friends of the Earth. http://www.wrm.org.uy/subjects/GMTrees/text.pdf 101 p.

LAMBRECHT B. 2001.*Dinner at the new gene café.* St. Martin's Press. NY,NY. 381 p.

LIBBY W.J. 2004. Clonal forestry: Yesterday, today, and tomorrow. *Abstract from:* 2004 IUFRO Joint Conference: Forest Genetics and Tree Breeding in the Age of Genomics: Progress and Future. Nov. 1-5,2004. Charleston, S.C. USA.

MARLIN C. B. 1978. A study of owners of small timber tracts in Louisiana. LA. State Univ. Agric. Exp. Sta. Bull. No. 710. August. 65 p.

MCLEAN M.A., AND CHAREST, P.J. 2000 The regulation of transgenic trees in North America. *Silvae Genet.* **49**: 233-239.

MULLIN T.J. AND BERTRAND S. 1998. Environmental release of transgenic trees in Canada- potential benefits and assessment of biosafety. For. Chron.74(2): 203-219.

MUTH R.M. AND J.C. HENDEE. 1980. Technology transfer and human behavior. J. For. 78(3): 141-144.

NAGUBADI V. K.T. MCNAMARA, W.L. HOOVER, AND W.L. MILLS, JR. 1996. Program participation behavior of nonindustrial forest landowners: a probit analysis. J. Agri. & Appl. Economics. 28(2): 323-336.

NWOS 2003.United States Department of Agriculture, Forest Inventory and Analysis National Woodland Owner Survey 2003 Preliminary Results http://www.fs.fed.us/woodlandowners/publications/nwos_draft_tables_july_2004.pdf

PENA L. AND SEGUIN A. 2001. Recent advances in the genetic transformation of trees. TRENDS in Biotechnology 19 (12): 500-506.http:tibtech.trends.com

POMEROY K.B. AND J.G. YOHO. 1964. North Carolina Lands: Ownership, use, and management of forests and related lands. Am. For. Assoc, Washington, 372 p.

RATHKE D.M. AND M.J. BAUGHMAN. 1996. Influencing nonindustrial private forest management through the property tax system. North. J. Appli. For. 13(1): 30-36.

ROGERS E.M. 2003. *Diffusion of Innovations.* Ed. 5. The Free Press, New York. 550 p.

ROSEN B.N. 1995. A longitudinal analysis of attitudes and marketing practices of non-industrial private forest landowners. North. J. Applied Forestry 12(4): 174-179.

ROYER J.P. AND H.F. KAISER. 1983. Reforestation decisions on harvested southern timberlands. J. For. 81(10):657-659.

ROYER J. P. AND R.J. MOULTON. 1987. Reforestation incentives: tax incentives and cost-sharing in the South. South. J. of Appl. For. 85(8): 45-47.

ROYER J.P. 1987. Determinants of reforestation behavior among southern landowners. For. Science 33(3): 654-667.

SALLEH N.M. 2001. Asia and the Pacific. *Unasylva* 204:52: 36-39.

SALTEIL J., J.W. BAUDER, AND S. PALAKOVICH. 1994. Adoption of sustainable agricultural practices: Diffusion, farm structure, and profitability. Rural Sociol. 59(2): 333-349.

SFN 2005. Southern Forest Network: Southern Sustainable Forests Website- Genetically Engineered Trees http://www.southernsustainableforests.org/ge%20trees.htm

SMITH W.B., P.D. MILES, J.S.VISSAGE, AND S.A.PUGH. 2004. Forest Resources of the United States, 2002. Gen.Tech. Rep. NC-241. St.Paul, MN. U.S.Dept. of Agri. 137 p.

STRUM L.S. 1992. Bolivian farmers and alternative crops: Some insights into innovation adoption. Unpub. Masters Thesis. N.C. State Univ, Raleigh. 80 p.

TAYLOR D.L. AND W.L. MILLER. 1978. The adoption process and environmental innovations: A case study of a government project. Rural Sociol. 43(3): 634-648.

THOMAS J.K., H . LADEWIG, AND W.A. MCINTOSH. 1990. The adoption of integrated pest management practices among Texas cotton growers. Rural Sociol. 55(3): 395-410.

VAN FRANKENHUYZEN K. AND BEARDMORE T. 2004. Current status and environmental impact of transgenic forest trees. Can. J. For. Res. 34: 1163-1180.

WEST P.C. J.M. FLY, D.J. BLAHNA, AND E.M. CARPENTER. 1988. The communication and diffusion of NIPF management strategies. North. J. Applied For. 5: 265-269.

ZOBEL B.J. AND SPRAGUE J.R. 1993. *A Forestry Revolution: The history of tree improvement in the southern United States.* Forest History Society, Carolina Academic Press.166p.

SECTION V

GOVERNMENT REGULATIONS AND BIOSAFETY

CHAPTER 13

CANADA'S REGULATORY APPROACH

ANNE-CHRISTINE BONFILS
Canadian Forest Service, Natural Resources Canada, Ottawa, Ontario, Canada

Abstract. Canada has adopted a cautious approach with its federal science-based regulatory framework, put in place in 1993 to require that the products of biotechnology meet high standards for human health and environmental safety. The framework is based on the development of regulations under existing legislation and using the *Canadian Environmental Protection Act* as a 'safety net' for products that would not be appropriately covered under other Acts. The *Seeds Act* allows the Canadian Food Inspection Agency to regulate the release of seeds (i.e. propagation material), including tree seeds. Regulatory review is triggered by the novelty of the plant material (plants with novel traits) rather than the specific means by which it was produced. In Canada, provincial and territorial governments own 77% of the forest and have legal authority to manage Crown land use. Sound regulations require policy developments that integrate the various national and international levels. Canadian regulatory principles and challenges will be described for trees with novel traits: terms and conditions for confined research field trials; evaluation criteria for unconfined release; biosafety issues; regulatory harmonization; pressures for change; and environmental safety research in support of regulations. Unconfined release of trees with novel traits is not expected soon in Canada, but is expected to occur. Before the first applications are submitted for regulatory approval, we have the opportunity to be proactive in research and regulatory developments.

1. INTRODUCTION

Genetic engineering offers a powerful research tool for advancing fundamental knowledge in genetics, particularly at the level of gene function. Applied research in this area pursues similar goals as conventional breeding, with the potential of more precise modifications, faster gains, and use of genes that would otherwise not be available for the development of specific desired characteristics. Together with strong propagation techniques, genetic engineering could thus allow the development of trees well suited for short-rotation intensive plantations for improved productivity from less land, and trees that could provide new options for addressing environmental threats or ecological restoration issues.

Although tree transformation and propagation techniques are continuously being improved, they are still at a proof-of-concept stage and many concerns are being raised by stakeholders related to: acceptation of plantation forestry; issues of genetic diversity and deployment; intellectual property and legal liabilities; uncertainties regarding long-term ecosystem effects; and lack of clear cost-benefit risk analysis. Uptake of these technologies by the forest sector will depend on many economical, social, and environmental factors that will need to be balanced before commercial applications can be envisioned.

C. G. Williams (ed.), Landscapes, Genomics and Transgenic Conifers, 229-243.
© 2006 *Her Majesty the Queen in right of Canada, Canadian Forest Service. Printed in the Netherlands.*

An integrated approach is required to fully benefit from these opportunities while minimizing the risks. In Canada, this is provided through the Canadian Biotechnology Strategy, an active horizontal framework that incorporates social, ethical, health, environmental and regulatory considerations. The implementation of the Canadian Biotechnology Strategy has resulted since its renewal in 1998 in the establishment of the Canadian Biotechnology Advisory Committee to advise the Government of Canada on emerging policy issues, as well as in significant new investments for genomics and regulatory research.

It is not the purpose of this paper to try and describe the complexities of the current debate around transgenic trees. Rather, it will focus on a description of the cautious approach provided through Canada's federal regulatory framework, put in place to require that the products of biotechnology meet high standards for human health and environmental safety.

2. FEDERAL REGULATORY FRAMEWORK

2.1. Approach

Canada approved the Federal Regulatory Framework for Biotechnology in 1993 (INDUSTRY CANADA 1993). Its basic premise is to provide a coordinated system to the regulations, with the consistent approach that safety concerns are posed primarily by the presence of a novel trait or substance in a product, rather than how the novelty was introduced.

The Federal Regulatory Framework of Canada is based on the development of regulations under existing legislation and using the *Canadian Environmental Protection Act* as a 'safety net' for products that would not be appropriately covered under other Acts, thus ensuring that there are no gaps in the framework. There is therefore no gene law in Canada.

As a result, the federal legislative authority for health and environmental assessment of biotechnology lies under several Acts of Parliament, which are administered by various departments and agencies responsible for regulating the development of equivalent products under more conventional techniques and processes. Examples applied to forest biotechnology include the *Seeds Act* for tree seeds, the *Plant Protection Act* for imports, the *Fertilizers Act* for bio-fertilizers and mycorrhizae, the *Pest Control Products Act* for microbial pest control agents and the *Canadian Environmental Protection Act* for microorganisms that would be released in the environment for purposes other than pest control (such as for pulp and paper processing). The Canadian Food Inspection Agency (CFIA) administers the *Seeds Act*, the *Plant Protection Act* and the *Fertilizers Act*, while the Pest Management Regulatory Agency administers the *Pest Control Products Act*. The *Canadian Environmental Protection Act* is co-administered by Environment Canada, for environmental safety, and Health Canada, for human health safety.

An interdepartmental committee meets regularly to ensure a coordinated approach in the development, ongoing improvements, and implementation of these regulations. The Canadian Forest Service participates as a member of this committee and contributes to the process by providing the above departments and agencies with scientific and technical expertise on forest sector biotechnology products.

2.2. Principles

The Framework provides a core set of common principles used by federal departments and agencies in the development of their detailed regulatory directives for specific classes of products derived through biotechnology. The principles of the regulatory framework are:

1. To maintain Canada's high standards for the protection of the health of workers, the general public, and the environment;
2. To use existing legislation and regulatory institutions to clarify responsibilities and avoid duplication;
3. To continue to develop clear guidelines for evaluating products of biotechnology which are in harmony with national priorities and international standards;
4. To provide for a sound scientific baseline on which to assess risks and evaluate products;
5. To ensure both the development and enforcement of Canadian biotechnology regulations are open and include consultations with stakeholders and the public; and,
6. To contribute to the prosperity and well-being of Canadians by fostering a favorable climate for investment, development, innovation, and adoption of sustainable Canadian biotechnology products and processes.

The federal Acts of most significance to trees with novel traits within the regulatory framework are the *Plant Protection Act* and the *Seeds Act*.

3. THE PLANT PROTECTION ACT

The purpose of the *Plant Protection Act* is to protect plant life and the agricultural and forestry sectors of the Canadian economy by preventing the importation, exportation and spread of pests, and by controlling or eradicating pests in Canada. Under this Act, the CFIA has developed Regulatory Directive D96-*13 Import permit requirements for plants with novel traits (including transgenic plants), and their products*, outlining the phytosanitary requirements to obtain import permits for plants with novel traits, including trees, and products derived from them such as fruit, tubers and grain (GOVERNMENT OF CANADA 1998).

4. THE SEEDS ACT

Under authority of the *Seeds Act* (GOVERNMENT OF CANADA 1985), the CFIA holds the mandate to regulate the required inspection and testing, the quality and sale of seeds in Canada, and the release of plants with novel traits in the environment (field tests and unconfined release).

4.1. Regulations and regulatory guidelines

Regulations under the *Seeds Act* were amended in early 1997 with the addition of Part V to meet the requirements of the *Canadian Environmental Protection Act* (CEPA). Part V of the *Seeds Regulations* requires the CFIA to evaluate environmental safety of PNTs, including trees, before approving research field tests or unconfined release. This is worded as the requirement to assess whether PNTs are toxic. According to the Seeds Regulations Part V, a PNT

> is toxic if it is entering or may enter the environment in a quantity or concentration, or under conditions that (a) have or may have an immediate or long-term harmful effect on the environment or its biological diversity; (b) constitute or may constitute a danger to the environment on which life depends; or (c) constitute or may constitute a danger in Canada to human life or health.

These amended regulations allowed the listing of the *Seeds Act* as exempted legislation under schedule 4 of CEPA 1999, indicating consistency with CEPA requirements and confirming that the CFIA has regulatory authority over the release of PNTs, including trees, in the environment.

Regulatory Directives outline the evaluation criteria to release PNTs and the information required from proponents to support their applications. Directives are more detailed than the regulations and much easier to update on a timely basis as new information becomes available from the scientific community or from practical experience in the field. Two Regulatory Directives guide the release of PNTs in the environment. Directive *Dir2000-07: Conducting Confined Research Field Trials of Plants with Novel Traits* (GOVERNMENT OF CANADA 2003) provides a guide on general and species-specific terms and conditions of confinement, including reproductive isolation, site monitoring and inspection, disposal of material, and post-harvest land use restrictions. Directive *Dir94-08: Assessment Criteria for Determining Environmental Safety of Plants with Novel Traits* (GOVERNMENT OF CANADA, 2004) describes the steps leading to authorization of unconfined release in Canada. A series of species-specific biology documents are used as regulatory companion documents to help identify potential risks associated with a given species. They describe the characteristics of the species, including distribution, life cycle, reproduction, and major interactions with other species in its range in Canada.

4.2. The concepts of novelty, familiarity, and substantial equivalence

Application of the regulations (commonly referred to as oversight) is triggered by the novelty of the plant material rather than the use of recombinant DNA technologies. This approach recognizes that the environmental impact of a new plant variety is determined by the characteristics of the plant itself, and not by the technique used to produce it and is largely supported by the scientific community. However, the introduction of new plant varieties (with the exception of exotic species) was not reflected as an environmental concern in Canada's regulatory system until 1988, when confined field trials of transgenic plants were first regulated (MCLEAN AND CHAREST 2000). Since 1995, about 85 percent of the authorized unconfined releases were for plants with novel traits produced through recombinant DNA techniques (CANADIAN FOOD INSPECTION AGENCY 2005).

Regulatory Directive Dir2000-07 defines the regulatory trigger as follows:

> A new variety of a species is subject to regulatory oversight when it possesses characteristic(s) or trait(s) novel to that species in Canada, i.e., the trait(s) itself/themselves, its/their presence in that plant species or its/their use are considered unfamiliar when compared with those of other plants of the same species that are grown in the Canadian environment and already regarded as safe, regardless of how the trait(s)

was/were introduced. As a result, plants developed through mutagenesis, somaclonal variation, wide cross, protoplast fusion or other techniques, as well as the plants developed through recombinant DNA technology may be considered PNTs.

Regulatory Directive Dir94-08 defines a PNT as:

> a plant containing a trait not present in plants of the same species already existing as stable, cultivated populations in Canada, or is present at a level significantly outside the range of that trait in stable, cultivated populations of that plant species in Canada.

In the concept of familiarity, the CFIA considers familiarity with the plant species, the trait, the introduction method, and cultivation practices. The introduction method through recombinant DNA techniques is currently considered unfamiliar, so transgenic plants normally trigger safety assessments. It is expected that, as experience with these materials increases, the techniques will become more trusted and the triggering factors will evolve.

The onus of determining novelty rests with developers. Environmental responsibility is expected to improve as a result, since developers, including those using conventional techniques, are required to consider issues of environmental impact that may not otherwise have been of concern to them. The determination of novelty for new forest tree cultivars could be quite challenging since they are largely undomesticated, and the concept of a 'stable, cultivated population' is largely inapplicable. Close communication between developers and regulators is encouraged by the CFIA and will be required on a case-by-case basis.

5. CONFINED RESEARCH FIELD TRIALS

Confined field trials are authorized by the CFIA subject to conditions designed to minimize environmental interactions and requiring close monitoring of the materials. These trials are necessary to evaluate the performance of the plants in a given environment and evaluate trait stability outside of the greenhouse setting. They also provide an important opportunity to study environmental safety and collect information required by the CFIA for unconfined environmental release.

The Canadian Forest Service was the first applicant to request approval by the CFIA to conduct a confined field trial of genetically engineered trees near Québec City, in 1997. The trial consisted of a 900 m^2 plot of poplar trees (*Populus alba* x *grandidentata*) genetically engineered to contain a marker gene attached to a wound-inducible promoter and unmodified trees used as controls, surrounded on all sides by two guard rows of unmodified hybrid poplar trees. Following are the CFIA conditions for research trials of poplar, set in consultation with authorities from the province of Quebec (K. THOMAS, pers. comm.):

1. The applicant must ensure that the trial seed and/or plant material are transported in clearly identified, secure containers and are kept separate from other seed and/or plant material.

2. Surplus transgenic trees from the trial must be clearly labeled and kept in a secure greenhouse facility or must be destroyed by mechanical means, heating or burning. Composting of this material is not an acceptable destruction method.

3. The trial term will be limited to eight (8) years from the date of commencement.

4. Seeding, transplanting and site maintenance machinery and equipment must be cleaned at the trial site to prevent dispersal of plant material.

5. In the case of accidental release, recoverable seeds or seedlings must be collected and destroyed, the site must be marked and monitored, and the Plant Biosafety Office notified immediately. Plants from unrecoverable seed or seedlings must be mechanically or chemically destroyed.

6. Two guard rows must be composed of non-transformed poplar (*Populus deltoides* x *trichocarpa*) producing no or very few suckers.

7. The trial material (including the guard rows) will be separated by a distance of at least 15 meters from other trees of the same or related species. The trial site and isolation distance must be monitored, at a minimum, twice a week during the period of flowering and budburst and, at a minimum, monthly during the growing season of the trial period to ensure that all suckers, precocious inflorescences and trees of same or related species that are not part of the trial are removed and destroyed.

8. During the trial period, all precocious inflorescences (if any) must be removed each year before the anthesis to prevent pollen dissemination and seed setting. Records must be kept of the date and number of flowering catkins removed from each genetic line.

9. Measurements from permanent surrounding landmarks must be provided for precise location of the site. Markers must also be placed at all corners of the trial site to identify the confined field trial boundaries. The markers must be obvious, identifiable and in place for the growing seasons of both the trial and the post-harvest restriction period.

10. Global Positioning System (GPS) coordinates must be taken precisely at all corners of each trial site. The GPS coordinates of each confined research field trial site location must be submitted to the CFIA within seven days after planting.

11. If a chemical treatment is used on the trial site location that requires a time until safe entry, a sign must be posted at the access to the trial indicating the date and time of spraying as well as the time until safe entry. This condition is intended to protect the health and safety of the CFIA inspection staff.

12. No plant material from these trials may enter the human food or livestock feed chain unless approved by Health Canada or the Feeds Section, CFIA, respectively.

13. Harvesting machinery and equipment will be cleaned of all residual plant material at the trial site prior to being moved to other locations. Plant material harvested (that is not to be retained) must be destroyed by burning, autoclaving, or burial to a depth of one meter. Composting of this material is not an acceptable destruction method.

14. Harvested seed and/or propagable plant material from the confined research field trial may only be retained if requested in the application and authorized by the CFIA. Any harvested seed and/or plant material must be clearly labeled, securely transported, and stored separately from other seed and/or plant material.

15. Applicants must provide the CFIA in writing within fifteen working days after harvest with information on quantity of seed and/or plant material harvested at the trial sites; date(s) of harvest; quantity of seed and/or plant material disposed of; location, method and date of disposal; quantity of seed and/or plant material retained and stored; storage location and method.

Disposal of plant material (propagable and/or non-propagable) includes harvested plant material as well as residual plant material on the trial site. If a trial is destroyed prior to harvest applicants must provide the CFIA in writing within fifteen working days after destruction with information on the trial's growth stage at the time of destruction, as well as the date and method of destruction.

16. A detailed trial log book must be kept. Records of the confined research field trial, including current season and post-harvest site monitoring, activities related to the trial site compliance, cleaning of machinery, transportation, disposition and storage of harvested seed and plant material, must be maintained by the applicant and made available to the CFIA upon request. A report summarizing the completed trial and experimental data, including any amendments to the original protocol, must also be made available to the CFIA upon request. Detailed records requirements can be found in section 3.8 of Regulatory Directive 2000-07.

17. The trees (including guard rows) must be cut down at the end of the trial period. Plant matter remaining at the end of the trial must be destroyed. Stumps and root systems must either be mechanically destroyed on site or removed and destroyed. The trial site must be tilled and any developing suckers after tillage must be destroyed.

18. The trial site must not be used to grow poplar trees from the date of termination of the trial until no suckers are observed for three (3) consecutive years. The site must be monitored, at a minimum, monthly during the post-harvest growing seasons to ensure that all volunteer plants, suckers and related species are removed and destroyed.

19. Applicants must notify the CFIA in writing of crop species planted on trial sites for each year the sites are subject to post-harvest restriction. This notification must be received every year by June 15.

The Canadian Forest Service initiated another field trial in 2000, again under strict conditions for confinement set by the CFIA, of white spruce lines carrying a gene from *Bacillus thuringiensis* for insect tolerance. These trials are planted for research purposes only and the tested materials are not targeted for future deployment.

6. UNCONFINED RELEASE: ENVIRONMENTAL SAFETY CRITERIA

Unconfined release of PNTs is authorized by the CFIA, with or without conditions, following thorough case-by-case environmental safety assessments. Included in the assessments are: precise evaluations of the novel proteins and the modified plant, stability of gene expression, comparisons with counterparts of the same species, considerations of weediness and invasiveness, ability to transfer genetic information to related species, potential to become a pest, potential to cause unwanted interactions with other organisms in the environment, and potential to cause negative impact on biodiversity. Evaluations also consider the unique combinations of species and traits, using biology companion documents as a baseline for comparison and for identifying species-specific considerations in relation to the novel trait. Unconfined release can be authorized when the testing and analyses indicate substantial equivalence, which means that the novel plant and its existing counterpart are found to be equally safe for the environment.

Unintended effects must also be evaluated. For example, modification of lignin is of importance to the pulping process for improving the ease of pulping and reducing the use of toxic chemicals, thereby reducing pollution. However, modification of this trait could change critical fitness properties of the forest tree, including tolerance to cold, rendering trees with modified lignin composition impractical for planting in cold-climate countries like Canada.

To date, no application for unconfined environmental release of a tree with novel traits has been submitted in Canada. The key biosafety considerations specific to forest trees are well covered in the literature (FRANKENHUYZEN AND BEARDMORE 2004; STRAUSS AND BRADSHAW 2001,2004; MULLIN AND BERTRAND 1998). Although forest trees differ significantly from agricultural crop plants, the biosafety issues are quite similar and could be adequately covered by the CFIA evaluation criteria described in Dir94-08.

Even so, addressing biosafety issues in forests are problematic. Due to the unique features of forest trees, forest tree populations and ecosystems, a sound impact assessment could potentially be very complex. Trees are undomesticated long-lived perennials interacting within complex ecosystems, forest tree populations are very diverse, and pollen movement can occur over enormous distances.

Several strategic questions arise: What would be the acceptable time frame to demonstrate safety? In the absence of stable, cultivated populations of the tree species in Canada, how would acceptable counterparts be defined? What would trees with novel traits be compared to for determining substantial equivalence? How would the natural range for a given trait in a species in Canada be determined? Do we have the tools and methods to predict and assess potential impacts? How much baseline knowledge of natural ecosystems will be required? How many interactions will need to be documented? Should the evaluation go beyond the characterization of trees with novel traits, and also cover impacts of plantations per se (e.g. impacts on biodiversity, on water resources and soil fertility, potential spread of pests and diseases, weed control)?

On the other hand, one could argue that environmental concerns regarding genetically engineered trees might be less than those associated with conventionally bred materials since the genes are extremely well characterized and can be tracked easily. For example, with the concern of gene flow. Gene movement from planted tree materials to wild relatives is probably inevitable, as well as the establishment of offspring outside plantations, for species like spruce, pine, or poplar. The probability of occurrence is very high. But the consequence, and therefore the environmental risk, may be negligible depending on the trait, and whether or not it would be favored by natural selection. With genetically engineered trees, the precise knowledge at the molecular level of the added trait facilitates this analysis greatly.

7. ENVIRONMENTAL SAFETY RESEARCH

The environmental impacts of transgenic trees and the research required to address them have been the specific subject of several international meetings. The Organization for Economic Co-operation and Development (OECD), with participation from six member countries including Canada, organized an International Workshop on Environmental Impacts of Transgenic Trees, 13-15 September 1999, Trondheim, Norway, to clarify tree specific questions and submit considerations to the OECD Working Group on Harmonization of Regulatory Oversight of Biotechnology. The full report can be found on the OECD Web site at http://www.oecd.org. The first International Symposium on Ecological and Societal

Aspects of Transgenic Plantations was organized in 2001 in Oregon, in parallel with a meeting of the International Union of Forest Research Organizations. Proceedings can be found at http://www.fsl.orst.edu/tgerc/iufro2001/eprocd.htm. The 2003 World Forestry Congress in Quebec hosted a side event dealing with regulatory challenges related to trees with novel traits.

While it is recognized that testing will not remove all scientific uncertainty, each of the above discussions resulted in the same definitive conclusion: more research is necessary, using a multidisciplinary approach. Research requirements include baseline understanding of the species biology and ecosystem interactions; thorough characterization of the transgenic tree, as well as of the transferred gene and its products; rigorous analysis of potential environmental consequences in the specific environments in which the trees might be deployed; and the development of appropriate tools, protocols, and criteria for risk assessment.

7.1. Snapshot of environmental safety research at the Canadian Forest Service

Research required to address environmental safety concerns constitutes an integral component of current CFS biotechnology research, although with limited resources. Following are a few examples of research areas currently being pursued by the CFS.

One of the key issues related to transgenic trees is gene flow. The direct study of gene flow from research trials of these trees is limited by the stringent conditions of reproductive isolation required by regulation. The use of model systems using exotic and hybrid tree species is an alternative that could help addressing some of the concerns. In collaboration with Laval University and *le Ministère des ressources naturelles du Québec*, the CFS is using a model system with exotic poplars and larches that can readily cross. Species-specific diagnostic markers were first developed for tracking purposes, then seed was sampled from maternal trees of natural populations, along perpendicular transects installed where native stands are located adjacent to mature clonal plantations. The DNA of 600 to 1000 seeds per site over two years will be analyzed for poplar and larch. With this, we will estimate the extent of introgression of genes from the exotic species into native species, and this information can be used to develop simulation models (N. ISABEL pers. comm.).

Pollen and seed have the greatest potential to facilitate gene movement. We have identified black spruce genes for cone initiation and development, the applied goal being to produce sterile trees (RUTLEDGE *et al.*, 1998). However, trees require an extensive time to flower so sterility is difficult to test, and engineering of early flowering in spruce presents major difficulties. Another potential approach is to use an embryo-specific promoter, which we have isolated, to drive a gene system that would stop embryo development (T. BEARDMORE pers. comm.). At this time, complete sterility in trees and its long-term stability cannot be guaranteed using the technology currently available.

Research field trials of poplar and spruce, which were carried out under confined conditions, were used to develop a protocol for tracking the fate of genetically modified DNA in forest soil and litter. As well, they were used to monitor changes in soil microbial populations and to enhance scientific understanding of the performance of the experimental trees (HAY et al., 2002).

7.2. Future research in Canada

Social acceptance of forest trees with novel traits will be highly dependent on our capacity to show that environmental issues have been properly addressed. Public opinion polling consistently reinforces public concern over long-term effects, along with the expectation of governments to monitor and report on these effects (EARNSCLIFFE RESEARCH AND COMMUNICATIONS 2000). The Royal Society of Canada expert panel on the future of food biotechnology and the Canadian Biotechnology Advisory Committee have both recommended the establishment of a national research program on long-term effects of genetically modified organisms on human and animal health and the environment (CANADIAN BIOTECHNOLOGY ADVISORY COMMITTEE; ROYAL SOCIETY OF CANADA).

In response, Environment Canada is leading an interdepartmental committee which is developing an initiative on Ecosystem Effects of Novel Living Organisms (EENLO). A series of workshops was held in February 2004, bringing together research scientists, science managers, policy specialists, regulators, and federal departmental advisory board members. Dovetailing nicely with recommendations made by previous biotechnology panels and workshops, seven areas of research were proposed:

1. Baseline data – accessing, generating, and maintaining baseline data on key ecosystems in order to be able to determine changes associated with the introduction of novel living organisms (NLOs);
2. Detection and monitoring – testing and developing new tools for detection of NLOs, genes, and gene products of concern;
3. Ecosystem impacts of novel living organisms – determining impacts of NLOs and associated production systems on biodiversity, nutrient cycling, water quality, etc.
4. Gene flow and its consequences – determining likelihood of, extent of, and consequences of gene flow on novel traits into other species and varieties;
5. Risk assessment method development – developing and implementing better techniques and approaches to predict risks associated with NLOs;
6. Containment and mitigation – testing and developing biological and physical containment mechanisms to restrict unwanted migration of NLOs or genes;
7. Stewardship of released products – advancing research to manage the long-term environmental impact of released products.

The EENLO initiative is planned to take the form of a multi-sectoral network of government and academic scientists, supported by a Community of Practice that allows identification of experts, coordination of research, information sharing, and knowledge transfer to regulators.

8. PROVINCIAL JURISDICTIONS

In Canada, provincial governments own 77% of the forest, federal and territorial governments own 16%, and only 7% of forest land is privately owned (NATURAL RESOURCES CANADA 2004). Provincial and territorial governments have legal authority to manage Crown land use: they set the rules for forest management and ultimately decide what materials can be planted. A formal federal/provincial committee was formed to discuss regulatory issues related to trees with novel traits, with representatives from all Canadian provinces and chaired by the Canadian Forest Service. A first meeting was called in January 1998. There was general agreement that the responsibility for environmental safety assessments of trees with novel traits was with the CFIA, with active input from the provinces where appropriate. The committee agreed that the Regulatory Directive Dir95-01 (since replaced with Dir2000-07) contained thorough enough information requirements for confined field tests of trees with novel traits. Because of the uniqueness of forestry and silviculture, the committee advised that the guidelines be adapted as a forest-specific document that would no longer make any reference to crop plants, providing a convenient and clear set of instructions for applicants. Another recommendation was to set up an expert committee to provide advice on the regulatory guidelines. An ad hoc expert committee was thus created, met in 1999 and 2002, and discussed regulatory requirements, the interpretation of the definition of tree with novel traits, and data to be collected. These discussions have lead to the development of a draft directive, which is intended to become a regulatory proposal subject to public consultation.

In essence, federal regulations can be viewed as the primary filter for assessing environmental safety of plants with novel traits, with a provincial role to review and advise. Provincial regulations are then applicable, especially on public land. To date, only two Canadian provinces have issued official statements on the use of genetically modified trees: British Columbia and Alberta.

8.1. British Columbia

In British Columbia, all Crown forestland must be reforested to specified standards following harvest. Approximately 27% of the area harvested is reforested naturally, and 73% is regenerated through tree planting. In 2000, around 200 million seedlings were requested for sowing, of which 63% were produced from seed obtained from wild stands, 34% from seed orchards, and 3% from superior provenances. British Columbia's tree improvement and reforestation programs are designed to maintain genetic diversity across the forestland base. By law, orchards producing seed for Crown land reforestation must be licensed, and all seedlots used on Crown land must be registered. The law also directs foresters to use seed of the highest genetic worth available (orchard or tested wild-stand seed), to use seed that is adapted to its environment, and to meet an acceptable level of genetic diversity (HADLEY *et al.,* 2001). The Chief Forester's Standards for Seed Use describe the requirements for using tree seed when reforesting Crown land in British Columbia, under the provincial *Forest and Range Practice Act*. Genetically engineered trees are not used in British Columbia and would not be permitted for use under the current registration policies and standards. These standards have come into effect on April 1, 2005 (GOVERNMENT OF BRITISH COLUMBIA 2005).

8.2. Alberta

In Alberta, the Forest Genetic Resources Council issued a Position Paper in November 2001, recognizing the potential of genetically modified trees in reforestation, recognizing also that their performance and impact on forest ecosystems are poorly understood, and therefore not recommending their use for reforestation at this time. This recommendation will be reviewed as knowledge evolves as a result of ongoing research (ALBERTA GOVERNMENT 2001).

9. INTERNATIONAL HARMONIZATION

9.1. Canada/USA

The CFIA, the United States Department of Agriculture, Animal and Plant Health Inspection Service (APHIS) and Health Canada met in 1998 to harmonize their respective regulatory requirements for the molecular characterization of transgenic plants. Then in September 2000, the United States Environmental Protection Agency (USEPA) and the CFIA met to harmonize the molecular characterization component of their respective evaluations, with observers from Argentina and Mexico. The text of the resulting Bilateral Agreement on Agricultural Biotechnology was finalized in December 2001, and is an important milestone on the way of mutual acceptance of assessments between Canada and the United States (CANADIAN FOOD INSPECTION AGENCY 2005).

9.2. Organization for Economic Co-operation and Development

The Organization for Economic Co-operation and Development (OECD) has representation from 30 industrialised member countries committed to democratic governments and market economies in North America, Europe and the Pacific. Observers from other countries and from international organizations attend many of the OECD meetings. The OECD Working Group on Harmonization of Regulatory Oversight of Biotechnology has made very useful contributions to international discussions on biotechnology regulations. It is an important regulatory forum because it allows countries to discuss issues of mutual concern, and share their scientific approach for addressing environmental issues. Harmonization with member countries of the OECD is a major consideration for Canada.

Canada proposed to use consensus documents to describe the biology of species and submitted the first document on canola in 1997. These are equivalent to the Canadian biology regulatory companion documents, with an international approach. There are now 28 species described in OECD consensus documents, including seven for tree species: European white birch (*Betula pendula*) submitted by Finland (2003); stone fruit (*Prunus* spp.) submitted by Austria (2002); eastern white pine (*Pinus strobus*) submitted by Canada (2002); Sitka spruce (*Picea sitchensis*) submitted by Canada (2002); poplar (*Populus* spp.) submitted by Canada (2000); white spruce (*Picea glauca*) submitted by Canada (1999); and Norway spruce (*Picea abies*) submitted by Norway (1999). Several more are currently under review by member countries. These documents aim at providing regulators detailed information about the biology of a specific plant species, distribution, mating systems, pollen flow distances, fecundity, seed dispersal and dormancy mechanisms. They are meant to be used for identifying potential issues associated with a particular species and for comparison purposes when assessing novel products.

9.3. Cartagena Protocol on Biosafety

The Cartagena Protocol on Biosafety, negotiated under the United Nations Convention on Biological Diversity, is designed to outline international rules to protect biological diversity from potential risks posed by transboundary movements of living modified organisms (LMOs). It entered into force on September 11, 2003. Canada has signed the Protocol, but has not ratified it yet. Still, Canadian delegations participate in Meetings of the Parties, as non-party, and are active in working groups and committees such as the Intergovernmental Committee on the Cartagena Protocol, which is responsible for preparing the entry into force of the Protocol.

The Cartagena Protocol only applies to living modified organisms, meaning those that can replicate or reproduce genetic material. It does not apply to non-living products derived from LMOs such as pulp, paper and dimensional lumber. The full text of the Protocol can be found at: http://www.biodiv.org/biosafety/protocol.asp.

10. CANADIAN REGULATORY STRATEGY FOR THE 21ST CENTURY

Canada's regulatory system is under constant pressure to evolve, especially in a field that develops at such a frantic pace as biotechnology. A catalyst for change is the current development of a regulatory governance framework designed for the 21st century, to modernize regulation in natural resources, environmental protection, biotechnology, health, food safety and transportation. It is called Smart Regulation, and aims at developing more efficient decision-making on policies, laws and regulations, consistent with the principles of sustainable development by protecting the environment and enabling a healthy economy and innovation. On September 23 2004, the External Advisory Committee on Smart Regulation issued a report entitled: *Smart Regulation: A Regulatory Strategy for Canada*. This Committee, with members representing industry, associations, academics, and international regulatory experts, was put in place in May 2003 with the mandate to provide an external view on how to redesign Canada's regulatory approach. The report was informed by consultations with non-governmental organizations, business associations and academics, as well as federal provincial and territorial government officials. Recommendations related to biotechnology regulation are as follows:

> The government should develop and implement a comprehensive biotechnology regulatory strategy to provide a more coherent government wide approach. Government should address legislative gaps and review relevant legislation regularly to ensure that it remains appropriate, implement a new approach to engage all stakeholders on policy issues with respect to biotechnology.

Highlights from the Committee's report to the Government of Canada can be found on http://www.pco-bcp.gc.ca/smartreg-regint/en/01/b-01.html.

11. CONCLUDING REMARKS

Unconfined release of trees with novel traits is not expected soon in Canada, but it is expected to occur. Before the first applications are submitted for regulatory approval, we have the opportunity to be proactive in research and regulatory developments. The case-by-case approach to product evaluation, as applied in Canada, is well suited to accommodate the assessment of trees with novel traits. The regulatory system is also flexible enough to respond to scientific developments.

The key challenge, clearly, is the level of knowledge required to answer regulatory questions related to environmental safety. There is a continuing need to pursue basic research to understand tree species at the molecular and cellular levels, the genetic makeup of Canadian tree populations, and their interactions within complex forest ecosystems. Environmental safety research needs to be effectively targeted to support regulations.

Currently some traits are more easily achieved using genetic engineering. Natural populations of forest trees present a rich pool of genes, and considerable progress is being achieved in the fundamental understanding of the genetics related to key systems such as wood formation and pest resistance mechanisms. In the near future, tree genomics, coupled with parentage analysis and marker-assisted breeding, will be capable of producing similar results using homologous genes within species and among close relatives, without the need to cross kingdom or even genus boundaries. It will be quite interesting to witness how regulatory, policy, ethical, and social questions evolve when this happens.

Acknowledgements: I am grateful to Bette Reid and Sylvie Richard at the Canadian Forest Service for their careful review of the manuscript, to Nathalie Isabel and Tannis Beardmore also at the Canadian Forest Service for an update on their environmental safety research, and to Phil MacDonald and Krista Thomas at the Canadian Food Inspection Agency for current information on the application of the Seeds Regulations.

12. REFERENCES

ALBERTA GOVERNMENT. 2001 Alberta Forest Genetics Research Council Position Paper – Genetically Modified Organisms. Alberta Sustainable Resource Development.

CANADIAN BIOTECHNOLOGY ADVISORY COMMITTEE. 2002. Regulation of Genetically Modified Foods. Report to the Government of Canada Biotechnology Ministerial Coordinating Committee. August 2002.

CANADIAN FOOD INSPECTION AGENCY. 2005 Plant Biosafety Office Main page http: www.inspection.gc.ca/english/plaveg/bio/pbobbve.shtml.

EARNSCLIFFE RESEARCH AND COMMUNICATIONS. 2000 Research Findings - Forest Biotechnology Focus Group Research 20pp. Report submitted to the Canadian Forest Service.

FRANKENHUYZEN, K.V., AND T. BEARDMORE, 2004 Current status and environmental impact of transgenic forest trees. Can. J. For. Res. **34**: 1163-1180.

GOVERNMENT OF BRITISH COLUMBIA. 2005 Chief Forester's Standards for Seed Use. British Columbia Ministry of Forests.

GOVERNMENT OF CANADA, CANADIAN FOOD INSPECTION AGENCY. 2004 Regulatory Directive 94-08 Assessment criteria for determining environmental safety of plants with novel traits.

GOVERNMENT OF CANADA, CANADIAN FOOD INSPECTION AGENCY. 2003 Regulatory Directive 2000-07 Conducting confined research field trials of plants with novel traits in Canada.

GOVERNMENT OF CANADA, CANADIAN FOOD INSPECTION AGENCY. 1998 Regulatory Directive D-96-13 Import permit requirements for plants with novel traits (including transgenic plants), and their products.

GOVERNMENT OF CANADA, DEPARTMENT OF JUSTICE. 1985 The *Seeds Act*, R.S., c. S-8. The Seeds Regulations, C.R.C., c. 1400, Part V. http://laws.justice.gc.ca/en/S-8/98571.html.

HADLEY, M.J., J.S. TANZ, AND J. FRASER, 2001 Biotechnology: potential applications in tree improvement. Forest Genetics Council of British Columbia Extension Note March 2001.

HAY, I., M.J. MORENCY, AND A. SÉGUIN, 2002 Assessing the persistence of DNA in decomposing leaves of genetically modified poplar trees. Can. J. For. Res. **32:** 977-982.

INDUSTRY CANADA 1993 Federal Regulatory Framework for Biotechnology. http://strategis.ic.gc.ca.

MCLEAN, M.A., AND P.J. CHAREST, 2000 The regulation of transgenic trees in North America. Silvae Genetica **49:** 233-239.

MULLIN, T.J., AND S. BERTRAND, 1998 Environmental release of transgenic trees in Canada – potential benefits and assessment of biosafety. For. Chron. **74:** 203-219.

NATURAL RESOURCES CANADA, 2004 The State of Canada's Forests. Canadian Forest Service. Ottawa, ON.

ROYAL SOCIETY OF CANADA, 2001 Elements of Precaution: Recommendations for the Regulation of Food Biotechnology in Canada http://www.rsc.ca/foodbiotechnology/indexEN.html.

RUTLEDGE, R.G., S. REGAN, O. NICOLAS, P. FOBERT, C. CÔTÉ, W. BOSNICH, C. KAUFFELDT, G. SUNOHARA, A. SÉGUIN, AND D. STEWART, 1998 Characterization of an AGAMOUS homologue from the conifer black spruce (*Picea mariana*) that produces floral homeotic conversions when expressed in *Arabidopsis*. Plant J. **15:** 625-634.

STRAUSS, S.H., AND H.D. BRADSHAW. 2001 Proceedings of the First International Symposium on Ecological and Societal Aspects of Transgenic Plantations. College of Forestry, Oregon State University WA. http://www.fsl.orst.edu/tgerc/iufro2001/eprocd.pdf

STRAUSS, S.H., AND H.D. BRADSHAW. 2004. The Bioengineered Forest. Challenges for Science and Society. Resources for the Future. Washington D.C.

CHAPTER 14

BIOSAFETY OF TRANSGENIC TREES IN THE UNITED STATES

RUTH IRWIN

Information Systems for Biotechnology, Virginia Tech University, Blacksburg, VA, USA

PHILLIP B. C. JONES

Freelance Writer, Spokane, Washington, USA

Abstract. The rapid advancement in transgenic forest technology has brought forth biosafety concerns and economic barriers to their commercialization. The primary ecological biosafety risk posed by transgenic organisms is the spread and establishment of transgenes in the environment. Transgenic trees pose unique ecological consideration due to their long life span, increasing the opportunity for transgene movement and integration into wild tree populations and prolonging selective pressure on affected ecosystem inhabitants. The challenges posed in fully assessing risks of introducing transgenic forest plantations are great, requiring multiple studies spanning numerous sites and successive generations of trees to evaluate invasiveness, weediness, and ecological performance. Due to the irreversible nature of introgressed transgenes into undomesticated populations, regulatory oversight and prudence is required. The US and Canada have robust systems for regulating field-testing of transgenic organisms prior to commercialization. To date, genetically-engineered trees have been regulated under rules designed for all GM plants. This is changing due to a growing awareness of the unique regulatory challenges posed by the production of transgenic forest plantations. On national and international levels, regulators are making new rules specific to the challenge posed by commercialization of genetically-engineered trees.

I. INTRODUCTION

The rapidly expanding development and production of transgenic trees has spurred the drive toward their commercialization. Major advances have been made in the development of efficient transformation, regeneration, and propagation protocols such that trees with engineered traits for herbicide tolerance, insect resistance, and altered lignin qualities have been extensively field-tested and show reasonable commercial potential. However, economic barriers to full production of transgenic plantations are significant, and the uncertainty about biosafety and regulation contributes greatly to those impediments. Concern over the impact and interaction of forest tree transgenes with the environment governs regulatory approval, and the empirical scientific knowledge needed to assess ecological effects to gain approval is lagging behind that of other transgenic crops.

The ecological risks associated with transgenic trees arise from genes affecting novel properties in trees. This applies to genes that are obtained from distant species, as well as native genes whose expression levels have been altered. The major issue of concern is whether transgenes will spread and establish themselves in the environment, either by inducing an increase in the invasiveness of the tree itself (weediness), or through the transfer of the transgene to wild relatives by either sexual (vertical gene flow) or non-sexual (horizontal gene flow) means. Other primary concerns interrelated to gene flow are, first, if the traits imparted by the transgene will negatively affect the ecology of the area or impact non-target organisms, and second, if the transgene will become unstable over time causing unpredictable negative effects on tree fitness (termed pleiotropic effects).

In addition to the ecological risks just mentioned, genetically engineered trees pose unique environmental concerns compared to GE agronomic crops. For example, the long life span of trees increases the likelihood that foreign transgene instability could be exhibited. Delayed flowering hampers the ability to examine transgene impacts over several generations. Long-term exposure to plant-expressed insecticides would prolong selective pressure on insect pests, improving the likelihood of pests evolving resistance to insecticidal proteins. Additionally, plantation trees, which have undergone less domestication than most agronomic crops, are more likely to survive and interbreed with native habitants in the wild. Wind-pollinated trees distribute pollen over very large distances, making pollination between GE and wild trees at great distances almost a certainty. The same applies to seeds of woody plants that are dispersed over large areas. Lastly, forest ecosystems support a complex array of organisms in multiple trophic levels, ranging from saprophytic soil fungi to mammal predators; therefore changes introduced by transgenic trees and any potential pleiotropic effects may have many, often unpredictable, consequences.

Historically, large-scale implementation of new methods of agriculture have altered the ecology of surrounding landscapes significantly, displacing native vegetation, creating weeds, or, in some cases, resulting in the extinction of species within restricted distribution areas. Risk assessment of genetically engineered forest plantations is therefore of profound importance. Assessing the risks of transgenic trees, however, poses many challenges, requiring copious time-consuming, complex, and multivariable studies. Multi-site and multi-generational studies to evaluate ecological performance are difficult for long-lived trees; therefore, predictive modelling approaches may be the only realistic method of assessing risks within a reasonable period required for regulatory approval.

The Animal and Plant Health Inspection Service (APHIS) of the US Department of Agriculture has issued more that 300 permits for transgenic tree field trials since 2000 and commercial production of GM trees may begin during 2005. In issuing these permits, the primary concern of regulators of transgenic trees is the environmental risks they may pose. In general, transgenic plants are assumed to have regulated status until they can be proved not harmful and are deregulated. Both the US and Canada have robust science-based regulatory systems in place for prevention of environmental risks of transgenic organisms. In the US, three basic steps are involved in the regulatory process (discussed in *Section 3* below). Anyone wanting to use a transgenic plant must first obtain a permit from APHIS to import, transport across state lines, or release into the plant into the environment. A notification must then be filed before field-testing the transgenic plant. Finally, a petition must be submitted for a determination of nonregulated status.

2. BIOSAFETY CONCERNS OF GENETICALLY ENGINEERED TREES

2.1. Invasiveness and introgression

Ecologists have been concerned that novel transgenes from genetically engineered trees may become permanently established in wild populations of native tree species, where they may exhibit a weedy or invasive character, spreading into the natural ecosystem and causing ecological or economic harm, or both. In order for this to happen, the transfer and establishment of transgene from tree plantation to wild population must occur, generally in a two-step process. First, the transgene must first "escape" the plantation, either by the spread of seeds containing the transgene or through vegetative propagation of trees containing the transgene, or the gene may be dispersed via pollen distribution. Second, the new gene must become permanently established or introgressed into the genome of a recipient population of trees, which occurs after transgenic trees have hybridized and backcrossed with native trees for many generations. Whether or not the "escaped" transgene will become permanently introgressed depends largely on selective pressures placed on it in from the environment. In general, new traits conferred by the transgene that give the tree a fitness advantage are more likely to support the persistence of the transgene in the recipient population genome.

A transformed tree may become invasive by three means. First, the tree itself may become "weedy" in character, as often exemplified by the uncontrolled spread of exotic species. However, some researchers have argued that the tendency for transgenic trees to become weedy is not likely (HANCOCK AND HOKANSON 2004). Hancock and Hokanson argue that weedy, exotic species are presented with few natural constraints when introduced into a new location; however, for transgenic trees, the addition of a novel transgene does little to alter the natural constraints already facing them. They further suggest that in order to make forestry crops more invasive, many traits, working in concert, would have to be genetically altered. The second means by which transgenic trees may become invasive is through pollination of sexually compatible nontransgenic members of the *same* species, and subsequent seed dispersal. Third, transgenes can be transferred by out-breeding to wild type, *related* tree species. The likelihood of the escape of a transgene from plantations into wild-type trees species is considered high. Trees are copious producers of pollen capable of travelling long distances. In addition, most trees easily interbreed with related species, so if transgenic forests are grown within pollen dispersal distance, the chance of hybridizing with wild relatives is great.

An important corollary to the question of whether or not gene flow occurs is what impact the genes might have on the recipient organism. In other words, does it matter? The ability to predict the evolutionary and ecological impact that transgenes may have is very difficult in long-lived species such as trees (JOHNSON AND KIRBY 2004; VAN FRANKENHUYZEN AND BEARDMORE 2004). Ultimately, gene flow and introgression depend on whether the transgene confers reproductive fitness on the recipient (MUIR AND HOWARD 2001; HANCOCK 2003). Deleterious genes, or those that, when passed on to offspring, decrease the likelihood of the transgenic species' survival to reproductive age, are likely to be lost from the population over time. Those genes that increase tree fitness, for example by conferring a selective advantage such as resistance to disease, insect pests, or harsh environmental conditions, will improve the likelihood that the altered species will survive and reproduce.

2.1.1. Horizontal gene flow

Horizontal gene flow is the transfer of new DNA sequences between organisms by means other than sexual reproduction. Genomics research has unearthed evidence that horizontal gene exchange among bacteria and between bacteria and plants or animals is a natural, though relatively rare, event. The possibility of transfer of foreign genes from transgenic trees to soil-dwelling organisms a regulatory concern largely because the ecological consequences of horizontal gene flow of exotic genes are unknown and because the length of time available for interaction between trees and soil microbes improves the likelihood of occurrence. However, although horizontal gene flow has been demonstrated under controlled, ideal laboratory conditions, it has not been conclusively shown to occur in the field. Therefore, horizontal gene flow is not considered a key concern at this time (see VAN FRANKENHUYZEN AND BEARDMORE 2004).

2.2. Unintended effects on organisms and ecosystems

Regardless of whether a transgene increases the invasiveness of the recipient tree species, it can still affect other organisms or the ecosystem, both directly and indirectly. Biotechnology, through the introduction of novel genes obtained from plants, animals, or microbes, alters traits (phenotypes) expressed by trees. Expression of these novel traits can affect commensal organisms exposed to the new proteins, whether those organisms are the intended or unintended targets. Newly introduced traits can also negatively alter critical ecosystem processes involving nutrient availability, litter decomposition, or plant health. Unintended targets include organisms closely associated with the trees, e.g., pollinators, herbivorous insects, and those that feed on decayed plant material, as well as those of higher feeding levels that may eat or otherwise rely on both the target and nontarget insects.

Therefore, transgenic trees should be carefully examined for their potential negative impacts on environmental organisms and processes.

2.2.1. Pest resistance

A primary environmental concern is that organisms will evolve resistance to an engineered trait. For example, selective pressure is increased on insects exposed to plant-incorporated protectants, such as *Bacillus thuringiensis* (*Bt*), in transgenic forests because of the sustained exposure insects receive over the life times of trees, compared to annual crops. Transgenic forests expressing herbicide tolerance traits may result in greater applications of herbicides, increasing selective pressure for resistance to develop in weeds and thereby making forest plantations less attractive to bird and animal species that rely on woodland understory and new forest growth habitats. Finally, long-term pest suppression could have effects on other, nonpest members of the food web community, some of which could become pests, thereby altering competitive interactions and biodiversity of the forest habitat.

Although herbicide resistance is a serious problem in agricultural weeds, no instances of herbicide resistance have been documented due to forestry applications (VAN FRANKENHUYZEN AND BEARDMORE 2004) to date. Therefore, selection for herbicide tolerant weeds in plantations is not considered a major environmental concern. Additionally, the risks of developing herbicide tolerant weeds should be weighed against the benefits of the technology. Herbicide-tolerant tree plantations require application of less toxic herbicides, such as glyphosate, and allow adoption of more environmentally beneficial, no-till methods of farming. Likewise, plantations comprised of pest-resistant trees can reduce quantity and frequency of pesticide applications and allow a shift from broad-spectrum to targeted sprays, decreasing ecological harm.

2.2.2. Quality traits

Any alteration in tree morphology or composition could have unintended ecological effects. Lignin modification for quantity and character is a desirable outcome of tree genetic engineering; however, these modifications may also pose undesirable consequences for tree fitness, as well as unintended environmental side effects. For example, reduced lignin content may result in increased winter mortality (CASLER *et al.*, 2002), increased numbers of insect defoliators, lowered resistance to viral disease, and altered soil structure and fertility due to changes in wood decomposition rates. Altered plant hormone levels could detrimentally affect formation of beneficial root-fungal partnerships (BARKER AND TAGU 2000). Woody plants exist in symbiosis with soil fungi and any change in the host plant, via either genetic engineering or other means, could alter the extent or composition of fungi colonization and hence the surrounding soil ecosystem.

2.3. Transgene instability and pleiotropic effects

Stable gene expression in which the transgene is passed on to subsequent generations is necessary for biosafety and confinement of inserted transgenes, especially for long-lived tree species. Expression of the transgene may be reduced or lost in two ways: through transgene silencing, a process through which foreign DNA is recognized and inactivated by the host plant, and by transgene loss. Reports of gene silencing triggered by environmental changes, such as cold-induced dormancy, and sexual reproduction in trees has increased concern of transgene stability (see MEILAN et al., 2002) in transgenic forests. The risk of transgene silencing is especially important when using sterility as a sole bioconfinement method—the engineered sterility system would need to be very stable if bioconfinement is required over the lifespan of the transgenic trees. Transformation techniques may lead to multiple copies of the transgene inserting as tandem repeats, which can be excised or silenced after several generations.

The process of genetic transformation itself can cause adverse affects on tree fitness. One example is somaclonal variation in which genetic instability can be introduced into plant body cells during cell culture. However, aberrant cell lines can be carefully screened during greenhouse trials, and STRAUSS et al. (2001) reported a very low rate (0.1%) of morphological abnormalities after years of field and greenhouse testing. In addition to somaclonal variation, the process of inserting the transgene itself can cause multiple effects, known as pleiotropic effects, arising from mutations in single genes or expression in nontarget genes. The effects may not be known for many years or may become apparent in response to various environmental conditions.

2.4. Risk assessment

How does one assess the likelihood of invasiveness of long-lived perennial trees and their effect on biodiversity? The answer appears to be, with great difficulty. Any potential ecological risk from transgenic forests needs to be assessed on a case-by-case basis, and considered within the ecological context of the particular organism, trait, and environs. Researchers must measure the rate of gene flow, and the impact of the transgene(s) on relevant characteristics of the recipients, including fitness effects. However, empirical data on the fitness costs and selective advantage of transgenic trees within particular habitats are largely lacking. In order to evaluate the ecological consequences of the introduction of transgenic trees, studies need to be conducted at many locations and across many generations. This requisite is nearly impossible to fulfil for long-lived species because any impacts would not be known for many years.

A realistic option for reasonable risk assessment of transgenic forests may be to use a predictive modelling approach. MUIR AND HOWARD (2001) have developed a model for predicting transgene impact on fitness parameters of Japanese medaka that could be extended to estimate fitness impacts of various transgenes inserted in trees, given that sufficient data can be gathered to estimate fitness components (JOHNSON AND KIRBY 2004). SLAVOV et al. (2002) introduced a spatial stimulation model called STEVE to estimate transgene flow and fate from poplar plantations to wild cottonwoods. The model "synthesizes data from a variety of ecological and genetic processes, and permits virtual experiments that investigate how diverse genetic, ecological, and management factors might influence the magnitude and variance of gene flow over a 50- to 100-year period." (SLAVOV et al., 2002) The authors suggest similar approaches could be used to set acceptable levels of gene flow from tree pollen and determine the effect of many variables on that level.

A modification of the case-by-case approach has also been proposed that includes ranking the relative risk posed by the transgenic organism based on existing knowledge about the biology of the crop, the proximity of wild relatives, and the phenotypic effects of the transgene (HANCOCK 2003). The assignment of a transgene to a risk category might not only speed regulatory approval by recognizing that not all transgenes pose the same risk, but could allow regulators to permit the long term testing of transgenic trees that pose little ecological risk of gene flow and nontarget effects.

2.5. Transgene containment options

Because it is highly likely that genes from engineered trees will flow to feral populations, biological confinement measures must be taken, except in those cases where transgenes pose no risk to wild populations, or where the risk of gene flow would be too great, regardless of precautions taken. The determination of biocontainment need will likely be based on the previous history of the gene or gene product, empirical risk assessment data already gathered, novel risks posed, and proximity to closely related relatives (NATIONAL RESEARCH COUNCIL 2004). Once an environmental risk is determined, investigators need to determine what degree of biocontrol is necessary, what techniques are available to achieve that control, and how the success of bioconfinement will be evaluated.

The use of sterility as a means of containment has wide support in the scientific community. The primary methods of achieving sterility are floral ablation, flowering control, and gene silencing techniques; detailed discussions of these techniques are beyond the scope of this chapter. The long life span of trees dictates that sterility must be complete in genetically engineered trees if the system is to be effective over their lifetime. Because long-term transgene stability and silencing cannot currently be guaranteed in trees, the use of sterility will likely need to be combined with other measures to ensure complete containment. Some other options for biocontainment available to researchers include use of the following: 1) triploid clones to achieve sterility; 2) mutant genotypes lacking a biosynthetic component that renders them less competitive; 3) tissue-specific, regulated expression of transgenes to restrict transgene expression; and 4) chloroplast engineering in angiosperms to prevent pollen transmission of transgenes.

3. REGULATION OF GENTICALLY ENGINEERED TREES

3.1. U.S. regulation of genetically engineered trees

In the United States, the federal government dominates the regulation of genetically modified (GM) plant field trials and the commercialization of GM plants and GM plant products. The U.S. government's influence over engineered plants extends to the earliest experimental stages through an agency of the Public Health Service, the National Institutes of Health (NIH).

3.1.1. Federal guidelines for genetically engineering life forms

The NIH manages an evolving set of guidelines for constructing and handling recombinant DNA molecules and organisms containing recombinant DNA molecules. The Guidelines cover experiments with genetically engineered plants and provide graded requirements for biological containment techniques based upon the level of hazard posed by experiments. The NIH requires certain GM plant experiments to be approved by an Institutional Biosafety Committee, which includes at least one expert in plant, plant pathogen, or plant pest containment principles.

The Guidelines apply not only to NIH-funded recombinant DNA research, but also to research conducted at or sponsored by an institution that receives any support for recombinant DNA research from the NIH. Now accepted as a standard for biotech research, the impact of the Guidelines has extended beyond even this very broad reach. The EPA, for example, expects that institutions subject to the Guidelines comply with its provisions, while the National Science Foundation requires grantees to comply with the Guidelines if the funded research falls within its scope. Similarly, institutions receiving USDA funding for a recombinant DNA research project must assure the agency that funded experiments are exempt from the NIH Guidelines or were approved by an Institutional Biosafety Committee.

3.1.2. The coordinated framework for regulation of biotechnology

As the development of a GM plant progresses to extensive field tests and commercialization, approval must be obtained from federal agencies that constitute the Coordinated Framework for Regulation of Biotechnology. In 1986, the federal government established the Coordinated Framework to provide a risk-based system for evaluating the products of biotechnology with the objective of ensuring that these new products would be safe for human health, animal health and the environment. Three agencies regulate biotech products according to their intended use: the U.S. Department of Agriculture's Animal and Plant Health Inspection Service (APHIS), the U.S. Environmental Protection Agency (EPA), and the Department of Health and Human Services' Food and Drug Administration (FDA). As discussed below, a product – a GM tree, for example – may be subject to review by all three of these agencies.

Table 1. APHIS field test permits issued for genetically modified (GM) trees.

Permits Issued	Tree	Illustrative Traits Affected by Genetic Modification	Locations
36	Pine	Growth rate, lignin levels, and fertility	SC
31	Poplar	Growth rate, fertility, flowering time, light response, uptake of heavy metals, *Coleopteran* resistance, and *Septoria* resistance	AL, CT, FL, GA, IN, KY, MO, MS, NY, OR, SC, TN, WA, WI, WV
11	Apple	Polyphenol oxidase levels, sugar alcohol levels, brown spot resistance, fruit ripening, and fire blight resistance	CA, NY, OR, WA, WV
11	Eucalyptus grandis	Lysine levels, fertility, lignin levels, and growth rate	FL, SC
11	Sweetgum	Fertility, phosphinothricin tolerance, and development	SC
6	Populus deltoides	Expression of a visual marker	SC
4	Eucalyptus camaldulensis	Fertility and growth rate	FL, SC
4	Persimmon	Cold tolerance, drought tolerance, and *Lepidopteran* resistance	CA
3	Pear	Fruit ripening	OR, WA, WV
2	Coffee	Ethylene production and caffeine levels	HI
1	American elm	Dutch elm disease resistance	NY
1	Citrange	*Xanthomonas campestris* resistance	TX
1	Grapefruit	*Closterovirus* resistance	TX
1	Papaya	Ringspot virus resistance	NY
1	Plum	Ethylene production	WV
1	Walnut	Adventitious root formation	CA

(Test permits in effect as of December 1, 2004. Information obtained from the Information Systems for Biotechnology website.)

3.1.2.1. Department of Agriculture's Animal and Plant Health Inspection Service

APHIS focuses on protecting agriculture from pests and diseases. The Plant Protection Act invests APHIS with supervisory authority over the import, handling, interstate movement, and release into the environment of certain "regulated articles." These regulated articles include organisms and products known or suspected to be plant pests or to pose a plant pest risk. Such organisms and products may be altered or produced through genetic engineering.

A field test of a genetically engineered plant requires permission from APHIS. The agency will approve a proposal for field-testing if the planned test conditions appear adequate to confine the regulated article within the field test site. In practice, an organization files either a notification or a permit application for a field test. The notification procedure is a streamlined approval process used for familiar crops and traits considered a low risk. On the other hand, a permit is required for field tests of plants that have an elevated risk, such as plants that synthesize a drug, industrial compound, or a plant pathogen. The agency's reviewers prepare an Environmental Assessment on the potential effect of the proposed release and issue a permit if they reach a Finding of No Significant Impact. By the end of 2004, APHIS had approved field tests for over 80 transgenic plant species, including trees. Table 1 provides an overview of field test permits issued for GM trees.

APHIS also provides a petition process for the determination of non-regulated status. To acquire deregulated status, a petitioner must supply information about the biology of the organism, experimental data, field test results and descriptions of the organism's genotypic characteristics. While evaluating an application, APHIS reviewers consider a number of issues, such as potential for plant pest risk, disease and pest susceptibilities, the expression of gene products or altered plant metabolism, risk of increasing weediness of sexually compatible plants, effects on non-target organisms, and potential for gene transfer to other types of organisms. The agency's fundamental standard is that a GM organism must not pose a greater plant pest risk than the unmodified organism from which it was derived.

APHIS publishes deregulation petitions for comment. After reviewing information and remarks provided by the public, the agency determines whether the plant should be granted deregulated status. If APHIS decides to grant a petition, then the organism will no longer be considered a regulated article subject to APHIS oversight. This means that the deregulated organism – and its progeny – no longer requires the agency's review for movement or release in the United States. The deregulated article may enter the marketplace unless other forms of approval must be obtained from the EPA, FDA or state agencies. As of November 26, 2004, APHIS had granted 63 petitions for nonregulated status, including a petition for ringspot virus resistant papaya, the only transgenic tree.

Some commentators have suggested that APHIS' regulatory authority over GM trees could be challenged. For example, a tree engineered for reduced lignin synthesis may not be considered a "plant pest," and therefore fall outside the literal scope of the Plant Protection Act. Commentators have also suggested that APHIS should adopt rules that treat GM trees as a separate category so that the regulations would cover all GM trees.

In January 2004, APHIS published its intent to prepare an environmental impact statement regarding potential changes to its regulations under the Plant Protection Act. Although the agency has not proposed separate rules for GM trees, it is considering an amendment of the regulations to include GM organisms that may pose a noxious weed risk. A noxious weed is a plant or plant product that can directly or indirectly injure or damage crops, livestock, irrigation, navigation, natural resources, public health or the environment. This definition is so broad that it may cover any transgenic tree and portions of transgenic trees.

3.1.2.2. Environmental Protection Agency

The Federal Insecticide, Fungicide, and Rodenticide Act (FIFRA) requires the EPA to evaluate a proposed pesticide before the product can be marketed and used in the United States. The objective of the statute is to assure that the use of a pesticide will not pose unreasonable risks of harm to human health and the environment

Using a registration process, the EPA regulates the testing, sale, distribution and use of pesticidal substances produced in plants and microbes, including pesticides produced by a GM organism. Typically, the agency issues Experimental Use Permits for field-testing. Before sale and distribution, an organization must register pesticidal products, providing the EPA with the opportunity to make the registration dependent upon conditions for use.

FIFRA covers biopesticides, such as naturally occurring substances that control pests and microorganisms that control pests. The EPA revised FIFRA regulations in 2001 by adding pesticidal substances produced by GM plants, otherwise known as Plant-Incorporated Protectants or PIPs. The agency regulates the pesticidal protein and the genetic material that encodes the protein, but not the plant that synthesizes the toxic protein. When assessing the risks posed by a PIP, the agency requires extensive studies on the potential for gene flow and risks to human health, non-target organisms and the environment.

In 2004 PIP registrations included a variety of *Bacillus thuringiensis* insecticidal proteins produced by GM corn, potato and cotton. Transgenic trees have also been engineered to express Bt toxin proteins. For example, a GM persimmon expresses CryIA(c) to render it Lepidopteran resistant, and a GM popular expresses CryIIIA for Coleopteran resistance.

The EPA also sets tolerance limits for pesticide residues associated with food and animal feed, or the agency can establish an exemption from the requirement for a tolerance. If a pesticide-synthesizing GM tree produces a food product, for example, then the EPA must determine whether there is a reasonable certainty that no harm will result from the aggregate exposure to the PIP. The agency derives its authority to render these pesticide tolerance decisions from the Federal Food, Drug, and Cosmetic Act, not the FIFRA.

Under the authority of the Toxic Substances Control Act (TSCA), the EPA identifies and regulates potentially hazardous chemicals not specifically covered by other statutes. TSCA, therefore, does not apply to pesticides, food, drugs or cosmetics. According to the EPA's interpretation, "new" microorganisms are chemical substances covered by TSCA if the microorganisms were created by combining genetic material from organisms of different genera. Consequently, the agency regulates GM microorganisms intended for industrial use and conducts a pre-market review of GM microorganisms. Commentators have suggested that engineered trees designed for bioremediation of hazardous waste may be regulated under TSCA. However, the EPA's policy in 2004 is that the regulation of such trees falls under APHIS' purview.

3.2.2.3. Food and Drug Administration

The FDA regulates GM food, feed, and food additives under the authority of the Federal Food, Drug, and Cosmetic Act (FFDCA). Over 20 years ago, the FDA took the position that its regulation must be based on the scientific evaluation of products, and not on any assumptions about processes used to make the products. The agency treats substances intentionally added to food through genetic engineering as food additives only if they differ significantly in structure, function, or amount than substances currently found in food.

During November 2004, the FDA proposed draft guidance for industry on new plant varieties intended for food use. According to the proposal, developers would provide the FDA with information about the food safety of a new non-pesticidal protein at a relatively early stage of development so that the agency can evaluate the protein's potential as a toxin or allergen. As noted above, a bioengineered food may contain a Plant-Incorporated Protectant – a pesticidal protein – that is subject to review by EPA.

In sum, federal agencies currently regulate GM trees following the same procedures used to regulate any other type of GM plant. APHIS determines whether GM plants are plant pests, EPA considers the environmental and food safety of GM plants with Plant-Incorporated Protectants, and the FDA looks at the safety of food produced from GM plants.

3.1.3. Regulatory oversight for genetically engineered papaya

Efforts to commercialize GM papaya illustrate how APHIS, the FDA and the EPA divide regulatory oversight. Cornell University and the University of Hawaii developed papaya lines resistant to papaya ringspot virus by genetically engineering the plants to express the viral coat protein. The universities petitioned APHIS for a determination of nonregulated status.

APHIS determined that the papaya lines exhibit no plant pathogenic properties, will not increase the likelihood of the emergence of new plant viruses, are no more likely to become weeds than papaya developed by traditional breeding techniques, will not increase the weediness potential for any other cultivated or wild species with which the GM papaya can breed, will not harm endangered species or other organisms beneficial to agriculture, and will not damage processed agricultural commodities. APHIS concluded that the papaya lines and any progeny will be as safe to grow as papaya in traditional breeding programs. Consequently, the GM papaya lines are no longer considered regulated articles under APHIS rules.

As part of a pre-market approval consultation, the universities submitted a safety and nutritional assessment of their GM papaya to the FDA. The agency acknowledged the developers' conclusion that the GM papaya does not materially differ, in terms of food safety and nutritional profile, from papaya varieties with a history of safe use and does not raise issues that would require pre-market review or approval by the FDA.

The EPA evaluated the safety of using papaya ringspot viral coat protein as a pesticide. In 1997, Cornell University filed a petition requesting an exemption from the requirement of tolerance for the plant-pesticide coat protein of papaya ringspot virus and the genetic material necessary for its production in agricultural commodities. The EPA granted the exemption, which eliminated the need to establish a maximum permissible level for residues of coat proteins and the genetic material that encoded the coat proteins. In effect, the agency had determined that there was a reasonable certainty that no harm will result from the aggregate exposure to the pesticide.

The EPA's 1997 decision fell under the authority of the FFDCA. In 2004, the FIFRA Scientific Advisory Panel held a meeting to consider risks associated with viral coat proteins used as a Plant-Incorporated Protectant. The EPA may require FIFRA registration for viral coat proteins.

3.1.4. Regulation of genetically engineered plants by states

At the state level, the regulation of GM plants can involve departments of agriculture, environmental protection and health – organizations that may administer plant health, pesticide, and food safety regulations similar to federal regulations. But federal law and federal actions strongly influence, and may control, the legal authority and activities of state programs. The Plant Protection Act, for instance, preempts states from regulating the movement in interstate commerce of plants and plant products designed to protect against plant pests if APHIS has acted to regulate the plant or plant product. In this case, federal law limits a state to implementing regulations consistent with APHIS requirements provided that the state's regulations do not exceed APHIS requirements. Only a few states issue their own authorization for a field trial, and these are based at least in part on APHIS' prior authorization of the trial.

FIFRA authorizes states to regulate pesticides, and states play an active enforcement role in overseeing the use of conventional chemical pesticides in fields. Yet most states do not perform experimental use permit inspections for Plant-Incorporated Protectants. This may be due to the EPA's position that, although the toxin protein and genetic material encoding the toxin are regulated as pesticides, plants that contain the protein and genetic material do not have to be regulated as pesticides. Since plants synthesize the PIPs, a state inspector cannot oversee the application of a pesticide.

3.2. International regulation of genetically engineered trees

International trade agreements on GM plants do not yet include provisions focused on the regulation of transgenic trees or wood from transgenic trees. Throughout the world, however, the growth of transgenic trees is generally regulated according to the same principles applied to transgenic crops.

For example, the European Parliament's Directive 2001/18/EC provides two regulatory regimes: one for research and development of a GM organism and one for marketing a GM organism within the European Union. In either case, the Deliberate Release Directive mandates notification to the appropriate authority in the Member State where the release is proposed to take place or where the GM organism is to be placed on the market. While decisions about experimental releases occur only at the national level, a dominating EU-wide authorization procedure goes into effect if a Member State approves a GM organism for marketing. By August 2004, the European Commission had logged over 1800 GM plant field trials in Member States. As shown in Table 2, these GM plants include genetically engineered trees.

Table 2. Genetically modified (GM) tree field trials within the European Union.

Tree	Number of field trials within Member State
Apple	Belgium (2), Germany (1), Netherlands (2), Sweden (1)
Eucalyptus	Spain (1), United Kingdom (1)
European aspen	Germany (2), Norway (1)
Norway spruce	Finland (2)
Poplar	Germany (2), Spain (1), France (8), United Kingdom (2)
Scotch pine	Finland (2)
Silver birch	Finland (4)

(Notifications available as of August 3, 2004. Information obtained from the European Commission's Biotechnology & GMO's Information website.)

The foremost international efforts relevant to GM plants are the Biosafety Protocol of the Convention on Biological Diversity, the International Plant Protection Convention, the Codex Alimentarius Commission, and the Agreement on the Application of Sanitary and Phytosanitary Measures. In brief, the Convention on Biological Diversity adopted an agreement in 2000 known as the Cartagena Protocol on Biosafety. The Protocol seeks to promote biosafety by establishing rules and procedures for the transfer, handling and use of GM organisms and focuses on regulating movements of these organisms from one country to another. Procedures govern the introduction GM organisms into the environment and GM organisms used as food, feed, or for processing. These procedures have been designed to provide information to importing countries so that they can determine whether or not to accept the imports. Governments access information about GM organisms through a Biosafety Clearing-House. The Protocol entered into force in September 2003; the United States is not a party to the agreement.

Administered by the United Nations' Food and Agriculture Organization, the International Plant Protection Convention is a multilateral standard setting body. The Convention aims to provide international standards for preventing the spread and introduction of plant pests.

Codex Alimentarius will impact regulations of produce grown by GM trees. This international body – created by the Food and Agriculture Organization and World Health Organization – sets standards for food safety, including limits for pesticide residues in foods and standards for contaminants and food additives. During 2004, the Codex Alimentarius Commission initiated an ad hoc Intergovernmental Task Force on Foods Derived from Biotechnology. The task force will develop standards, guidelines or recommendations for food derived from modern biotechnology or traits introduced into foods by biotechnology.

The Sanitary and Phytosanitary Measures Agreement, which arose from a General Agreement on Tariffs and Trade meeting, aspires to protect human, animal, or plant life or health from risks arising from the spread of pests, diseases, additives, toxins, or contaminants found in food, beverages, or feed. To avoid disguised trade protectionism, health and safety measures must be based on either a scientific risk assessment or on international standards, such as those established by the International Plant Protection Convention for plant health or by the Codex Alimentarius Commission for food safety.

4. CONCLUSIONS

Relatively little is known about the ecological risks and biosafety issues surrounding transgenic forest plantations compared with agronomic crops. Detailed, long term studies on the invasiveness, weediness and ecological performance of transgenic trees are required. Because the escape of transgenes into feral populations cannot be reversed, regulatory prudence is required. Current regulations that restrict the field testing of trees past sexual maturity hinder the collection of data that would address many of the biosafety concerns expressed by the ecologists and regulators. Short term testing can be used to assess many deleterious ecological effects but cannot reliably be used to predict ecological consequences of transgenic plantations over the time and space scale required.

However, the increasing safety, effectiveness, and number of biocontainment options available to researchers creates optimism that the biosafety of genetically engineered trees can be ensured. The full acceptance of transgenic plantations will only occur, however, when biocontainment methods are used by the research and industry communities. As with other genetically engineered crops, the public needs to concede that a certain level of risk is acceptable, especially when viewed in light of the benefits to be gained though this technology. There also needs to be a commitment to increased funding available for risk assessment research, as well as for development of new engineering technologies. Finally, regulations for field testing transgenic trees that recognize the efficacy of assigning risk categories and corresponding confinement measures would enable risk assessment research to advance by allowing the collection of data from transgenic systems that pose little risk.

By the end of 2004, the regulation of GM trees under U.S. law and international agreements relied upon measures for controlling GM plants in general. That may change, however. APHIS and the Organization for Economic Co-operation and

Development have held public meetings to address concerns about genetically engineered trees. According to the OECD's Working Group on Harmonization of Regulatory Oversight in Biotechnology, a growing awareness of the differences between crop plants and trees triggered its discussion about tree-specific regulation. This awareness should increase as the products of GM trees enter the marketplace at national and international levels.

5. REFERENCES

FEDERAL INSECTICIDE, FUNGICIDE, AND RODENTICIDE ACT, 7 U.S.C. §136 et seq. (2004).
FEDERAL FOOD DRUG AND COSMETIC ACT, 21 U.S.C. §301 et seq. (2004).
TOXIC SUBSTANCES CONTROL ACT, 15 U.S.C. §2601 et seq. (2004).
PLANT PROTECTION ACT, 7 U.S.C. §7701 et seq. (2004).
APHIS, 1996 Cornell University and University of Hawaii; Availability of determination of nonregulated status for papaya lines genetically engineered for virus resistance. *Federal Register* **61**: 48663-48664.
APHIS, 2003 Genetically engineered forest and fruit trees; Public meeting. *Federal Register* **68**: 28193.
APHIS, 2004 Environmental Impact Statement; Introduction of genetically engineered organisms. *Federal Register* **69**: 3271-3272.
BARKER, S. J., AND D. TAGU, 2000 The roles of auxins and cytokinins in mycorrhizal symbioses. *J Plant Growth Regul.* 19: 144-154.
CASLER, M. D., D. R. BUXTON, AND K. P. VOGEL, 2002 Genetic modification of lignin concentration affects fitness of perennial herbaceous plants. *Theor Appl Genet.* **104**: 127-31.
EPA, 2001 Regulations under the Federal Insecticide, Fungicide, and Rodenticide Act for Plant-Incorporated Protectants (formerly Plant-Pesticides). *Federal Register* **66**: 37772-37817.
FDA, 2004 Draft guidance for industry: Recommendations for the early food safety evaluation of new non-pesticidal proteins produced by new plant varieties intended for food use. Federal Register **69**: 68381-68383.
VAN FRANKENHUYZEN, K., AND T. BEARDMORE, 2004 Current status and environmental impact of transgenic forest trees. Can. J. For. Res. 34: 1163-180.
FRIEDMAN, S., P. JENKINS AND C.D. INGRAM. 2001. Landscaping a regulatory system, pp. 24-28 in *Biotech Branches Out: A Look at the Opportunities and Impacts of Forest Biotechnology*. Pew Initiative on Food and Biotechnology, Society of American Foresters and the Ecological Society of America, Atlanta, GA. http://pewagbiotech.org/.
HANCOCK, J. F., 2003 A framework for assessing the risk of transgenic crops. *Bioscience* **53**: 512-519.
HERON D.S. AND J.L. KOUGH, 2001 Regulation of transgenic plants in the United States, pp. 101-104 in *Tree Biotechnology in the New Millennium*, edited by S. H. Strauss and H. D. Bradshaw. http://www.fsl.orst.edu/tgerc/iufro2001/eprocd.htm.
JOHNSON, B., AND K. KIRBY, 2004. Genetically Modified Trees and Biodiversity of Forestry Plantations: A Global Perspecitve. In The Bioengineered Forest: Challenges for Science and Society, S. H. Strauss and H. D. Bradshaw (Editors), pp 190-207. RFF Press, Washington, DC.
MAYER, S. 2001 International regulation and public acceptance of GM trees: Developing a new approach to risk evaluation, pp. 101-104 in *Tree Biotechnology in the New Millennium*, edited by S. H. STRAUSS and H. D. BRADSHAW. http://www.fsl.orst.edu/tgerc/iufro2001/eprocd.htm.
MEILAN, R., D. J. AUERBACH, C. MA, S. P. DIFAZIO, AND S. H. STRAUSS, 2002. Expression in transgenic hybrid poplars (*Populus* sp.) during four years of field trials and vegetative propagation. HortScience 37: 277-280.
MUIR, W. M., AND R. D. HOWARD, 2001 Fitness components and ecological risk of transgenic release: A model using Japanese Medaka (*Oryzias latipes*). American Naturalist **158**: 1-15.
NATIONAL RESEARCH COUNCIL, 2004. *Biological Confinement of Genetically Engineered Organisms.* National Academies Press, Washington, DC.
OFFICE OF SCIENCE AND TECHNOLOGY POLICY. 1986 Coordinated Framework for Regulation of Biotechnology. *Federal Register* **51**: 23302-23393.

ORGANISATION FOR ECONOMIC CO-OPERATION AND DEVELOPMENT, 2001 *Report of the Workshop on the Environmental Considerations of Genetically Modified Trees*. OECD, Norway. http://www.oecd.org/.

SEDJO, R.A. 2004 Transgenic trees: Implementation and outcomes of the Plant Protection Act (Discussion Paper 04-10). Resources for the Future, Washington, D.C. http://www.rff.org.

SLAVOV, G. T., S. P. DEFAZIO, AND S. H. STRAUSS, 2002. Gene flow in forest trees: From empirical estimates to transgenic risk assessment. Gene Flow Workshop, The Ohio State University, March 5 and 6, 2002. http://www.biosci.ohio-state.edu/~asnowlab/gene_flow.htm.

STRAUSS, S. H., S. P. DIFAZIO, AND R. MEILAN, 2001. Genetically modified poplars in context. For. Chron. 77: 271-279.

TAYLOR, M.R., J.S. TICK AND D.M. SHERMAN. 2004 *Tending the Fields: State & Federal Roles in the Oversight of Genetically Modified Crops*. Pew Initiative on Food and Biotechnology, Washington, D.C. http://pewagbiotech.org/.

USDA, EPA, AND FDA, 2004 U.S. database of completed regulatory agency reviews. http://usbiotechreg.nbii.gov/database_pub.asp.

HANCOCK, J. F., AND K. HOKANSON. 2004. Invasiveness of Transgenic versus Exotic Plant Species. How Useful is the Analogy? In *The Bioengineered Forest: Challenges for Science and Society*, S. H. Strauss and H. D. Bradshaw (Editors), pp 181-189. RFF Press, Washington, DC.